美国数学会经典影印系列

出版者的话

近年来，我国的科学技术取得了长足进步，特别是在数学等自然科学基础领域不断涌现出一流的研究成果。与此同时，国内的科研队伍与国外的交流合作也越来越密切，越来越多的科研工作者可以熟练地阅读英文文献，并在国际顶级期刊发表英文学术文章，在国外出版社出版英文学术著作。

然而，在国内阅读海外原版英文图书仍不是非常便捷。一方面，这些原版图书主要集中在科技、教育比较发达的大中城市的大型综合图书馆以及科研院所的资料室中，普通读者借阅不甚容易；另一方面，原版书价格昂贵，动辄上百美元，购买也很不方便。这极大地限制了科技工作者对于国外先进科学技术知识的获取，间接阻碍了我国科技的发展。

高等教育出版社本着植根教育、弘扬学术的宗旨服务我国广大科技和教育工作者，同美国数学会（American Mathematical Society）合作，在征求海内外众多专家学者意见的基础上，精选该学会近年出版的数十种专业著作，组织出版了"美国数学会经典影印系列"丛书。美国数学会创建于1888年，是国际上极具影响力的专业学术组织，目前拥有近30000会员和580余个机构成员，出版图书3500多种，冯·诺依曼、莱夫谢茨、陶哲轩等世界级数学大家都是其作者。本影印系列涵盖了代数、几何、分析、方程、拓扑、概率、动力系统等所有主要数学分支以及新近发展的数学主题。

我们希望这套书的出版，能够对国内的科研工作者、教育工作者以及青年学生起到重要的学术引领作用，也希望今后能有更多的海外优秀英文著作被介绍到中国。

高等教育出版社
2016年12月

美国数学会经典影印系列

Frobenius Manifolds, Quantum Cohomology, and Moduli Spaces

Frobenius 流形、量子上同调和模空间

Yuri I. Manin

高等教育出版社·北京

To Xenia, with love and gratitude

Contents

Preface	xi
Chapter 0. Introduction: What Is Quantum Cohomology?	1
Chapter I. Introduction to Frobenius Manifolds	17
§1. Definition of Frobenius manifolds and the structure connection	17
§2. Identity, Euler field, and the extended structure connection	22
§3. Semisimple Frobenius manifolds	27
§4. Examples	35
§5. Weak Frobenius manifolds	42
Chapter II. Frobenius Manifolds and Isomonodromic Deformations	49
§1. The second structure connection	49
§2. Isomonodromic deformations	55
§3. Semisimple Frobenius manifolds as special solutions to the Schlesinger equations	60
§4. Quantum cohomology of projective spaces	66
§5. Dimension three and Painlevé VI	71
Chapter III. Frobenius Manifolds and Moduli Spaces of Curves	83
§1. Formal Frobenius manifolds and $Comm_\infty$-algebras	83
§2. Pointed curves and their graphs	88
§3. Moduli spaces of genus 0	91
§4. Formal Frobenius manifolds and Cohomological Field Theories	99
§5. Gromov–Witten invariants and quantum cohomology: Axiomatic theory	115
§6. Formal Frobenius manifolds of rank one and Weil–Petersson volumes of moduli spaces	121
§7. Tensor product of analytic Frobenius manifolds	134
§8. K. Saito's frameworks and singularities	146
§9. Maurer–Cartan equations and Gerstenhaber–Batalin–Vilkovysky algebras	157
§10. From dGBV–algebras to Frobenius manifolds	168
Chapter IV. Operads, Graphs, and Perturbation Series	175
§1. Classical linear operads	175
§2. Operads and graphs	184
§3. Sums over graphs	187
§4. Generating functions	193

Chapter V. Stable Maps, Stacks, and Chow Groups ... 201
 §1. Prestable curves and prestable maps ... 201
 §2. Flat families of curves and maps ... 206
 §3. Groupoids and moduli groupoids ... 210
 §4. Morphisms of groupoids and moduli groupoids ... 213
 §5. Stacks ... 219
 §6. Homological Chow groups of schemes ... 226
 §7. Homological Chow groups of DM–stacks ... 232
 §8. Operational Chow groups of schemes and DM–stacks ... 236

Chapter VI. Algebraic Geometric Introduction to the Gravitational Quantum Cohomology ... 245
 §1. Virtual fundamental classes ... 245
 §2. Gravitational descendants and Virasoro constraints ... 250
 §3. Correlators and forgetful maps ... 256
 §4. Correlators and boundary maps ... 266
 §5. The simplest Virasoro constraints ... 270
 §6. Generalized correlators ... 272
 §7. Generating functions on the large phase space ... 278

Bibliography ... 285

Subject Index ... 299

Preface

The mathematical language of classical physics is based upon real numbers. Configuration spaces and phase spaces of classical systems are differentiable manifolds, and physical laws are expressed by differential equations in the real domain.

The mathematical language of quantum physics is based upon complex numbers, and it would be natural to expect that the complex analytic and the algebraic geometry should replace the differential geometry of the classical period. In a sense, this is what has been happening during the last two or three decades, with the advent of scattering matrices, twistors, strings propagating in the ten–dimensional space–time, quantum cohomology, and M–theory. The mathematical physics of the dawning New Age sets as its ultimate goal construction of the universal quantum theory of all interactions including gravity. In the meantime it distanced itself from the traditional preoccupations of experimental particle physics and cosmology and did not just become heavily mathematicized, but in fact almost merged with mathematics. What made this development so exciting for mathematicians was that physicists brought not only a wealth of fresh insights, ideas, and problems, but also heuristic tools of great power and a certain freedom of expression which supplanted a rather strait–laced mood in the mathematical community of the fifties and sixties.

This book summarizes some of the mathematical developments that took place in the last decade or so and that focus on the notion of Quantum Cohomology, introduced by Cumrun Vafa (see [Va]) and Edward Witten. However, this is a mathematical monograph, and the reader who is interested in physical motivation and history will have to refer to other sources: see [MirS1], [MirS2], and the references therein.

Quantum Cohomology is a construction which endows with an additional highly non–linear structure the usual cohomology space $H = H^*(V)$ with complex coefficients of any projective algebraic (or symplectic) manifold V. The resulting structure, suitably axiomatized by B. Dubrovin, is called the *Frobenius manifold*. Interest in this axiomatization depends on the fact that there exist several general constructions of Frobenius manifolds, seemingly quite different, and unexpected isomorphisms between Frobenius manifolds of various classes (dualities, including Mirror duality). The first part of the book, Chapters I–IV, is dedicated to this notion and its multiple interconnections with geometry, differential equations, operads, and perturbation formalism. A more detailed summary can be found in the Introduction.

Although Quantum Cohomology in the proper sense of the word is invoked in several places in the first part of the book (Introduction, examples in Chapter II, axiomatic exposition in Chapter III), its systematic treatment is postponed until

Chapters V and VI. But whereas Chapters I–IV are reasonably self–contained and provide complete proofs of the main results, the final part of the book is meant as an introduction to the original papers and cannot replace them. In fact, the construction of Quantum Cohomology requires considerable algebraic geometric technique: the machinery of the Deligne–Mumford and Artin stacks, including intersection theory and the deformation theory for them. Already for schemes, this machinery takes hundreds of pages in standard sources: see [Ful] for intersection theory and [Il] for the deformation formalism. A monograph exhaustively treating the algebraic–geometric background for Quantum Cohomology is highly desirable. Hopefully, this book might stimulate its appearance.

A word of warning is in order: although the Mirror Conjecture initially provided the main stimulus for studying Quantum Cohomology, it is not treated in this book. On the one hand, this subject is still in a state of flux and rapid change. On the other hand, the body of firmly established facts, among which Givental's proof of the Mirror Identity of [COGP] for quintics occupies the prominent position (see [Giv2], [BiCPP], [Pa3], and the further development in [LiLY]), still constitutes only a fraction of the extremely varied and fascinating insights into what might be called the Mirror Phenomenon, which is an ambitious collective project bridging the physical and the mathematical communities.

Acknowledgements. Work on this book started in 1992–93, when Iz Singer and I led a seminar on the Mirror Conjecture at MIT. Contacts with Cumrun Vafa and Ed Witten were crucial at this stage.

The book took its present form after several lecture courses given at the Max–Planck–Institut für Mathematik in Bonn in 1994–98, and many shorter lecture courses delivered at various summer schools and conferences.

The vision of Quantum Cohomology expounded here was greatly influenced by Maxim Kontsevich, with whom I collaborated at the Max–Planck–Institut in 1994 and later. A part of the results in this book, including the axiomatic treatment of Gromov–Witten invariants, the theory of operadic tensor products in Chapter III, and the treatment of gravitational descendants in Chapter VI, is based on our joint work. Boris Dubrovin's papers, in particular his lecture notes [D2], provide the basic source of information about Frobenius manifolds, and most of the key definitions and theorems of Chapters I–II are due to him. The notion of weak Frobenius manifolds was introduced in my joint paper with Claus Hertling. Ralph Kaufmann's study of tensor products in the categories of local and global (as opposed to the operadic and formal) Frobenius manifolds is also incorporated in Chapter III. Chapter IV can serve as a brief introduction to operads and perturbation series. Our presentation owes much to the work of Misha Kapranov and Ezra Getzler. The final part of the book prepares and presents the construction of Gromov–Witten invariants which in genus zero are the coefficients of the formal series (potential) embodying Quantum Cohomology, and in higher genus provide a far–reaching extension of this theory in which much work remains to be done. This construction is due to Kai Behrend and Barbara Fantechi: see [Beh] and [BehF]. It was motivated by the earlier construction of the Gromov–Witten invariants in the symplectic and complex–analytic context due to J. Li and G. Tian: see [LiT1] and [LiT2]. The Behrend–Fantechi theory uses in essential ways stacks and their intersection theory, which are reviewed in Chapter V of this book. It is based on the work of Pierre Deligne, David Mumford, Mike Artin, Vistoli, and many others.

During the course of the work, I profited from many enlightening conversations and/or correspondence with my colleagues, friends, and collaborators mentioned above, and with Victor Batyrev, Sergei Barannikov, Alexander Givental, Vadim Schechtman, Sergey Merkulov, Markus Rosellen, and Don Zagier. Their contributions are gratefully acknowledged.

CHAPTER 0

Introduction: What Is Quantum Cohomology?

0.1. An overview. We start with a rather detailed overview of the two central themes of this book: Quantum Cohomology and Frobenius Manifolds.

Let $H = H^*(V, k)$ be the cohomology space of a projective algebraic manifold V with coefficients in a field k of characteristic zero.

The quantum cohomology $H^*_{quant}(V)$ consists of H *plus* an additional piece of data which can be described in at least three seemingly unrelated ways:

i) As a formal series ("potential") Φ in coordinates on H whose third derivatives can be used to define on $K \otimes H$ the structure of a $\mathbf{Z}_2$–graded commutative associative algebra, K being the ring of all formal series in the coordinates.

ii) As a family of polylinear cohomological operations $[m] : H^{\otimes n} \to H$, $n \geq 2$, indexed by all homology classes $m \in H_*(\overline{M}_{0,n+1}, k)$. Here $\overline{M}_{0,n+1}$ denotes the moduli space of stable $(n+1)$-marked algebraic curves of genus zero (cf. 0.2, Chapter III, §3, and [Ke]).

iii) As a "completely integrable system" on the tangent sheaf of the formal spectrum $\mathrm{Spf}(K)$ (i.e. a formal completion of H at the origin considered as a linear supermanifold). In this context, the system itself consists of a one–parametric family of flat connections on the tangent bundle of $\mathrm{Spf}(K)$.

The structures i)–iii) can and must first be described abstractly. We will do it in more detail in 0.2–0.4, and then discuss in what sense they are equivalent in 0.5. In the main body of the book and in 0.4.1 below, they are called *formal Frobenius manifolds*: cf. Chapter III, §4. Chapters I and II introduce and study Frobenius manifolds in more geometric categories (differentiable, analytic, algebraic).

A constructive description of these structures on cohomology spaces, i.e. quantum cohomology of V in the proper sense, involves counting (parametrized) rational curves on V (Gromov–Witten invariants) and is thus related to some classical problems of enumerative algebraic geometry. In 0.6 and 0.7, we will give two examples of the potential Φ constructed in this way, for $V = \mathbf{P}^2$ and for $V =$ a quintic hypersurface in $\mathbf{P}^4$. For more systematic treatment, see Chapter III, §5, and Chapter VI. The geometry underlying these constructions leads naturally to the descriptions of types i) and ii). Quantum cohomology is a functor from the category of smooth projective algebraic manifolds *and their isomorphisms* to the category of formal Frobenius manifolds. The study of its properties with respect to the more general morphisms has not been carried out systematically as yet, and remains an important problem. However, the quantum Künneth formula is reasonably well understood via the general construction of the tensor product of Frobenius manifolds.

In the language of physicists, quantum cohomology is a mathematical theory of the topological quantum sigma–model with target space V (in the tree approximation). In the context of the Mirror Conjecture (cf. below) it is also referred to as the *A-model*.

Algebraic geometry also furnishes several constructions of (formal and non-formal) Frobenius manifolds of different nature. They are supported by the moduli spaces of various kinds: versal deformations of an isolated singularity (Kyoji Saito's theory, physicist's Landau–Ginzburg models), Hurwitz spaces, and moduli spaces of Calabi–Yau manifolds and their extended (formal) versions constructed by S. Barannikov and M. Kontsevich. The relevant potentials and metrics are constructed via periods of algebraic integrals and variations of Hodge structure. In the context of the Mirror Conjecture, we call such constructions *B-models*. We will discuss an example in 0.8, cf. also 0.9. For more general constructions see I.4.5 (A_n–singularities), III.8 (an axiomatic version of K. Saito's construction), III.9 and III.10 (Barannikov–Kontsevich's theory.)

If a potential Φ obtained by counting curves on a manifold can be identified with another potential Ψ related to the periods on another manifold, this gives a strong hold on the analytical properties of Φ and on the behavior of its coefficients. Existence of such an identification first suggested for quintics in [COGP] and proved in [Giv2] is (a part of) the famous Mirror Conjecture for Calabi–Yau manifolds. It is already clear that it constitutes a part of a much vaster mirror pattern, whose formulation suggested by M. Kontsevich might involve identification of a triangulated category related to coherent sheaves (*A*–model) with another triangulated category related to the Lagrange and Kähler geometry (*B*–model.)

We will now fix notation for the remaining part of the Introduction. Denote by (H,g) a $\mathbf{Z}_2$–graded finite–dimensional k–linear space H endowed with an even non–degenerate graded symmetric bilinear form g. Let $\{\Delta_a\}$ be a basis of H, $g_{ab} = g(\Delta_a, \Delta_b)$, $(g^{ab}) = (g_{ab})^{-1}$, and $\Delta = \sum \Delta_a g^{ab} \otimes \Delta_b \in H \otimes H$. Denote by $\{x^a\}$ the dual basis of the dual space of H. We will consider x^a as formal independent graded commuting variables of the same parity as Δ_a. Put $K = k[[x^a]]$; this is the same as the completed symmetric algebra of the dual space. Put $\partial_a = \partial/\partial x^a : K \to K$. We will write Φ_a instead of $\partial_a \Phi$, etc.

0.2. Definition. *A formal solution Φ to the associativity equations on (H,g), or simply a potential, is a formal series $\Phi \in K$ satisfying the following differential equations:*

$$(0.1) \qquad \forall a,b,c,d: \sum_{ef} \Phi_{abe} g^{ef} \Phi_{fcd} = (-1)^{\tilde{x}_a(\tilde{x}_b+\tilde{x}_c)} \sum_{ef} \Phi_{bce} g^{ef} \Phi_{fad}$$

where generally $\tilde{x}$ denotes the $\mathbf{Z}_2$-parity of x.

Define a K–linear multiplication $\circ$ on $H_K := K \otimes_k H$ by the rule

$$(0.2) \qquad \Delta_a \circ \Delta_b = \sum_{cd} \Phi_{abc} g^{cd} \Delta_d.$$

Clearly, it is supercommutative.

0.2.1. Proposition. *a) $(H_K, \circ)$ is associative iff Φ is a potential. Multiplication $\circ$ does not change if one adds to Φ a polynomial of degree ≤ 2 in x^a.*

b) An element Δ_0 of the basis is a unit with respect to $\circ$ iff it is even and $\Phi_{0bc} = g_{bc}$ for all b, c. Equivalently:

$$(0.3) \qquad \Phi = \frac{1}{6} g_{00}(x^0)^3 + \frac{1}{2} \sum_{c \neq 0} x^0 x^b x^c g_{bc} + \text{terms independent of } x^0.$$

If $H = H^*(V, k), g =$ Poincaré pairing $(g_{ab} = \int_V \Delta_a \wedge \Delta_b)$, and Φ is obtained via the Gromov–Witten counting of rational curves on V, then $(H_K, \circ)$ is called *the quantum cohomology ring of V*.

0.3. Moduli spaces $\overline{M}_{0n}$. Before giving the next definition, we recall some basic facts about stable curves of genus 0 with $n \geq 3$ labeled pairwise distinct non–singular points $(x_1, \ldots, x_n)$ (cf. [Kn1], [Ke]). Such a curve is a tree of $\mathbf{P}^1$'s: any two irreducible components are either disjoint or intersect transversely at one point. Each component must contain at least three special (singular or labeled) points.

The space $\overline{M}_{0n}$ is a smooth projective algebraic manifold of dimension $n - 3$ supporting a universal family $X_n \to \overline{M}_{0n}$ of stable curves whose labeled points are given by n structure sections $x_i : \overline{M}_{0n} \to X_n$. An open subset ("big cell") parametrizes $\mathbf{P}^1$ with n pairwise distinct points on it. The boundary, or infinity, of $\overline{M}_{0n}$ is stratified according to the degeneration type of fibers of X_n: the combinatorics of the incidence tree of the curve and the distribution of labeled points among the components. The number of the components diminished by one is the codimension of the stratum. Of course, the closure of such a stratum includes its own boundary corresponding to further degeneration.

In particular, the irreducible boundary divisors D_σ of $\overline{M}_{0n}$ correspond to the stable (unordered) 2–partitions $\sigma : \{1, \ldots, n\} = S_1 \coprod S_2, |S_i| \geq 2$, describing the distribution of the labeled points among the two $\mathbf{P}^1$'s at the generic point of D_σ. A choice of the ordering of the partition defines an identification of D_σ with $\overline{M}_{0,n_1+1} \times \overline{M}_{0.n_2+1}, n_i = |S_i|$: on each $\mathbf{P}^1$ add to the labeled points the intersection point of the two components. Thus we have a family of closed embeddings

$$(0.4) \qquad \varphi_\sigma : \overline{M}_{0,n_1+1} \times \overline{M}_{0,n_2+1} \to \overline{M}_{0n}$$

inducing the restriction morphisms of the cohomology groups with coefficients in k,

$$(0.5) \qquad \varphi_\sigma^* : H^*(\overline{M}_{0n}) \to H^*(\overline{M}_{0,n_1+1}) \otimes H^*(\overline{M}_{0,n_2+1}).$$

Besides, S_n acts on $\overline{M}_{0n}, H^*(\overline{M}_{0n})$ and on the partitions σ by renumbering the labeled points, and (0.5) is compatible with this action.

0.3.1. Definition. *A structure of the Cohomological Field Theory (CohFT) (or an algebra over the operad $H_*\overline{M}_0$, cf. Chapter IV and [GeK2]) on (H, g) consists of a family of S_n-equivariant $\mathbf{Z}_2$-even polylinear maps*

$$(0.6) \qquad I_n : H^{\otimes n} \to H^*(\overline{M}_{0n}, k), \ n \geq 3,$$

satisfying the following conditions. For every stable 2-partition σ of $\{1, \ldots, n\}$ and all homogeneous $\gamma_1, \ldots, \gamma_n \in H$ we have

$$(0.7) \qquad \varphi_\sigma^*(I_n(\gamma_1 \otimes \cdots \otimes \gamma_n)) = \epsilon(\sigma)(I_{n_1+1} \otimes I_{n_2+1}) \left(\bigotimes_{i \in S_1} \gamma_i \otimes \Delta \otimes \left(\bigotimes_{i \in S_2} \gamma_i \right) \right),$$

where $\epsilon(\sigma)$ is the sign of the permutation induced by σ on the odd-dimensional classes γ_i.

Another way of looking at such a structure is to make a partial dualization with the help of the Poincaré pairing on $\overline{M}_{0,n+1}$ and g on H. Then one can rewrite $(0.6)_{n+1}$ as

(0.8) $$H_*(\overline{M}_{0,n+1}) \otimes H^{\otimes n} \to H, \ n \geq 2,$$

that is, to interpret every class $m \in H_*(\overline{M}_{0,n+1})$ as an n-ary multiplication $[m]$ on H linearly depending on $[m]$. Then (0.7) gives a complex system of quadratic identities between these multiplications which are best described in the operadic formalism.

However, the situation simplifies considerably if we restrict ourselves to looking only at those multiplications that correspond to the fundamental classes $[\overline{M}_{0,n+1}] \in H_*(\overline{M}_{0,n+1})$ and denote them simply by

(0.9) $$[\overline{M}_{0,n+1}] \otimes (\gamma_1 \otimes \cdots \otimes \gamma_n) \mapsto (\gamma_1, \ldots, \gamma_n), \ n \geq 2.$$

These multiplications are supercommutative. Moreover:

0.3.2. Proposition. *The identities (0.7) imply the following generalized associativity equations for these multiplications: for any $\alpha, \beta, \gamma, \delta_1, \ldots, \delta_n \in H$, $n \geq 0$, we have*

(0.10) $$\sum_\sigma \epsilon'(\sigma)((\alpha, \beta, \delta_i \,|\, i \in S_1), \gamma, \delta_j \,|\, j \in S_2)$$
$$= \sum_\sigma \epsilon''(\sigma)(\alpha, (\beta, \gamma, \delta_i \,|\, i \in S_1), \delta_j \,|\, j \in S_2)$$

where σ runs over 2-partitions $\sigma : \{1, \ldots, n\} = S_1 \coprod S_2$ (non-necessarily stable), and ϵ are the standard signs.

In particular, for $n = 0, 1$ we get respectively

$$((\alpha, \beta), \gamma) = (\alpha, (\beta, \gamma)),$$

(0.11) $$((\alpha, \beta), \gamma, \delta) + (-1)^{\tilde\gamma\tilde\delta}((\alpha, \beta, \delta), \gamma) = (\alpha, (\beta, \gamma, \delta)) + (\alpha, (\beta, \gamma), \delta).$$

Remarkably, this family of n-ary multiplications is actually equivalent to the whole structure described in 0.3.1: cf. sketch of the proof of Theorem 0.5 below and its complete version in III.4.

In conclusion, let us formally compare the system of operations (0.8) on $H = H^*(V, k)$ (in the situation of quantum cohomology) with the more traditional Steenrod operations.

i) Steenrod powers are defined on the cohomology with coefficients in $\mathbf{F}_p$ whereas we allow characteristic zero coefficients.

ii) Steenrod powers generate an *algebra* whereas $[m]$, $m \in H_*(\overline{M}_{0,n+1})$, are elements of an *operad*.

iii) Steenrod powers are defined solely in terms of topology of V, whereas to construct $[m]$ we need additionally the structure of algebraic (or symplectic) manifold, in order to be able to define holomorphic curves on V.

0. INTRODUCTION: WHAT IS QUANTUM COHOMOLOGY?

0.4. Frobenius manifolds. The term "completely integrable system" is used rather indiscriminately in a wide variety of contexts. The notion relevant here was introduced by B. Dubrovin (cf. [D1], [D2]) under the name of Frobenius manifold. We start with the formal version.

0.4.1. Definition. *a) The structure of a formal Frobenius manifold on (H,g) is a one–parametric system of flat connections on the module of derivations of K/k given by its covariant derivatives*

$$(0.12) \qquad \nabla_{\lambda,\partial_a}(\partial_b) := \lambda \sum_{cd} A_{abc} g^{cd} \partial_d = \lambda \sum_d A_{ab}{}^d \partial_d,$$

where $A_{abc} \in K$ is a symmetric tensor, λ an even parameter.

b) This structure is called a potential one if the tensor $\partial_d A_{abc}$ is totally symmetric.

More generally, a Frobenius manifold (M, g, A) (in any of the standard geometric categories: smooth, analytic, algebraic (super)manifolds) is a manifold M endowed with a flat metric g and a tensor field A of rank 3 such that if we write the components of A in local g–flat coordinates, the conditions of 0.4.1 a) and eventually b) are satisfied.

0.5. Theorem. *For a given (H,g), there exists a natural bijection between the sets of the additional structures described above:*

i) Formal solutions of the associativity equations on (H,g), modulo terms of degree ≤ 2.

ii) Structures of the CohFT on (H,g).

iii) Structures of the formal potential Frobenius manifold on (H,g).

Easy part of the proof (sketch). We will first describe maps $ii) \to i) \to iii)$.

$ii) \to i)$. Assume that we have on (H,g) the structure of CohFT given by some maps I_n as in (0.6). Construct first the symmetric polynomials

$$(0.13) \qquad Y_n : H^{\otimes n} \to k, \; Y_n(\gamma_1 \otimes \cdots \otimes \gamma_n) := \int_{\overline{M}_{0n}} I_n(\gamma_1 \otimes \cdots \otimes \gamma_n)$$

and form the series

$$(0.14) \qquad \Phi(x) := \sum_{n \geq 3} \frac{1}{n!} Y_n \left((\sum_a x^a \Delta_a)^{\otimes n} \right).$$

Keel ([Ke]) has described the linear relations between the cohomology classes of the boundary divisors D_σ defined in 0.3. Namely, choose a quadruple of pairwise distinct indices $i, j, k, l \in \{1, \ldots, n\}$, $n \geq 4$. For a stable 2–partition $\sigma = \{S_1, S_2\}$ write $ij\sigma kl$ if $i, j \in S_1$, $k, l \in S_2$ for some ordering of the parts. Then the $\{ijkl\}$–th Keel's relation is

$$(0.15) \qquad \sum_{\sigma: ij\sigma kl} D_\sigma \cong \sum_{\sigma: ik\sigma jl} D_\sigma \quad \text{in } H^*(\overline{M}_{0n}).$$

Geometrically, it follows from the fact that the two sides of (0.15) are the two fibers of the projection

$$\overline{M}_{0n} \to \overline{M}_{0,\{ijkl\}} \cong \overline{M}_{0,4} = \mathbf{P}^1$$

forgetting all the labeled points except for x_i, x_j, x_k, x_l. The space $\overline{M}_{0.4}$ has exactly three boundary points corresponding to the three stable partitions of $\{i,j,k,l\}$. In (0.15) we use two of them.

Notice that the existence of the forgetful morphism is a non–trivial geometric fact, because on the level of fibers of X_n (i.e. geometric points of the moduli) it involves contracting those components that become unstable: cf. [Kn1] and V.4.4.

If we restrict $I_n(\gamma_1 \otimes \cdots \otimes \gamma_n)$ to D_σ using (0.7) then integrate over D_σ and take into account (0.15), we will get a series of bilinear identities: $\forall i,j,k,l$

$$\sum_{\sigma:ij\sigma kl} \epsilon(\sigma)(Y_{|S_1|+1} \otimes Y_{|S_2|+1}) \left(\bigotimes_{p \in S_1} \gamma_p \otimes \Delta \otimes \left(\bigotimes_{q \in S_2} \gamma_q \right) \right)$$

$$(0.16) \qquad = \sum_{\sigma:ik\sigma jl} \epsilon(\sigma)(Y_{|S_1|+1} \otimes Y_{|S_2|+1}) \left(\bigotimes_{p \in S_1} \gamma_p \otimes \Delta \otimes \left(\bigotimes_{q \in S_2} \gamma_q \right) \right).$$

On the other hand, writing the associativity equations (0.1) for the series (0.14), one can directly show that they reduce to a subfamily of the relations (0.16), which implies the whole family by the standard polarization argument. Thus Φ encodes the same amount of information as $\{Y_n\}$ and (0.16).

$i) \to iii)$. Given a potential Φ, we simply put $A_{abc} = \partial_a \partial_b \partial_c \Phi$. This is in fact a bijection, because given (H, g, A), the symmetry of A_{abc} and $\partial_d A_{abc}$ implies the existence of Φ with $A_{abc} = \partial_a \partial_b \partial_c \Phi$, and the curvature vanishing equation $\nabla_\lambda^2 = 0$ implies the associativity equations for Φ.

Difficult part of the proof. It remains to show that nothing is lost or gained in the passage from I_n to Y_n, i.e., that the arrow $ii) \to i)$ is both injective and surjective. Injectivity is again easy, because using (0.7) consecutively one sees that the knowledge of Y_n allows us to reconstruct integrals of I_n along all the boundary strata, whose classes span $H^*(\overline{M}_{0n})$. But surjectivity requires considerable work. Basically, it reduces to showing that the ad hoc formulas for the integrals over the boundary strata do define a cohomology class, i.e., satisfy all the linear relations between the classes. A remarkable reformulation of this property asserts that the homology of moduli spaces forms a Koszul operad. For details, see the main text.

0.5.1. Remark. What this last argument additionally shows is that the structure of a CohFT on (H, g) can be replaced by the structure of a $Comm_\infty$-algebra given by a family of n–ary operations, one for each $n \geq 2$, satisfying the generalized associativity relations (0.10). This structure looks simpler because it does not involve the moduli spaces $\overline{M}_{0n}$ which look completely irrelevant also for the remaining two descriptions. However, there are at least three reasons not to eliminate the moduli spaces, and even to consider $ii)$ as the most important structure.

a) In the applications to quantum cohomology, the geometry of the Gromov–Witten invariants naturally involves total maps I_n, not just their top–dimensional terms Y_n describing the physicists' correlation functions.

b) The higher genus theory of Gromov–Witten invariants furnishes cohomological operations parametrized by all homology classes of the moduli spaces of stable curves $\overline{M}_{gn}$, and unlike the genus zero case, they cannot be reconstructed from the operations corresponding to the fundamental classes, because there exist cohomology classes vanishing on the boundary.

c) The whole theory can be extended to include the so-called gravitational descendants. The respective correlation functions can be calculated, if we know the complete Gromov–Witten invariants, but not if we know only its top degree parts. For details, see Chapter VI.

d) Returning to the genus zero case, in the abstract framework of $Comm_\infty$–algebras, there exists an operation of their tensor product. It can be defined as follows:
$$(H', g', I'_n) \otimes (H'', g'', I''_n) = (H' \otimes H'', g' \otimes g'', I_n),$$
where I_n are given by
$$I_n(\gamma'_1 \otimes \gamma''_1 \otimes \ldots \otimes \gamma'_n \otimes \gamma''_n) := \epsilon(\gamma', \gamma'') I'_n(\gamma'_1 \otimes \ldots \otimes \gamma'_n) \wedge I''_n(\gamma''_1 \otimes \ldots \otimes \gamma''_n).$$

This is an important and natural operation necessary e.g. for the formulation of the quantum Künneth formula. However, it seems impossible to construct this product without invoking $\overline{M}_{0n}$. In fact, its existence is a reflection of the fact that $H_*(\overline{M}_{0n})$ forms an operad of coalgebras, and not just linear spaces.

In particular, consider C_∞–algebras of rank 1 (i.e. $\dim(H)=1$). In terms of potentials, they correspond to arbitrary power series in one variable $\Phi(x) = \sum_{n \geq 3} \frac{C_n}{n!} x^n$ because the associativity equations in one variable are satisfied identically. Hence we can define a tensor multiplication of such series. It turns out to be given by quite non–trivial polynomials in the coefficients involving a generalization of the Petersson–Weil volumes of $\overline{M}_{0n}$ (see Theorem III.6.5.)

We will now describe some examples.

0.6. Quantum cohomology of $\mathbf{P}^2$. First, we have
$$H^{2i}(\mathbf{P}^2, k) = k\Delta_i, \quad \Delta_i = c_1(\mathcal{O}(1))^i, \quad i = 0, 1, 2.$$

Denote by $N(d)$ (for $d \geq 1$) the number of rational curves of degree d in $\mathbf{P}^2$ passing through $3d-1$ points in general position. The first few values of $N(d)$ starting with $d = 1$ are
$$1, \ 1, \ 12, \ 620, \ 87304, \ 26312976, \ 14616808192.$$
The potential $\Phi^{\mathbf{P}^2}$, by definition, is
$$\Phi^{\mathbf{P}^2}(x\Delta_0 + y\Delta_1 + z\Delta_2) = \frac{1}{2}(xy^2 + x^2 z) + \sum_{d=1}^\infty N(d) \frac{z^{3d-1}}{(3d-1)!} e^{dy}$$

(0.17)
$$:= \frac{1}{2}(xy^2 + x^2 z) + \varphi(y, z).$$

A direct computation shows:

0.6.1. Proposition. *The associativity equations (0.1) for the potential (0.17) are equivalent to one differential equation for φ:*

(0.18)
$$\varphi_{zzz} = \varphi_{yyz}^2 - \varphi_{yyy}\varphi_{yzz}$$

which is in turn equivalent to the family of recursive relations uniquely defining $N(d)$ starting with $N(1) = 1$:

$$(0.19) \qquad N(d) = \sum_{k+l=d} N(k)N(l)k^2 l \left[l \binom{3d-4}{3k-2} - k \binom{3d-4}{3k-1} \right], \ d \geq 2.$$

0.6.2. Geometry. The identities (0.19) showing that $(H^*(\mathbf{P}^2, \mathbf{Q}), g, \Phi^{\mathbf{P}^2})$ is actually an instance of the structure described above were first proved by M. Kontsevich. He skillfully applied an old trick of enumerative geometry: in order to understand the number of solutions of a numerical problem, try to devise a degenerate case of the problem where it becomes easier. In this setting, Kontsevich starts with a new problem having *one-dimensional space of solutions* and looks at two different degeneration points in the line of solutions.

More precisely, fix $d \geq 2$ and consider a generic configuration in $\mathbf{P}^2$ consisting of two labeled points y_1, y_2, two labeled lines l_1, l_2, and a set of $3d - 4$ unlabeled points Y. Look at the space of quintuples $(\mathbf{P}^1, x_1, x_2, x_3, x_4, f)$ where $x_i \in \mathbf{P}^1$ are pairwise distinct points, $f : \mathbf{P}^1 \to \mathbf{P}^2$ is a map of degree d such that $f(x_i) = y_i$ for $i = 1, 2$, $f(x_i) \in l_i$ for $i = 3, 4$, and $Y \subset f(\mathbf{P}^1)$. We identify such diagrams if they are isomorphic (identically on $\mathbf{P}^2$). Then we can assume that $(x_1, x_2, x_3, x_4) = (1, 0, \infty, \lambda)$. If λ is fixed and generic, the number of maps does not depend on it. Kontsevich counts it by first letting $\lambda \to \infty$, and then letting $\lambda \to 1$. In the stable limit, $\mathbf{P}^1$ degenerates into two projective lines, and we must sum over all possible distributions of $\{x_i\} \cup f^{-1}(Y)$ on these components. Comparison of the two limits furnishes (0.19).

To make all of this rigorous, one must introduce not only the moduli spaces of stable curves, but also the moduli spaces of stable maps $\overline{M}_{0n}(\mathbf{P}^2)$ parametrizing Kontsevich-stable maps to $\mathbf{P}^2$. Then it will become clear that the calculation we sketched above furnishes a particular case of the identities (0.16).

Chapter V is a systematic introduction to the study of stable maps.

0.7. Quantum cohomology of a three–dimensional quintic. Let $V \subset \mathbf{P}^4$ be a smooth quintic hypersurface. Its even cohomology has rank four and is spanned by the powers of a hyperplane section, the odd cohomology has rank 204 and consists of three–dimensional classes. For a generic even element $\gamma = \sum x^a \Delta_a \in H^*(V)$, denote by y the coefficient at $\Delta_1 := c_1(\mathcal{O}(1))$ and put

$$(0.20) \qquad \Phi^V(\gamma) = \frac{1}{6}(\gamma^3) + \sum_{d \geq 1} n(d) Li_3(e^{dy})$$

where (γ^3) means the triple self–intersection index, $Li_3(z) = \sum_{m \geq 1} z^m/m^3$, and $n(d)$ is the appropriately defined number of rational curves of degree d on V.

Before we turn to the definition of $n(d)$, let us notice that in this case the associativity equations are satisfied with whatever choice of these coefficients! This can be checked by a direct calculation. An arguably more enlightening argument runs as follows: in quantum cohomology of any V, the associativity equations must reflect the degeneration properties of rational curves on V as was the case with $\mathbf{P}^2$. Now, on a quintic, the rational curves are typically rigid so that there is nothing to degenerate. (See, however, the discussion in 0.7.3.)

Algebraically, the quantum cohomology ring of the projective plane with ∘-multiplication (cf. 0.2 above) is semisimple whereas that of the quintic is nilpotent. B. Dubrovin has developed a rich theory of the Frobenius manifolds with pointwise semisimple multiplication in a tangent sheaf: see I.3, II.4 below, and the recent preprint [DZh3]. This should eventually provide analytic tools for the numerical theory of rational curves on Fano varieties. On the contrary, potentials of the Calabi–Yau threefolds are conjecturally constrained by the Mirror identities rather than associativity equations. Nevertheless, there are at least two contexts in which the Calabi–Yau quantum cohomology can be understood as a limiting case of the semisimple situation. First, Givental's approach via equivariant cohomology produces a family of Frobenius manifolds, whose generic fiber is semisimple, and a special fiber is the relevant quantum cohomology. Second, Gepner's approach via Landau–Ginzburg models conjecturally realizes quantum cohomology of certain Calabi–Yau hypersurfáces as a closed Frobenius submanifold of a generically semisimple Frobenius manifold: see III.8.7.2.

0.7.1. A definition of the numbers $n(d)$. A naive argument showing that the number of rational curves of degree d on V must be finite runs as follows. The space of maps $f: \mathbf{P}^1 \to \mathbf{P}^4$, $(t_0, t_1) \mapsto (f_0(t_0, t_1), \ldots, f_4(t_0, t_1))$ of degree d is a Zariski open subset in the space $\mathbf{P}^{5d+4}$ of the coefficients of forms f_i. The condition $F(f_0(t_0, t_1), \ldots, f_4(t_0, t_1)) = 0$ where $F = 0$ is the equation of V furnishes $5d + 1$ equations on these coefficients. If these equations were independent, the space of solutions would be 3-dimensional. It is acted upon effectively by $\mathrm{Aut}(\mathbf{P}^1)$ (linear reparametrizations) which leaves us with finitely many equivalence classes of unparametrized curves.

Unfortunately, it is unknown whether there exists a sufficiently generic V for which these equations actually are independent after deleting degenerating maps. The symplectic approach to this problem going back to M. Gromov uses a drastic deformation of the complex structure of V destroying its integrability. In this way the problem is put into general position. More precisely, only isolated non–singular pseudoholomorphic spheres in V with normal sheaf $\mathcal{O}(-1) + \mathcal{O}(-1)$ survive; they can be counted directly, and their number is stable.

Another strategy which we will sketch below does not leave the algebraic geometric framework and even allows one to calculate $n(d)$ using the same degeneration philosophy as in example 0.6, although in a rather different setting. This construction is also due to M. Kontsevich ([Ko7]).

Consider a pair (C, f) where C is a connected curve of genus 0 (a tree of $\mathbf{P}^1$'s), and $f: C \to \mathbf{P}^4$ is a map of degree d such that the inverse image of any point in $f(C)$ is either 0–dimensional, or a stable curve of genus zero whose labeled points are intersection points with non–contracted components. Such pairs (C, f) are called (Kontsevich–)stable maps (of genus zero, to $\mathbf{P}^4$). There exists a diagram

$$\overline{M}(\mathbf{P}^4, d) \leftarrow \overline{C}_d \to \mathbf{P}^4$$

where $\overline{M}(\mathbf{P}^4, d)$ is the moduli space (or rather stack) of stable maps of degree d, and $\overline{C}_d$ is the universal curve on it. Denote the right arrow (the universal map) by φ_d, and the left arrow by π. Put $\mathcal{E}_d = \varphi_d^*(\mathcal{O}(5))$, $E_d = \pi_*(\mathcal{E}_d)$.

0.7.2. a) *$\overline{M}(\mathbf{P}^4, d)$ is a smooth orbifold of dimension $5d + 1$.*

b) E_d *is a locally free sheaf on it of rank* $5d + 1$.

0.7.3. Definition. $n(d) := c_{5d+1}(E_d)$.

Motivation for this definition is simple: if a quintic V is defined by $s = 0$, $s \in \Gamma(\mathbf{P}^4, \mathcal{O}(5))$, then s produces a section $\bar{s} \in \Gamma(\overline{M}(\mathbf{P}^4, d), E_d)$, and

$$c_{5d+1}(E_d) = \text{the number of zeroes of } \bar{s}$$

calculated with appropriated multiplicities. But $\bar{s}([\varphi]) = 0$ for $[\varphi] \in \overline{M}(\mathbf{P}^4, d)$ iff $\varphi_d(\overline{C}_{d,[\varphi]}) \subset V$. Thus we simply avoided the problem of assigning ad hoc multiplicities to actual rational curves on V (which may have a "wrong" normal sheaf, singularities, or come in families) by reducing it to a calculation of Chern numbers on orbifolds.

Moreover, we simultaneously created a setting in which degeneration can easily occur. In fact, instead of considering curves in a fixed quintic V, we are now looking at curves in $\mathbf{P}^4$ lying in V, i.e., treat V as an "incidence condition", similar to $3d-1$ points in $\mathbf{P}^2$ in 0.6 above. We may now freely change the equation $s = 0$ for V and can take, e.g., $s = \prod_{i=0}^{4} s_i$ where $s_i \in \Gamma(\mathbf{P}^4, \mathcal{O}(1))$ are coordinates in $\mathbf{P}^4$.

To make sense of the problem of "counting rational curves on the algebraic symplex $V_\infty := \bigcup_{i=0}^{4} \{s_i = 0\}$" Kontsevich proceeds as follows. Consider the G_m-action on the whole setting $(\mathbf{P}^4, \mathcal{O}(5), \overline{M}(\mathbf{P}^4, d))$ given by $s_i \mapsto e^{\lambda_i t} s_i$, $i = 0, \ldots, 4$, where λ_i are the parameters of this action considered as independent variables.

0.7.4. Claim. a) V_∞ *is the only reduced quintic fixed with respect to this action.*

b) *Fixed points of this action in* $\overline{M}(\mathbf{P}^4, d)$ *consist of stable pairs* (C, f) *where C is a tree of* $\mathbf{P}^1$*'s mapped by* f *to the 1-skeleton of* V_∞ *(consisting of 10 projective lines).*

Each such (C, f) has a combinatorial invariant (τ, λ) which is, roughly speaking, the dual tree τ of C, each vertex of which is labeled either by zero (if the respective component of C is contracted by f), or by the name of the line in the skeleton to which it is mapped and the degree of this map.

Bott's formula for Chern numbers of a bundle E in a situation where G_m acts upon the whole setting involves a sum of local contributions over the connected components of the set of G_m-fixed points, each contribution depending on the weights of G_m on the normal sheaf of the component and on the restriction of E upon it.

Kontsevich shows that in our case we get a sum

$$(0.21) \qquad n(d) = \sum w(\tau, \lambda)$$

where the Bott multiplicities $w(\tau, \lambda)$ of the parametrized curves in the 1-skeleton of V_∞ are explicit but complex rational functions on the parameters λ of the G_m-action. Since $n(d)$ must be a rational or even integral *number*, miraculous cancellations must take place in the r.h.s. of (0.21) which are not at all evident algebraically.

Computer calculations furnish the following values for the first four $n(d)$'s:

(0.22) \qquad\qquad 2875, \ 609250, \ 317206375, \ 242467530000.

More direct methods of counting rational curves lead to the same numbers.

Although in a sense the potential (0.20) is now explicitly known, it is still difficult to identify it with its conjectural mirror image which we will shortly describe.

0.8. Moduli spaces of Calabi–Yau threefolds as weak Frobenius manifolds. As the discussion in 0.4 and 0.5 shows, the geometry of a Frobenius manifold on M is basically defined by a flat structure and a symmetric cubic tensor which is the third Taylor differential of a potential in flat coordinates. A flat metric is then used in order to raise indices and write the associativity equations.

If we are interested in a class of potentials for which the associativity equations are trivial, like (0.20), we may as well forget about the metric, and call the resulting structure *weak Frobenius*. For a precise definition of weak Frobenius manifolds, see I.5. This geometry naturally arises from the theory of variation of Hodge structure of Calabi–Yau threefolds.

Let $\pi : W \to Z$ be a complete local family of Calabi–Yau threefolds. Recall that each fiber W_z is a projective algebraic manifold with trivial canonical bundle and $h^{i,0} = 0$ for $i = 1, 2$. Denote by $\mathcal{L} = \pi_* \Omega^3_{W/Z}$ the invertible sheaf of holomorphic volume forms on the fibers of π. We will construct an $\mathcal{L}^{-2}$-valued cubic differential form $G : S^3(\mathcal{T}_Z) \to \mathcal{L}^{-2}$ in the following way. First, according to Bogomolov–Todorov–Tian, the Kodaira–Spencer map (following from $0 \to \mathcal{T}_{W/Z} \to \mathcal{T}_W \to \pi^*(\mathcal{T}_Z) \to 0$)

$$KS : \mathcal{T}_Z \to R^1\pi_*\mathcal{T}_{W/Z}$$

is actually an isomorphism so that the tangent space at $z \in Z$ can be identified with $H^1(W_z, \mathcal{T}_{W_z}) \cong H^1(W_z, \Omega^2_z) \otimes \mathcal{L}(z)^{-1}$. Second, the convolution $i : \mathcal{T}_{W/Z} \times \Omega^p_{W/Z} \to \Omega^{p-1}_{W/Z}$ induces the pairings

$$R^1\pi_*(i) : R^1\pi_*\mathcal{T}_{W/Z} \times R^q\pi_*\Omega^p_{W/Z} \to R^{q+1}\pi_*\Omega^{p-1}_{W/Z}$$

or else

$$R^1\pi_*\mathcal{T}_{W/Z} \to \mathcal{E}nd^{(-1,1)}\left(\bigoplus_{p,q} R^q\pi_*\Omega^p_{W/Z}\right)$$

which is essentially the graded symbol of the Gauss–Manin connection defined thanks to the Griffiths' transversality condition. Iterating it three times and using Serre's duality we get finally:

$$G : S^3(\mathcal{T}_Z) \cong S^3(R^1\pi_*\mathcal{T}_{W/Z}) \to \mathcal{H}om(\pi_*\Omega^3_{W/Z}, \pi_*\mathcal{O}_W) \cong \mathcal{L}^{-2}.$$

In order to identify $\mathcal{L}^{-2}$ with $\mathcal{O}_Z$ (which we need to define a weak Frobenius structure) we must choose a trivialization of the volume form sheaf. In the context of the Mirror Conjecture, this is achieved by postulating that Z can be partially compactified by $\dim(Z)$ divisors with normal intersection in such a way that the family W can be extended to a family of "degenerate Calabi–Yau's" and the zero-dimensional stratum of the boundary W_∞ becomes a maximally degenerate manifold, like the simplex V_∞ in the family of quintics. A precise description of this condition is fairly technical, and we omit it here; but see Deligne's paper [De2], [Mo1], and [Pea].

Then the monodromy invariant part of $H_3(W_z, \mathbf{Z})/(tors)$ around zero will be generated by one cycle γ defined up to sign (more or less by the definition of maximal degeneration), and we locally trivialize $\mathcal{L}$ by choosing a volume form ω_z on W_z in such a way that $\int_{\gamma_z} \omega_z = (2\pi i)^3$.

The flat coordinates in which G is the third Taylor differential of a potential Ψ can be constructed in the same context as the action variables of the algebraically completely integrable system whose phase space is the family of Griffiths Jacobians of W_z: cf. [DoM].

A family W is called the mirror family for V if one can identify the weak Frobenius manifold structure on $H^2(V)$ obtained via curve counting on V (A-model) with that corresponding to the variation of Hodge structure for W (B-model).

For the particular case of quintics considered in 0.7 the mirror family depends on one parameter z, and W_z is obtained by resolving singularities of the spaces $\widetilde{W}_z/(\mathbf{Z}/5\mathbf{Z})^3$ where $\widetilde{W}_z \subset \mathbf{P}^4$ is given by the equation $\sum_{j=1}^5 x_j^5 = z \prod_{j=1}^5 x_j$, and $(\mathbf{Z}/5\mathbf{Z})^3$ acts by $x_j \mapsto \xi_j x_j$, $\xi_j^5 = 1$, $\prod_{j=1}^5 \xi_j = 1$.

All the periods $\psi(z) := \int_{\gamma_z} \nu_z$ of an explicit algebraic volume form along $\gamma_z \in H_3(W_z, \mathbf{Z})$ (any horizontal cycle) satisfy the Picard–Fuchs differential equation $\partial := zd/dz$:
$$[\partial^4 - 5z(5\partial + 1)(5\partial + 2)(5\partial + 3)(5\partial + 4)]\psi(z) = 0.$$
It has four linearly independent solutions near $z = 0$:
$$\psi_0(z) = \sum_{n=0}^\infty \frac{(5n)!}{(n!)^5} z^n,$$
$$\psi_1(z) = \log(z)\psi_0(z) + 5\sum_{n=0}^\infty \frac{(5n)!}{(n!)^5}\left(\sum_{k=n+1}^{5n} k^{-1}\right) z^n,$$
and two more for which we give only the top terms:
$$\psi_2(z) = \frac{1}{2}(\log z)^2 \psi_0(z) + \ldots, \psi_3(z) = \frac{1}{6}(\log z)^3 \psi_0(z) + \ldots.$$

An appropriate flat coordinate on the z–line by definition is $\frac{\psi_1}{\psi_0}(z)$. Under the mirror correspondence, it becomes y in (0.20), thus locally identifying $H^2(V, \mathbf{C})$ (where V is a generic quintic) to the moduli space of the dual family W. Putting

(0.23) $$F(y) := \Phi^V(y) = \frac{5}{6}y^3 + \sum_{d=1}^\infty n(d) Li_3(e^{dy})$$

we have the following mirror identity:

(0.24) $$F'''\left(\frac{\psi_1}{\psi_0}\right) = \frac{5}{2}\frac{\psi_1 \psi_2 - \psi_0 \psi_3}{\psi_0^2}.$$

Since ψ_i are explicitly known, one can check that the first coefficients agree with (0.22).

However, conceptually (0.24) looks baffling. In order to reduce our problem to the proof of an explicit identity, we have oversimplified the geometry. In particular, the mirror pattern must involve some operator of parity change or an odd scalar product on the full Frobenius supermanifold, because an even part of $H^*(V)$ becomes identified with an odd part of $H^*(W)$. E. Witten and M. Kontsevich suggested that generally one should extend the moduli space of the model B rather than restrict (to H^2) the moduli of the problem A. This is crucially important for understanding the mirror picture for the higher–dimensional Calabi–Yau manifolds where rational curves cease to be isolated and a considerably larger (depending on $\dim(V)$) portion of $H^*(V)$ becomes affected by the instanton corrections. According to one suggestion due to M. Kontsevich, one should construct deformations of a Calabi–Yau manifold in a mysterious universe of non–commutative and/or non–associative objects like A_∞–categories (cf. [Ko4]). A less ambitious construction due to S. Barannikov and M. Kontsevich produces extended *formal* moduli spaces with Frobenius structure using solutions of formal Maurer–Cartan equations: see III.9 and III.10.

A. Givental proved the mirror identity (0.24) in [Giv2] by refining Kontsevich's approach, passing to the equivariant cohomology, and completing the geometric picture by extremely ingenious calculations (cf. also [BiCPP], [Pa3] and the further extension in [LiLY]).

V. Batyrev developed a theory of mirror correspondence between complete Calabi–Yau intersections in toric varieties which conjecturally should serve as the background for a multitude of mirror identities: see [Ba1], [BaBo1], [Babo2]. It is also expected that mirror identities reflect only a part of a much richer geometric picture, which still remains rather mysterious. For some additional insights, see [Va], [Ko4], [Ko5], [Bor2].

0.9. Weil–Petersson volumes as rank 1 Cohomological Field Theory. The rank of the CohFT on (H, g) is, by definition, $\dim(H)$. Let it be 1. Assume for simplicity that $g(\Delta_0, \Delta_0) = 1$ for a basis vector $\Delta_0 \in H$ and fix it. Then the whole structure boils down to a sequence of (non–necessarily homogeneous) cohomology classes

$$(0.25) \qquad c_n := I_n(\Delta_0^{\otimes n}) \in H^*(\overline{M}_{0n})^{S_n}, \ n \geq 3,$$

satisfying the identities

$$(0.26) \qquad \phi_\sigma^*(c_n) = c_{n_1+1} \otimes c_{n_2+1}, \ n = n_1 + n_2, \ n_i \geq 2$$

(cf. (0.6) and (0.7)).

By Theorem 0.5, we see that each such theory is uniquely determined by the coefficients of its potential

$$(0.27) \qquad \Phi(x) := \sum_{n \geq 3} \frac{C_n}{n!}, \ C_n = \int_{\overline{M}_{0n}} c_n$$

(cf. (0.14)), which can be totally arbitrary because any series in one variable satisfies the associativity equations. Therefore, rank one theories seem to be rather trivial objects. However, this is not so for at least two reasons: first, there are quite interesting specific theories of algebro–geometric origin; second, the behavior of $\Phi(x)$ with respect to the tensor product of theories is non–trivial.

Here we give an example (the first term of a hierarchy) of algebro–geometric theories.

There is a standard Weil–Petersson hermitian metric on the non–compact moduli spaces M_{0n} parametrizing irreducible curves. On the boundary this metric becomes singular. Nevertheless, its Kähler form extends to a closed L^2–current on $\overline{M}_{0n}$, thus defining a real cohomology class $\omega_n^{WP} \in H^2(\overline{M}_{0n})^{S_n}$. There is also a purely algebro–geometric definition of this class (see [AC1]). Consider the universal curve $p_n : X_n \to \overline{M}_{0n}$. Let $x_i \subset X_n$ be the divisors corresponding to the structure sections, and $\omega = \omega_{X_n/\overline{M}_{0n}}$ the relative dualizing sheaf. Then

$$(0.28) \qquad \omega_n^{WP} = 2\pi^2 p_{n*}\left(c_1(\omega(\sum_{i=1}^n x_i))^2\right).$$

The main property of ω_n^{WP} is

$$(0.29) \qquad \varphi_\sigma^*(\omega_n^{WP}) = \omega_{n_1+1}^{WP} \otimes 1 + 1 \otimes \omega_{n_2+1}^{WP}.$$

Comparing this with (0.26) one sees that

$$(0.30) \qquad c_n := \exp(\omega_n^{WP}/2\pi^2) \in H^*(\overline{M}_{0n}, \mathbf{Q})$$

is a rank one CohFT. Its potential is a generating function for the Weil–Petersson volumes considered in [Zo1]:

$$(0.31) \qquad \Phi^{WP}(x) := \sum_{n=3}^\infty \frac{v_n}{n!(n-3)!} x^n,$$

$$(0.32) \qquad \frac{v_n}{(n-3)!} := \frac{1}{\pi^{2(n-3)}} \int_{\overline{M}_{0n}} \frac{(\omega_n^{WP})^{n-3}}{(n-3)!}.$$

P. Zograf proved that $v_4 = 1$, $v_5 = 5$, $v_6 = 61$, $v_7 = 1379$, and generally

$$(0.33) \qquad v_n = \frac{1}{2}\sum_{i=1}^{n-3} \frac{i(n-i-2)}{n-1}\binom{n-4}{i-1}\binom{n}{i+1} v_{i+2}v_{n-i},\ n \geq 4.$$

This is equivalent to a non-linear differential equation for $\Phi^{WP}(x)$. What is more remarkable, the inverse function for the second derivative of the potential satisfies a linear (modified Bessel) equation:

$$(0.34) \qquad y = \sum_{n=3}^\infty \frac{v_n}{(n-2)!(n-3)!} x^{n-2} \iff x = \sum_{m=1}^\infty \frac{(-1)^{m-1}}{m!(m-1)!} y^m.$$

This can be considerably generalized to the complete description of the tensor product of invertible rank one CohFT's: see III.6. Thus, in addition to the associativity equations for the quantum cohomology of plane (and other Fano manifolds) and the hypergeometric equations for Calabi–Yau (made non-linear by a coordinate change) we have one more differential equation of a seemingly different origin. In fact, this is a reincarnation of the (partly conjectural) Virasoro constraints of a fuller theory, involving correlators of all genera with gravitational descendants: see Chapter VI.

0.10. Plan of the book. From this sketchy overview, it must be clear that the quantum cohomology is an exceptionally rich and tightly woven structure.

Chapters I and II develop the local and global geometric and analytic theory of Frobenius manifolds. Chapter III introduces the more algebraic aspects: formal Frobenius manifolds, moduli spaces and their homology operads. Besides, Chapter III contains the theory of tensor products and several constructions of large classes of Frobenius manifolds of algebraic geometric origin.

Chapters V and VI focus on the algebraic geometric constructions of the Gromov–Witten invariants. In the first part they figure only as examples or in axiomatic form. The theory of quantum cohomology is thereby considerably enriched, but its relationship to the basic substructure of Frobenius manifolds needs further clarification. To be more precise, it is clear that a considerable part of the higher genus theory with gravitational descendants can be extended to more general Frobenius manifolds than actual quantum cohomology (cf. the notion of Frobenius manifolds of qc–type, III.5.4.) However, the exact scope of such an extension remains unclear.

There is one more structure that keeps appearing in all the ramifications of this subject: trees and more general graphs, eventually with labels. They enumerate the strata and cells of $\overline{M}_{g,n}$, help to visualize the composition laws of operads and operadic algebras, and govern the counting of curves on quintics via Kontsevich's construction. Many generating functions and potentials Φ, when they can be explicitly calculated, often appear in the guise of sums over labeled graphs of rather special type, perturbation series, which are well known in statistical physics and quantum field theory.

One can look at graphs as a mere book–keeping device and treat them in an *ad hoc* manner whenever they appear. However, I thought it worthwhile to pay them more respect as a combinatorial skeleton of the theory. Chapter IV summarizes some of their applications.

CHAPTER I

Introduction to Frobenius Manifolds

§1. Definition of Frobenius manifolds and the structure connection

1.1. Supermanifolds. We will work throughout this section and the next one in the superextension of one of the classical categories of manifolds $\mathcal{M}an$: C^∞, real analytic, or complex analytic. Whenever integration can be avoided, $\mathcal{M}an$ may even be a category of smooth algebraic manifolds over a field of characteristic zero. To fix notation, we briefly recall the basic framework of [Ma2], Chs. 4 and 5.

1.1.1. Definition. *A supermanifold is a locally ringed space $(M, \mathcal{O}_M)$ with the following properties.*

a) $\mathcal{O}_M = \mathcal{O}_{M,0} \oplus \mathcal{O}_{M,1}$ is the structure sheaf of $\mathbf{Z}_2$-graded supercommutative rings.

b) $M_{\text{red}} = (M, \mathcal{O}_{M,\text{red}} := \mathcal{O}_M/(\mathcal{O}_{M,1}))$ is a classical manifold, object of the respective classical category.

c) $\mathcal{O}_M$ is locally isomorphic to the exterior algebra $\wedge(E)$ of a free $\mathcal{O}_{M,\text{red}}$-module E.

A morphism of supermanifolds is a morphism of locally ringed spaces extending a classical morphism of underlying reduced manifolds.

We denote $(M, \mathcal{O}_M)$ simply by M, when there is no risk of confusion.

1.1.2. Conventions. By $\widetilde{x}$ we denote the $\mathbf{Z}_2$-degree, or parity, of a homogeneous object x (local function, vector field, scalar product, etc.).

If M is a supermanifold, local coordinates in a neighborhood of a point form a family of sections of the structure sheaf which can be obtained as follows. Choose a local isomorphism $\varphi : \wedge(E) \to \mathcal{O}_M$ as above, local coordinates $(\overline{x}^1, \ldots, \overline{x}^m)$ on M_{red}, and free local generators $(\overline{x}^{m+1}, \ldots, \overline{x}^{m+n})$ of E. Put $x^i = \varphi(\overline{x}^i)$. Then $(x^1, \ldots, x^{m+n})$ are local coordinates on M. Any local function on M can be expressed as a polynomial in anticommuting odd coordinates $x^{m+1}, \ldots, x^{m+n}$ whose coefficients are classical (C^∞, analytic, etc.) functions of the commuting even coordinates $x^1, \ldots, x^m$. Odd coordinates are sometimes denoted by Greek letters.

If M is connected, the pair $m|n$ is an invariant of M called its (super)dimension. When $n = 0$, we say that M is pure even. Transition functions between various local coordinate systems, of course, need not be linear in odd coordinates, e.g. $(x, \xi, \eta) \mapsto (x + \xi\eta, x\xi, x^{-1}\eta)$ is a transition function outside $x = 0$.

The de Rham complex of sheaves on M is the universal $(\mathbf{Z}_2, \mathbf{Z})$-graded differential $\mathcal{O}_M$-algebra (Ω_M^*, d) with *odd* differential d. This means that $\widetilde{dx} = \widetilde{x} + 1$, and the Leibniz formula reads
$$d(fg) = df\, g + (-1)^{\tilde{f}} f\, dg.$$
Notice that as an $\mathcal{O}_M$-algebra, Ω_M^* is the *symmetric* algebra of the $\mathcal{O}_M$-module Ω_M^1 rather than the exterior one. This is the combined effect of our choice of odd d and the rule of signs defining the action of S_n upon $P^{\otimes n}$:
$$\sigma(p_1 \otimes \cdots \otimes p_n) = \epsilon(\sigma, p) p_{\sigma^{-1}(1)} \otimes \cdots \otimes p_{\sigma^{-1}(n)},$$
where $\epsilon(\sigma, p)$ is the sign of the permutation induced on odd p_i (i.e. when even p_i are simply disregarded).

Given local coordinates (x_a) on M, they determine the local vector fields $\partial_a = \partial/\partial x^a$ by the rule
$$df = \sum dx^a \partial_a f$$
for any f in $\mathcal{O}_M$. Notice that $\widetilde{\partial}_a = \widetilde{x}_a$ and $\partial_a \partial_b = (-1)^{\tilde{x}_a \tilde{x}_b} \partial_b \partial_a$ so that the supercommutator, which we denote by the usual square brackets $[\partial_a, \partial_b]$, vanishes. To shorten notation, a sign of the type $(-1)^{\tilde{x}_a(\tilde{x}_b + \tilde{x}_c)}$ will be denoted $(-1)^{a(b+c)}$.

The tangent sheaf $\mathcal{T}_M$ (resp. cotangent sheaf $\mathcal{T}_M^*$) is locally freely generated by (∂_a) (resp. by (dx_a) with *reverse* parity).

A Riemannian metric on M is an even symmetric pairing $g : S^2(\mathcal{T}_M) \to \mathcal{O}_M$, inducing an isomorphism $g' : \mathcal{T}_M \to \mathcal{T}_M^*$. We put $g_{a,b} := g(\partial_a, \partial_b)$. Clearly, $\widetilde{g}_{ab} = \widetilde{x}_a + \widetilde{x}_b$. No positivity condition is imposed, even in the pure even case over $\mathbf{R}$.

A warning: in many situations it is necessary to consider the relative versions of all these notions, that is, to work with submersions of supermanifolds $M \to S$ considered as a family parametrized by the base S. Functions on S are "constants", and since there are no odd constants in $\mathbf{R}$ or $\mathbf{C}$, the need for a base extension arises in supergeometry more often than in the pure even setting. The necessary changes are routine.

The following structure is important in the theory of Frobenius manifolds.

1.2. Definition. *a) An affine flat structure on the supermanifold M is a subsheaf $\mathcal{T}_M^f \subset \mathcal{T}_M$ of linear spaces of pairwise (super)commuting vector fields, such that $\mathcal{T}_M = \mathcal{O}_M \otimes \mathcal{T}_M^f$ (tensor product over the ground field).*

Sections of $\mathcal{T}_M^f$ are called flat vector fields.

b) The metric g is compatible with the structure $\mathcal{T}_M^f$, if $g(X, Y)$ is constant for flat X, Y.

In the smooth or analytic case, an affine flat structure can also be equivalently described by a complete atlas whose transition functions are affine linear, because for a maximal commuting set of linearly independent vector fields (X_a) one can find local coordinates such that $X_a = \partial/\partial x^a$, and they are defined up to a constant shift.

If a metric g is compatible with an affine flat structure, it is flat in the sense of the straightforward (not involving spinors) superextension of Riemannian geometry. The parallel transport endows $\mathcal{T}_M^f$ with the structure of a local system.

We now introduce the central definition of this book, due to B. Dubrovin.

1.3. Definition. *Let M be a supermanifold. Consider a triple $(\mathcal{T}_M^f, g, A)$ consisting of an affine flat structure, a compatible metric, and an even symmetric tensor $A: S^3(\mathcal{T}_M) \to \mathcal{O}_M$.*

Define an $\mathcal{O}_M$-bilinear symmetric multiplication $\circ = \circ_{A,g}$ on $\mathcal{T}_M$:

(1.1) $$\mathcal{T}_M \otimes \mathcal{T}_M \to S^2(\mathcal{T}_M) \xrightarrow{A'} \mathcal{T}_M^* \xrightarrow{g'} \mathcal{T}_M : X \otimes Y \to X \circ Y$$

where prime denotes a partial dualization, or equivalently,

(1.2) $$A(X, Y, Z) = g(X \circ Y, Z) = g(X, Y \circ Z).$$

This means that the metric is invariant with respect to the multiplication.

a) M endowed with this structure is called a pre-Frobenius manifold.

b) A local potential Φ for $(\mathcal{T}_M^f, A)$ is a local even function such that for any flat local tangent fields X, Y, Z

(1.3) $$A(X, Y, Z) = (XYZ)\Phi.$$

A pre-Frobenius manifold is called potential, if A everywhere locally admits a potential.

c) A pre-Frobenius manifold is called associative, if the multiplication $\circ$ is associative.

d) A pre-Frobenius manifold is called Frobenius, if it is simultaneously potential and associative.

1.3.1. Remarks. a) Denote by $\mathcal{O}_M^f$ the sheaf of local functions x on M such that Xx is constant for all local flat vector fields X. Put $\Omega_M^{1f} := d\mathcal{O}_M^f$. The sections of $\mathcal{O}_M^f$ are called *flat functions*, and the sections of Ω_M^{1f} are called *flat 1-forms*. Clearly, Ω_M^{1f} is dual to $\mathcal{T}_M^f$ as the sheaf of linear spaces.

b) If a potential Φ exists, it is unique up to a polynomial in flat local coordinates of degree ≤ 2.

c) In flat local coordinates (x^a) (1.3) becomes $A_{abc} = \partial_a \partial_b \partial_c \Phi$, and (1.2) can be rewritten as

(1.4) $$\partial_a \circ \partial_b = \sum_c A_{ab}{}^c \partial_c,$$

where

$$A_{ab}{}^c := \sum_e A_{abe} g^{ec}, \quad (g^{ab}) := (g_{ab})^{-1}.$$

Furthermore,

$$(\partial_a \circ \partial_b) \circ \partial_c = \left(\sum_e A_{ab}{}^e \partial_e\right) \circ \partial_c = \sum_{ef} A_{ab}{}^e A_{ec}{}^f \partial_f,$$

$$\partial_a \circ (\partial_b \circ \partial_c) = \partial_a \circ \sum_e A_{bc}{}^e \partial_e = (-1)^{a(b+c+e)} \sum_{ef} A_{bc}{}^e A_{ae}{}^f \partial_f$$

(1.5) $$= (-1)^{a(b+c)} \sum_{ef} A_{bc}{}^e A_{ea}{}^f \partial_f$$

(notice our abbreviated notation for signs).

Comparing the coefficients of ∂_f in (1.5), lowering the superscripts, and expressing A_{abc} through a potential, we finally see that the notion of the Frobenius manifold is a geometrization of the following highly non–linear and overdetermined system of PDE:

(1.6) $\quad \forall a,b,c,d: \quad \sum_{ef} \Phi_{abe} g^{ef} \Phi_{fcd} = (-1)^{a(b+c)} \sum_{ef} \Phi_{bce} g^{ef} \Phi_{fad}.$

They are called Associativity Equations, or WDVV (Witten–Dijkgraaf–Verlinde–Verlinde) equations.

We will now express (1.6) as a flatness condition.

1.4. Definition. *Let (M, g, A) be a pre–Frobenius manifold (we omit T_M^f in the notation, since it can be reconstructed from g). Define the following objects:*

a) The connection $\nabla_0 : T_M \to \Omega_M^1 \otimes T_M$ well determined by the condition that flat vector fields are ∇_0–horizontal.

Denote its covariant derivative along a vector field X by

$$\nabla_{0,X}(Y) = i_X(\nabla_0(Y)), \ i_X(df \otimes Z) = Xf \otimes Z.$$

b) A pencil of connections depending on an even parameter λ:

(1.7) $\quad \nabla_\lambda : T_M \to \Omega_M^1 \otimes T_M : \ \nabla_{\lambda,X}(Y) := \nabla_{0,X}(Y) + \lambda X \circ Y.$

We will call ∇_λ the structure connection of (M, g, A).

1.4.1. Remark. In flat coordinates (1.7) reads:

(1.8) $\quad \nabla_{\lambda,\partial_a}(\partial_b) = \lambda \sum_c A_{ab}{}^c \partial_c = \lambda \partial_a \circ \partial_b = (-1)^{ab} \lambda \partial_b \circ \partial_a = (-1)^{ab} \nabla_{\lambda,\partial_b}(\partial_a).$

Therefore ∇_λ has vanishing torsion for any λ. In particular, ∇_0 is the Levi–Civita (super)connection for g.

Notice that the covariant differential ∇_λ is odd. As in the pure even case, it can be naturally extended to all Ω_M^*.

1.5. Theorem. *Let ∇_λ be the structure connection of the pre–Frobenius manifold (M, g, A). Put $\nabla_\lambda^2 = \lambda^2 R_2 + \lambda R_1$ (there is no constant term since $\nabla_0^2 = 0$). Then*

a) $R_1 = 0 \iff (M, g, A)$ is potential.

b) $R_2 = 0 \iff (M, g, A)$ is associative.

Therefore (M, g, A) is Frobenius, iff ∇_λ is flat.

Proof. a) Calculating the λ–terms in

$$[\nabla_{0,\partial_a} + \lambda \partial_a \circ, \nabla_{0,\partial_b} + \lambda \partial_b \circ](\partial_c)$$

we see that $R_1 = 0$ iff $\forall a, b, c, e, \ \partial_a A_{bc}{}^e = (-1)^{ab} \partial_b A_{ac}{}^e$, or better

(1.9) $\quad \forall a, b, c, d, \quad \partial_a A_{bcd} = (-1)^{ab} \partial_b A_{acd}.$

If A is potential, this follows from (1.3). Conversely, assume (1.9). Then for all c, d, the form $\sum_b dx^b A_{bcd}$ is closed, hence locally exact by the superversion of the Poincaré lemma. Thus we can find local functions $B_{cd} = (-1)^{cd} B_{dc}$ such that

$$A_{bcd} = \partial_b B_{cd} = (-1)^{bc} \partial_c B_{bd} = (-1)^{bc} A_{cbd},$$

because A is symmetric. It follows that for all d, $\sum_c dx^c B_{cd}$ is closed. By the same reasoning, we have locally $B_{cd} = \partial_c C_d$ and finally $C_d = \partial_d \Phi$, so that $A_{bcd} = \partial_b \partial_c \partial_d \Phi$.

b) Calculating the λ^2 terms in $[\nabla_{\lambda,X}, \nabla_{\lambda,Y}](Z)$, we find that

$$R_{2,XY}(Z) = X \circ (Y \circ Z) - (-1)^{\widetilde{X}\widetilde{Y}} Y \circ (X \circ Z).$$

Hence if $\circ$ is associative, $R_2 = 0$, because $\circ$ is always (super)commutative. Conversely, if $R_2 \equiv 0$,

$$X \circ (Y \circ Z) = (-1)^{\widetilde{X}\widetilde{Y}} Y \circ (X \circ Z) = (-1)^{\widetilde{X}(\widetilde{Y}+\widetilde{Z})} Y \circ (Z \circ X)$$
$$= (-1)^{\widetilde{X}\widetilde{Y}+\widetilde{X}\widetilde{Z}+\widetilde{Y}\widetilde{Z}} Z \circ (Y \circ X) = (X \circ Y) \circ Z.$$

This finishes the proof.

The potentiality criterion $R_1 = 0$ can be somewhat rewritten. I learned the following statement from C. Hertling. Write for brevity ∇_X instead of $\nabla_{0,X}$.

1.6. Proposition. *The pre-Frobenius manifold (M, g, A) is potential iff for any local vector fields X, Y, Z*

(1.10) $\quad \nabla_X(Y \circ Z) - \nabla_Y(X \circ Z) + X \circ \nabla_Y Z - Y \circ \nabla_X Z - [X, Y] \circ Z = 0.$

Proof. The left hand side of (1.10) is a tensor. So it vanishes iff it vanishes on flat vector fields, which is equivalent to the family of identities

$$\nabla_{\partial_a}(\partial_b \circ \partial_c) = (-1)^{ab} \nabla_{\partial_b}(\partial_a \circ \partial_c).$$

But the reasoning after formula (1.9) shows that this is the same as potentiality.

1.7. Induced structures. Let $M' \to M$ be any morphism of supermanifolds which is an isomorphism locally at any point of M', for instance, an open embedding, or an unramified covering of an open submanifold. Then all structures on M described above induce the respective structures on M'.

Induction on closed submanifolds is less common. However, it is well defined in two important situations.

(i) One can always induce a (pre-) Frobenius structure from M to M_{red}. Functions on M_{red} are obtained by factoring out all nilpotents (their ideal is generated by odd local coordinates). In the de Rham complex, the differentials of odd coordinates are factored out as well. Under this reduction, the flat even coordinates by definition remain flat; the even–even part of the metric form remains the same; the new potential is the reduction of the old one. It is not difficult to check that (1.6) after reduction will become the Associativity Equations for the reduced potential.

In Quantum Cohomology, this allows us to restrict attention to the pure even-dimensional subspace if need be. However some information will be lost thereby.

(ii) In the presence of the flat identity and the Euler vector field (cf. the next section), and under certain restrictive assumptions, one can show that the induced Frobenius manifold structure exists on the closed submanifold defined by the vanishing of local coordinates corresponding to the non-integral points of spectrum: see Chapter III, Proposition 8.7.1.

This construction arises in the physical context when one compares Landau–Ginzburg models (unfolding of singularities) to the Calabi–Yau models.

1.8. Example: Cubic potentials. The simplest examples of Frobenius manifolds are furnished by potentials which are cubic polynomials in flat coordinates with constant coefficients. The algebra of tangent vectors at any point is just a commutative Frobenius (super)algebra with invariant scalar product, locally independent on the point (flat local fields identify two algebras at a neighborhood of any point). For more sophisticated examples, see §4 below and the next Chapter.

1.9. Notes and references. The WDVV–equations in flat coordinates (1.6) were introduced around 1990: see [DijVV2] and [W1]. The flat coordinates in this context parametrize certain families of two–dimensional Topological Field Theories. B. Dubrovin axiomatized the physical setting and extracted its geometric content in a series of publications, amply summarized in [D2], where the interested reader can also find the physical background.

§2. Identity, Euler field, and the extended structure connection

2.1. Definition. *Let (M, g, A) be a pre–Frobenius manifold. An even vector field e on M is called the identity, if $e \circ X = X$ for all X.*

If e exists at all, it is uniquely defined by $\circ$, hence by g and A.

Conversely, given A and e, there can exist at most one metric g making (M, g, A) a pre–Frobenius manifold with this identity:

$$g(X, Y) = A(e, X, Y).$$

This follows from (1.2). If A has a potential Φ, identity $g(X,Y) = A(e,X,Y)$ translates into a non–homogeneous linear differential equation for Φ supplementing the Associativity Equations (1.6):

(2.1) $$\forall \text{ flat } X, Y, \quad eXY\Phi = g(X, Y).$$

In fact, if $e = \sum_a e^a \partial_a$, ∂_a flat, we have from (1.3):

$$A(e, X, Y) = \sum_a e^a \partial_a XY\Phi = eXY\Phi.$$

Notice that if not A but $\circ$–multiplication with identity is given, which can be completed to a Frobenius structure, then the compatible flat metric is not necessarily unique, even up to a constant multiple. Counterexamples arise in the context of Schlesinger's equations (cf. Chapter II, Theorem 3.4.3) and in the K. Saito theory (when one encounters non–uniqueness of primitive forms). The structure obtained by forgetting the metric is more systematically studied in §5 of this Chapter.

In most (although not all) important examples e itself is *flat*. If this is the case, one can everywhere locally find a flat coordinate system $(x^0, \ldots, x^n)$ such that $e = \partial/\partial x^0 = \partial_0$, and (2.1) becomes

(2.2) $$\forall a, b, \quad \Phi_{0ab} = g_{ab}.$$

Since all g_{ab} are constants, we get

2.1.1. Corollary. *On a potential pre-Frobenius manifold with flat identity $e = \partial_0$ (in a flat coordinate system) we have modulo terms of degree ≤ 2:*
(2.3)
$$\Phi(x^0, \ldots, x^n) = \frac{1}{2} x^0 \left(\sum_{a,b \neq 0} g_{ab} x^a x^b + \sum_{a \neq 0} g_{0a} x^0 x^a + \frac{1}{3} g_{00} (x^0)^2 \right) + \Psi(x^1, \ldots, x^n).$$

2.1.2. Flat function x^0. In some situations, we can choose "the best" flat coordinate x^0 or rather its differential $\xi = dx^0$ dual to e: see 2.4.2 below. Even when such a choice is not fully constrained, it can be a useful part of the structure, and we give it a special name.

2.1.3. Definition. *A (pre-)Frobenius manifold with flat identity endowed with a flat 1-form ξ with $i_e(\xi) = 1$, or equivalently, a splitting $\mathcal{T}_M^f = \mathcal{T}_M^{f0} \oplus \langle e \rangle$, is called the manifold with split identity.*

Whenever such a ξ is chosen, the flat coordinate x^0 with $dx^0 = \xi$ is locally defined up to an additive constant. If M is simply connected, we obtain a morphism $x^0 : M \to \mathbf{A}^1$. Its relative tangent sheaf is generated by its flat subsheaf $\mathcal{T}_M^{f0}$.

If M_0 is a domain where $(x^1, \ldots, x^n)$ are local coordinates and the function Ψ from (2.3) is defined, the potential Φ from (2.3) produces the structure of the (pre-)Frobenius manifold on $M_0 \times \mathbf{A}^1$, where the two factors are the leaves of the distributions $\mathcal{T}_M^{f0}, \langle e \rangle$.

When we are speaking of flat local coordinates (x^a), we reserve whenever possible the superscript 0 to denote a flat function x^0 with properties described above.

2.1.4. Co-identity and flat function $x_0 = \eta$. The metric g identifies $\mathcal{T}_M$ and $\mathcal{T}_M^*$. We will call the 1-form which is the image of e (with reverse parity) *the co-identity* and denote it by ε. More precisely, ε is defined by

$$\forall X \in \mathcal{T}_M, \quad i_X(\varepsilon) = g(X, e).$$

If (x^a) is a local coordinate system, then

$$\varepsilon = \sum_a dx^a g(\partial_a, e).$$

Finally, if e and (x^a) are flat, then $g(\partial_a, e)$ are constant, and

(2.4) $$\varepsilon = d\eta, \quad \eta = \sum_a x^a g(\partial_a, e).$$

Thus η is obtained by lowering the superscript of x^0 in the classical tensor formalism. Notice however that η is defined independently of the choice of splitting. This function plays a very important role in the theory of semisimple Frobenius manifolds where, together with Dubrovin's *canonical coordinates*, it replaces the potential: cf. §3 below.

2.2. Euler field. We will say that an even vector field E on a manifold with flat metric (M, g) is *conformal*, if $\text{Lie}_E(g) = Dg$ for some constant D. In other words, for all vector fields X, Y we have

(2.5) $$E(g(X, Y)) - g([E, X], Y) - g(X, [E, Y]) = Dg(X, Y).$$

It follows that in flat coordinates we have $E = \sum_a E^a(x)\partial_a$ where $E^a(x)$ are polynomials of degree ≤ 1. In fact, E is a sum of infinitesimal rotation, dilation and constant shift. Hence $[E, \mathcal{T}_M^f] \subset \mathcal{T}_M^f$. Moreover, the operator

$$\mathcal{V}: \mathcal{T}_M^f \to \mathcal{T}_M^f, \quad \mathcal{V}(X) := [X, E] - \frac{D}{2}X$$

is skew symmetric:

$$\forall \text{ flat } X, Y: \quad g(\mathcal{V}(X), Y) + g(X, \mathcal{V}(Y)) = 0.$$

2.2.1. Definition. *Let E be an even vector field on a pre–Frobenius manifold (M, g, A). It is called an Euler field if it is conformal and $\mathrm{Lie}_E(\circ) = d_0 \circ$ for some constant d_0; that is, for all vector fields X, Y,*

(2.6) $\qquad [E, X \circ Y] - [E, X] \circ Y - X \circ [E, Y] = d_0 X \circ Y.$

Notice that it suffices to check (2.5) and (2.6) for X, Y in any (local) basis of $\mathcal{T}_M$, because both sides are $\mathcal{O}_M$–bilinear.

Clearly, any scalar multiple of an Euler field is also an Euler field. One can use this remark in order to normalize E by requiring that some non–vanishing eigenvalue becomes one. A convenient choice is often $d_0 = 1$, if we have reason to restrict ourselves to the $d_0 \neq 0$ case.

2.2.2. Proposition. *Let E be a conformal vector field on a Frobenius manifold (M, g, Φ). Then E is Euler, iff*

(2.7) $\qquad E\Phi = (d_0 + D)\Phi + \text{a quadratic polynomial in flat coordinates.}$

Proof. Clearly, (2.7) is equivalent to the following statement: for all flat X, Y, Z

(2.8) $\qquad XYZE\Phi = (d_0 + D)XYZ\Phi.$

Now

(2.9) $\qquad XYZE\Phi = EXYZ\Phi - XY[E, Z]\Phi - X[E, Y]Z\Phi - [E, X]YZ\Phi.$

Using (1.3), (1.2), and the fact that $[E, \mathcal{T}_M^f] \subset \mathcal{T}_M^f$, we can rewrite the right hand side of (2.9) as

$$Eg(X \circ Y, Z) - g(X \circ Y, [E, Z]) - g([E, X \circ Y], Z)$$
$$+ g([E, X \circ Y], Z) - g(X \circ [E, Y], Z) - g([E, X] \circ Y, Z).$$

The first three terms add up to $Dg(X \circ Y, Z) = DXYZ\Phi$. The last three terms add up to $d_0 g(X \circ Y, Z) = d_0 XYZ\Phi$ precisely if E is Euler.

2.3. Gradings induced by E. Now put

(2.10) $\qquad \mathcal{T}_M(r) := \{X \in \mathcal{T}_M \mid [E, X] = (r - d_0)X\}, \quad \mathcal{T}_M(*) := \bigoplus_{r \in \mathbf{C}} \mathcal{T}_M(r).$

Notice that we are considering not necessarily flat fields, and shift the eigenvalues by d_0. Similarly, put

(2.11) $\qquad \mathcal{O}_M(s) := \{f \in \mathcal{O}_M \mid Ef = sf\}, \quad \mathcal{O}_M(*) := \bigoplus_{s \in \mathbf{C}} \mathcal{O}_M(s).$

This is a graded sheaf of algebras.

2.3.1. Proposition. *On any pre-Frobenius manifold M with Euler field E, the sheaf $\mathcal{T}_M(*)$ is*

a) *A graded $\mathcal{O}_M(*)$-module.*

b) *A graded supercommutative algebra with multiplication $\circ$.*

c) *A graded Lie superalgebra with the bracket of degree $-d_0$.*

This is proved by a straightforward calculation which is left to the reader.

As a corollary, since $[E,E] = 0$, we have $E \in \mathcal{T}_M(d_0)$, so that $E^{\circ n} \in \mathcal{T}_M(nd_0)$, or

$$(2.12) \qquad [E, E^{\circ n}] = (n-1)d_0 E^{\circ n}.$$

Below we will extend this to the commutation relations between arbitrary $E^{\circ m}$ and $E^{\circ n}$ and find (for $d_0 = 1$) the algebra of vector fields on a line: see Proposition 3.6.2 for the semisimple case, and Theorem 5.6 in general.

2.4. Case of semisimple ad E. We will call the set of eigenvalues of $-\text{ad}\, E$ on $\mathcal{T}_M^f$, together with d_0 and D, *the spectrum* of E. We will say that E is *semisimple*, if ad E, acting on flat fields, is. For semisimple E we can construct many homogeneous elements of $\mathcal{O}_M(*)$ and $\mathcal{T}_M(*)$ explicitly.

Let (∂_a) be a local basis of $\mathcal{T}_M^f$ such that

$$(2.13) \qquad [\partial_a, E] = d_a \partial_a,$$

where (d_a) forms a part of the spectrum of E. (We assume here that the ground field is $\mathbf{C}$ or else complexify the tangent sheaf.) Putting $E = \sum E^a(x)\partial_a$, we find from (2.13) that $\partial_a E^b = \delta_a^b d_a$. Hence if $\partial_a = \partial/\partial x^a$, we have

$$E = \sum_{a:\, d_a \neq 0} (d_a x^a + r^a)\partial_a + \sum_{b:\, d_b = 0} r^b \partial_b.$$

By shifting x^a, we can make $r^a = 0$ for $d_a \neq 0$. Multiplying x^b by a constant, we can make $r^b = 0$ or 1 for $d_b = 0$. So finally we can choose local flat coordinates in such a way that

$$(2.14) \qquad E = \sum_{a:\, d_a \neq 0} d_a x^a \partial_a + \sum_{\text{some } b:\, d_b = 0} \partial_b.$$

Clearly, E assigns definite degrees to the following local functions:

$$(2.15) \qquad Ex^a = d_a x^a \text{ for } d_a \neq 0; \; E\exp x^b = \exp x^b \text{ or } 0 \text{ for } d_b = 0.$$

Assume now that M has an identity e. From (2.6) we get

$$(2.16) \qquad [e, E] = d_0 e.$$

Hence our notation for the spectrum will be consistent, if in the case of flat e we put $e = \partial_0$, and otherwise do not use 0 as one of the subscripts in (2.13). We have already agreed on this in 2.1.4.

In a more invariant form (2.14) can be written as

$$E = \sum_{s \in \mathbf{C}} E[s],$$

where $E[s]$ is the part of (2.13) consisting of summands with $d_a = s$ for $s \neq 0$, and the remaining summands for $s = 0$. This decomposition does not depend on the remaining arbitrariness in the choice of local coordinates.

We can now present some of our previous remarks in more concrete form. Put
$$\mathcal{T}_M^f[r] := \{X \in \mathcal{T}_M^f \mid [X, E] = rX\}.$$
(Notice the difference with (2.10).) Then condition (2.5) is equivalent to the following one:

$\mathcal{T}_M^f[d_a]$ and $\mathcal{T}_M^f[d_b]$ are orthogonal unless $d_a + d_b = D$.

In fact, (2.5) in the basis (2.13) becomes
$$\forall a, b: \; g(d_a \partial_a, \partial_b) + g(\partial_a, d_b \partial_b) = D g_{ab},$$
that is,

(2.17) $$(d_a + d_b - D) g_{ab} = 0.$$

In particular, $g(e, e) = 0$ unless $D = 2d_0$.

2.4.1. Proposition. *(2.6) is equivalent to any one of the following sets of equations written in the basis (2.13):*

(2.18) $$\forall a, b, c: \; E A_{ab}{}^c = (d_0 - d_a - d_b + d_c) A_{ab}{}^c,$$

(2.19) $$\forall a, b, c: \; E A_{abc} = (d_0 + D - d_a - d_b - d_c) A_{abc}.$$

This follows from the homogeneity of multiplication.

We now have the following supply of homogeneous functions: components of A and mixed monomials in local functions (2.15):

(2.20) $$\prod_{a: d_a \neq 0} (x^a)^{m_a} \prod_{b: d_b = 0} \exp(n_b x^b) \in \mathcal{O}_M\left(\sum_{a: d_a \neq 0} m_a d_a + \sum_{b: d_b = 0} n_b r^b\right),$$

where $m_a \in \mathbf{Z}$ and $n_b \in \mathbf{R}$ (or $\mathbf{C}$).

2.4.2. Additional constraint on x^0. When E is semisimple, we find in the notation of (2.14) for $\xi = dx^0$ that $\text{Lie}_E(\xi) = d_0 \xi$.

(In fact, $\text{Lie}_E(\xi) = d\,i_E \xi + i_E d\xi$.)

This can be used in addition to 2.1.2 to define ξ uniquely if d_0 has multiplicity one.

2.5. Extended structure connection. Let M be a pre–Frobenius manifold with a conformal vector field E. Put $\widehat{M} := M \times (\mathbf{P}_\lambda^1 \setminus \{0, \infty\})$, where $\mathbf{P}_\lambda^1$ is the completion of $\text{Spec } \mathbf{C}[\lambda, \lambda^{-1}]$. Furthermore, put $\widehat{\mathcal{T}} = \text{pr}_M^*(\mathcal{T}_M)$. If X is a vector field on M, it may be lifted to $\widehat{M}$ in two different guises: as a vector field annihilating λ, denoted again X, and as a section of $\widehat{\mathcal{T}}$, then denoted $\widehat{X}$.

Choose a constant d_0 and put $\mathcal{E} := E - d_0 \lambda \dfrac{\partial}{\partial \lambda} \in \mathcal{T}_{\widehat{M}}$. Clearly, $\widehat{X}$ for flat X span $\widehat{\mathcal{T}}$, whereas flat X and $\mathcal{E}$ span $\mathcal{T}_{\widehat{M}}$, provided $d_0 \neq 0$, which we will assume.

2.5.1. Definition. *Let M be a pre-Frobenius manifold with a conformal field E, and d_0 a non-zero constant. The extended structure connection for M is the connection $\widehat{\nabla}$ on the sheaf $\widehat{\mathcal{T}}$ on $\widehat{M}$, defined by the following formulas for its covariant derivatives: for any local vector fields $X \in \mathcal{T}_M$, $Y \in \mathcal{T}_M^f$,*

(2.21) $$\widehat{\nabla}_X(\widehat{Y}) := \lambda \widehat{X \circ Y},$$

(2.22) $$\widehat{\nabla}_{\mathcal{E}}(\widehat{Y}) := \widehat{[E, Y]}.$$

2.5.2. Theorem. *The extended structure connection is flat iff M is Frobenius and E is Euler with $\mathrm{Lie}_E (\circ) = d_0 \circ$.*

Proof. From (2.21) it follows that the vanishing of the XY-components of the curvature of $\widehat{\nabla}$ for all flat X, Y is equivalent to the flatness of the structure connection of M.

It remains to calculate the $\mathcal{E}X$-components, i.e. the expression

(2.23) $$\widehat{\nabla}_{[\mathcal{E},X]}(\widehat{Y}) - [\widehat{\nabla}_{\mathcal{E}}, \widehat{\nabla}_X](\widehat{Y})$$

for all flat X, Y. Since $[\mathcal{E}, X]$ is the lift of the flat field $[E, X]$, from (2.21) and (2.22) it follows that the first term of (2.23) is $\lambda(\widehat{[E, X] \circ Y})$. Furthermore, $\widehat{\nabla}_X(\widehat{Y}) = \lambda \widehat{X \circ Y}$, so that

$$\widehat{\nabla}_{\mathcal{E}}\widehat{\nabla}_X(\widehat{Y}) = \lambda \widehat{[E, X \circ Y]} - d_0 \lambda \widehat{X \circ Y}, \quad \widehat{\nabla}_X \widehat{\nabla}_{\mathcal{E}}(\widehat{Y}) = \lambda \widehat{X \circ [E, Y]}.$$

We see that the vanishing of this part of the curvature is equivalent to (2.6). This finishes the proof.

From (2.21) and (2.22) one can derive a formula for the covariant derivative in the λ-direction: if Y is flat, we have

$$\widehat{[E, Y]} = \widehat{\nabla}_{E - d_0 \lambda \partial/\partial \lambda}(\widehat{Y}) = \widehat{\nabla}_E(\widehat{Y}) - d_0 \lambda \widehat{\nabla}_{\partial/\partial \lambda}(\widehat{Y}) = \lambda \widehat{E \circ Y} - d_0 \lambda \widehat{\nabla}_{\partial/\partial \lambda}(\widehat{Y})$$

so that

(2.24) $$d_0 \widehat{\nabla}_{\partial/\partial \lambda}(\widehat{Y}) = \widehat{E \circ Y} - \frac{1}{\lambda} \widehat{[E, Y]}.$$

§3. Semisimple Frobenius manifolds

Let (M, g, A) be an associative pre-Frobenius manifold of dimension n. In this section and the next one we will assume that M is classical, that is, pure even.

3.1. Definition. *M is called semisimple (resp. split semisimple) if an isomorphism of the sheaves of $\mathcal{O}_M$-algebras*

(3.1) $$(\mathcal{T}_M, \circ) \overset{\sim}{\to} (\mathcal{O}_M^n, \text{componentwise multiplication})$$

exists everywhere locally (resp. globally.)

This means that in a local (resp. global) basis $(e_1, \ldots, e_n)$ of $\mathcal{T}_M$ the multiplication takes the form

$$\left(\sum f_i e_i\right) \circ \left(\sum g_j e_j\right) = \sum f_i g_i e_i,$$

and, in particular,

(3.2) $$e_i \circ e_j = \delta_{ij} e_j.$$

Such a family of idempotents is well defined up to renumbering. Another way of saying this is that a semisimple manifold comes with the structure group of $\mathcal{T}_M$ reduced to S_n. Notice that e_i are generally not flat, so that this reduction is not compatible with that induced by $\mathcal{T}_M^f$, with the structure group $GL(n)$.

Hence if M is semisimple, there exists an unramified covering of degree $\leq n!$, upon which the induced pre–Frobenius structure is split.

Denote by (ν^i) the basis of 1–forms dual to (e_i). From (1.2) and (3.2) we find

$$g(e_i, e_k) = g(e_i \circ e_i, e_k) = g(e_i, e_i \circ e_k) = \delta_{ik} g_{ii}.$$

We will denote g_{ii} by η_i. We see that the symmetric 2–form representing g is diagonal in the basis (ν^i):

(3.3) $$g = \sum_i \eta_i (\nu^i)^2.$$

Moreover, according to (1.2), $A(e_i, e_j, e_k) = \delta_{ij} \delta_{ik} \eta_i$, so that the symmetric 3–form representing A is diagonal with the same coefficients:

(3.4) $$A = \sum_i \eta_i (\nu^i)^3.$$

Finally, $e := \sum_i e_i$ is the identity in $(\mathcal{T}_M, \circ)$, and the co–identity, defined in 2.1.2, nicely complements (3.3) and (3.4):

(3.5) $$\varepsilon = \sum_i \eta_i \nu^i.$$

Thus Definition 3.1 can be restated as follows:

3.2. Definition. *The structure of the semisimple pre–Frobenius manifold on M is determined by the following data:*

a) A reduction of the structure group of $\mathcal{T}_M$ to S_n, specified by a choice of local bases (e_i) and dual bases (ν^i).

b) A flat metric g, diagonal in (e_i), (ν^i).

c) A diagonal cubic tensor A with the same coefficients as g.

Associativity of $(\mathcal{T}_M, \circ)$ is automatic in both descriptions. However, potentiality (and the flatness of g which we postulated) are non–trivial conditions.

3.3. Theorem. *The structure described in Definition 3.2 is Frobenius iff the following conditions are satisfied:*

a) $[e_i, e_j] = 0$, or equivalently, $e_i = \partial/\partial u^i$, $\nu^i = du^i$ for a local coordinate system (u^i), called canonical.

b) $\eta_i = e_i \eta$ for a local function η defined up to addition of a constant. Equivalently, ε is closed.

We will call η the *metric potential* of this structure. (Sometimes this term refers to h such that $g_{ab} = \partial_a \partial_b h$; our meaning is different.)

Canonical coordinates are defined up to renumbering and constant shifts.

Proof. Let ∇_λ be the structure connection of the pre–Frobenius manifold M. According to Theorem 1.5, M is Frobenius iff the curvature ∇_λ^2 vanishes, i.e. iff

(3.6) $$\forall i,j,k: \quad [\nabla_{\lambda,e_i},\nabla_{\lambda,e_j}](e_k) = \nabla_{\lambda,[e_i,e_j]}(e_k).$$

Since M is associative, and since we assumed that g is flat, we have to worry only about the λ–linear terms in (3.6). Let us start by introducing the Riemannian connection coefficients of g for the basis e_k:

(3.7) $$\nabla_{0,e_i}(e_k) = \sum_q \Gamma_{ik}{}^q e_q.$$

Since $\nabla_{\lambda,X} = \nabla_{0,X} + \lambda X\circ$ (cf. (1.7)), the left hand side of (3.6) produces the λ–terms

$$(\nabla_{0,e_i} + \lambda e_i \circ)(\nabla_{0,e_j} + \lambda e_j \circ)(e_k) - (i \leftrightarrow j)$$
$$= \lambda e_i \circ \sum_q \Gamma_{jk}{}^q e_q + \lambda \sum_q \delta_{jk}\Gamma_{ik}{}^q e_q - (i \leftrightarrow j) + \ldots$$

(3.8) $$= \lambda \sum_q (\delta_{iq}\Gamma_{jk}^q + \delta_{jk}\Gamma_{ik}^q - \delta_{jq}\Gamma_{ik}^q - \delta_{ik}\Gamma_{jk}^q)e_q + \cdots.$$

Now introduce the Lie coefficients

$$[e_i,e_j] = \sum_q f_{ij}{}^q e_q.$$

The λ–terms in the right hand side of (3.6) amount to

$$\nabla_{\lambda,[e_i,e_j]}(e_k) = \lambda \sum_q f_{ij}{}^q e_q \circ e_k + \cdots = \lambda f_{ij}{}^k e_k + \cdots.$$

But the coefficient of e_k in (3.8) vanishes. Therefore, if M is Frobenius, then (3.6) is satisfied, so that $f_{ij}{}^k = 0$. Hence e_i pairwise commute, and local canonical coordinates u^i do exist.

Moreover, the left hand side of (3.6) vanishes. Again, it suffices to investigate the meaning of this, looking only at λ–linear terms.

Recall that for any metric $g = \sum g_{ij}du^i du^j$ the coefficients of the Levi–Civita connection are given by the formulas

$$\Gamma_{ij}{}^k = \sum_l \Gamma_{ijl}g^{lk}, \quad \Gamma_{ijk} = \frac{1}{2}(e_i g_{jk} - e_k g_{ij} + e_j g_{ki}).$$

The non–vanishing connection coefficients of $g = \sum \eta_i (du^i)^2$ are $(i \neq j)$:

$$\Gamma_{ii}{}^i = \frac{1}{2}\eta_i^{-1}e_i\eta_i, \quad \Gamma_{ii}{}^j = -\frac{1}{2}\eta_j^{-1}e_j\eta_i,$$

(3.9) $$\Gamma_{ij}{}^i = \Gamma_{ji}{}^i = \frac{1}{2}\eta_i^{-1}e_j\eta_i.$$

Hence putting $\nabla := \nabla_0$ (the Levi–Civita connection), $\nabla_i := \nabla_{0,e_i}$, we have

$$\nabla_i(e_i) = \frac{1}{2}\eta_i^{-1}e_i\eta_i \cdot e_i - \sum_{j\neq i}\frac{1}{2}\eta_j^{-1}e_j\eta_i \cdot e_j,$$

(3.10) $$\nabla_i(e_j) = \frac{1}{2}\eta_i^{-1} e_j \eta_i \cdot e_i + \frac{1}{2}\eta_j^{-1} e_i \eta_j \cdot e_j.$$

Now, the vanishing of the λ–terms in the left hand side of (3.6) means that

(3.11) $$\forall i,j,k: \quad e_i \circ \nabla_j(e_k) + \nabla_i(e_j \circ e_k) = (i \leftrightarrow j).$$

Using (3.10), one checks that (3.11) is identically satisfied for $i = j$ and for $i \neq j \neq k \neq i$, whereas the case $i \neq j = k$ gives

(3.12) $$e_i \eta_j = e_j \eta_i.$$

The same condition is obtained for $k = i \neq j$. It follows that $\eta_i = e_i \eta$ for some η, defined at least locally.

Reading this argument in the reverse direction, we see that if a) and b) are satisfied, then ∇_λ is flat, and M is Frobenius.

3.4. The Darboux–Egoroff equations. Theorem 3.3 establishes a (not very explicit) equivalence between the following functional spaces on M (modulo self–evident equivalence):

a) Flat coordinates $(x^1, \ldots, x^n)$, flat metric g_{ab}, function $\Phi(x)$ satisfying the Associativity Equations (1.6), and semisimplicity.

b) Canonical coordinates $(u^1, \ldots, u^n)$, function $\eta(u)$ such that the metric $g = \sum e_i \eta (du^i)^2$ is flat, where $e_i = \partial/\partial u^i$.

The constraints on η, implicit in b), are called the Darboux–Egoroff equations. In order to write them down explicitly, let us introduce the rotation coefficients of the potential metric:

(3.13) $$\gamma_{ij} := \frac{1}{2} \frac{\eta_{ij}}{\sqrt{\eta_i \eta_j}},$$

where, as before, $\eta_i = e_i \eta$, $\eta_{ij} = e_i e_j \eta$.

3.4.1. Proposition. *The diagonal potential metric $g = \sum e_i \eta (du^i)^2$ is flat iff $\forall k \neq i \neq j \neq k$:*

(3.14) $$e_k \gamma_{ij} = \gamma_{ik} \gamma_{kj}$$

and

(3.15) $$e \gamma_{ij} = 0.$$

Proof. This is established by a straightforward calculation, complementing that in the proof of Theorem 3.3. In fact, we now want to make explicit the condition $\nabla^2 = 0$, where ∇ is the Levi–Civita connection. So we return to (3.6) at $\lambda = 0$, i.e.

$$\nabla_i \nabla_j(e_k) = \nabla_j \nabla_i(e_k).$$

Non–vanishing curvature components can occur only for $i \neq j$. Calculating them directly we arrive at (3.14) and (3.15).

3.5. Proposition. *Let e be the identity, and ε the co-identity of the semisimple Frobenius manifold. Then*

a) $\varepsilon = d\eta$, *where η is the metric potential.*

b) e is flat iff for all i, $e\eta_i = 0$, or equivalently, $e\eta = g(e,e) = $ const. This condition is satisfied in the presence of an Euler field with $D \neq 2d_0$ (see (2.5), (2.16), (2.17)).

c) If e is flat and (x^a) is a flat coordinate system, then

$$\eta = \sum_a x^a g(\partial_a, e) + \text{const.} \tag{3.16}$$

The formula (3.16) shows that in the passage from the (x^a, Φ)-description to the (u^i, η)-description the main information is encoded in the transition formulas $u^i = u^i(x)$, at least in the presence of flat identity.

Proof. The first claim follows from (3.5) and Theorem 3.3 b).

The second claim can be obtained directly from (3.10). We have

$$\nabla_i(e) = \nabla_i(e_i + \sum_{j \neq i} e_j) = \frac{1}{2}\frac{e\eta_i}{\eta_i} \cdot e_i.$$

These derivatives vanish iff $e\eta = $ const. But $e\eta = \sum \eta_i = g(e,e)$. From (2.17) it follows that $g(e,e) = 0$ if $D \neq 2d_0$.

Finally, (3.16) is (2.4).

Notice that the equations $e\eta_i = 0$ imply (3.15).

We will now see that, like the identity, the Euler field is almost uniquely defined by the canonical coordinates, if it exists at all.

3.6. Theorem. *Let E be a vector field on the semisimple Frobenius manifold M, and d_0 a constant.*

a) We have $\text{Lie}_E(\circ) = d_0(\circ)$, iff

$$E = d_0 \sum_i (u^i + c^i)e_i, \tag{3.17}$$

where c^i are some constants.

b) For the field of the form (3.17) and a constant D, we have $\text{Lie}_E(g) = Dg$ iff for all i, $E\eta_i = (D - 2d_0)\eta_i$, or equivalently

$$E\eta = (D - d_0)\eta + \text{const.} \tag{3.18}$$

Thus in the presence of a non-vanishing Euler field we may and will normalize the canonical coordinates so that $E = d_0 \sum u^i e_i$.

Proof. a) Put $E = \sum_i E^i e_i$ and write (2.6) for $X = e_k, Y = e_l$. Since $[E, e_k] = -\sum_i e_k(E^i) \cdot e_i$, we get $e_k(E^i) = d_0 \delta_{ik}$, so that $E^i = d_0(u^i + c^i)$.

b) Likewise, (2.5) for $X = e_i, Y = e_j$ is identically satisfied for $i \neq j$, and is equivalent to $E\eta_i = (D - 2d_0)\eta_i$ for $i = j$. Since $\eta_i = e_i\eta$ and $Ee_i = e_iE - d_0e_i$, this is the same as (3.18).

3.6.1. Grading. The semisimplicity of $\text{ad } E$ on T_M^f does not seem to have a good alternate formulation. However, if it holds, then the grading of functions and vector fields defined in 2.3 becomes especially simple in the canonical coordinates. For instance, let $d_0 = 1$; then $Ef = sf$ iff $f(\lambda u^1, \ldots, \lambda u^n) = \lambda^s f(u^1, \ldots, u^n)$.

Finally, we can complete the commutation relations (2.12).

3.6.2. Proposition. *If $d_0 = 1$, then*

$$[E^{\circ m}, E^{\circ n}] = (n - m) E^{\circ (m+n-1)} \tag{3.19}$$

for $m, n \geq 0$ everywhere on M, and for arbitrary integral m, n outside of $\bigcup_i (u^i = 0)$, that is, exactly where E is $\circ$-invertible.

In fact, from (3.2) one sees that $E^{\circ m} = \sum u_i^m e_i$.

3.7. A pencil of flat metrics. Equations (3.14) are stable with respect to a semigroup of coordinate changes. Namely, let f_i be arbitrary functions of one variable such that $\check{u}^i := f_i(u^i)$ form a local coordinate system, $\check{e}_i = \partial/\partial \check{u}^i$, $\check{\eta}_i = \check{e}_i \eta$, etc.

3.7.1. Proposition. *If (e_i, γ_{ij}) satisfy (3.14), then $(\check{e}_i, \check{\gamma}_{ij})$ satisfy (3.14) as well.*

Proof. The rotation coefficients of $\check{g} := \sum_i \check{e}_i \eta (d\check{u}^i)^2$ are (cf. (3.13))

$$\check{\gamma}_{ij} = \frac{1}{2} \check{e}_i \check{e}_j \eta (\check{e}_i \eta \, \check{e}_j \eta)^{-\frac{1}{2}} = \gamma_{ij} (f_i'(u^i) f_j'(u^j))^{-\frac{1}{2}}.$$

Hence for $k \neq i \neq j \neq k$ we have, in view of (3.14):

$$\check{e}_k \check{\gamma}_{ij} = \gamma_{ik} \gamma_{kj} (f_i'(u^i) f_j'(u^j))^{-\frac{1}{2}} f_k(u^k)^{-1}$$

and

$$\check{\gamma}_{ik} \check{\gamma}_{kj} = \gamma_{ik} \gamma_{kj} (f_i'(u^i) f_j'(u^j))^{-\frac{1}{2}} f_k(u^k)^{-1}$$

so that $\check{\gamma}_{ij}$ satisfy (3.14).

In order to satisfy (3.15) as well, we will have to restrict ourselves to the one-parametric family of local coordinate changes

$$\check{u}^i = \log(u^i - \lambda), \quad \check{e}_i = (u^i - \lambda) e_i, \quad \check{g}_\lambda = \sum_i (u^i - \lambda)^{-1} e_i \eta (du^i)^2 \tag{3.20}$$

which make sense on $M_\lambda := \{x \in M \mid \forall i, \, u^i \neq \lambda\}$.

3.7.2. Theorem. *Let M be a semisimple Frobenius manifold with canonical coordinates (u^i) and metric potential η. Then the following statements are equivalent.*

a) For all λ, the structure (3.20) is semisimple Frobenius on M_λ.

b) The same for a particular value of λ.

c) For all $i \neq j$,

$$\sum_k u^k e_k \gamma_{ij} = -\gamma_{ij}. \tag{3.21}$$

Moreover, (3.21) is satisfied if $E = \sum_k u^k e_k$ is the Euler field on M with $d_0 = 1$.

Notice that generally $\check{e} = \sum \check{e}_k$ is not flat for $\check{g}_\lambda$ and $\check{E} = \sum \check{u}^k \check{e}_k$ is not an Euler field.

§3. SEMISIMPLE FROBENIUS MANIFOLDS

Proof. Let us start by deducing (3.21). If E is the Euler field with $d_0 = 1$, we have $\sum_k u^k \eta_{ik} = (D-2)\eta_i$ (see Theorem 3.6 b)). Applying e_j we obtain $\sum_k u^k \eta_{ijk} = (D-3)\eta_{ij}$. Hence

$$E\gamma_{ij} = \sum_k u^k e_k \gamma_{ij} = \sum_k u^k \left[\frac{1}{2} \frac{\eta_{ijk}}{\sqrt{\eta_i \eta_j}} - \frac{1}{4} \frac{\eta_{ij}\eta_{ik}}{\eta_i \sqrt{\eta_i \eta_j}} - \frac{1}{4} \frac{\eta_{ij}\eta_{jk}}{\eta_j \sqrt{\eta_i \eta_j}} \right]$$

$$= -\frac{1}{2} \frac{\eta_{ij}}{\sqrt{\eta_i \eta_j}} = -\gamma_{ij}.$$

Now we turn to the Darboux–Egoroff equations. We know from the assumptions and Proposition 3.7.1 that (3.14) is satisfied both for g and $\check{g}_\lambda$. The second half (3.15) in this situation is equivalent to

$$(3.22) \qquad \forall i \neq j, \quad \sum_{k \neq i,j} \gamma_{ik}\gamma_{kj} = -(e_i + e_j)\gamma_{ij},$$

so that it remains to see the meaning of (3.22) now written for $\check{\gamma}_{ij}, \check{e}_j$.

We have, using (3.20), $\check{\gamma}_{ij} = \gamma_{ij}(u^i - \lambda)^{1/2}(u^j - \lambda)^{1/2}$. Hence for $i \neq j$

$$\sum_{k \neq i,j} \check{\gamma}_{ik}\check{\gamma}_{kj} = \left[\sum_{k \neq i,j} \gamma_{ik}\gamma_{kj}(u^k - \lambda) \right] (u^i - \lambda)^{1/2}(u^j - \lambda)^{1/2}$$

$$= \left[\sum_{k \neq i,j} u^k \gamma_{ik}\gamma_{kj} + \lambda(e_i + e_j)\gamma_{ij} \right] (u^i - \lambda)^{1/2}(u^j - \lambda)^{1/2}$$

$$= \left[\sum_{k \neq i,j} u^k e_k \gamma_{ij} + \lambda(e_i + e_j)\gamma_{ij} \right] (u^i - \lambda)^{1/2}(u^j - \lambda)^{1/2}$$

$$(3.23) \qquad = \left[E\gamma_{ij} - (u^i - \lambda)e_i\gamma_{ij} - (u^j - \lambda)e_j\gamma_{ij} \right] (u^i - \lambda)^{1/2}(u^j - \lambda)^{1/2}.$$

On the other hand,

$$-(\check{e}_i + \check{e}_j)\check{\gamma}_{ij} = -\left[(u^i - \lambda)e_i + (u^j - \lambda)e_j \right] \left[\gamma_{ij}(u^i - \lambda)^{1/2}(u^j - \lambda)^{1/2} \right]$$

$$(3.24) \qquad = -\left[\gamma_{ij} + (u^i - \lambda)e_i\gamma_{ij} + (u^j - \lambda)e_j\gamma_{ij} \right] (u^i - \lambda)^{1/2}(u^j - \lambda)^{1/2}.$$

Comparing (3.23) and (3.24) one sees that their coincidence for one or for all values of λ is equivalent to (3.21). This finishes the proof.

3.7.3. Remarks. a) If E is Euler, the metric $\check{g}_\lambda$ in (3.20) can be written in coordinate free form:

$$(3.25) \qquad \check{g}_\lambda(X, Y) = g((E - \lambda)^{-1} \circ X, Y).$$

In fact (3.25) is flat on any Frobenius manifold with semisimple Euler field on it, non–necessarily semisimple: cf. [D2].

b) The inverse metrics $\check{g}_\lambda^t$ on the cotangent sheaf form a pencil of flat metrics with two marked points. Conversely, given such a pencil and two metrics g, h in it, we can define the spectrum of such data: zeroes of $\det(g - uh)$. If the spectrum (u^i) forms a local coordinate system, the pair (u^i, h) has a chance to define the Frobenius structure: we have to check the potentiality of h written in u^i-coordinates, which is equivalent to the flatness of the structural connection: see Theorem 3.3.

3.8. Summary. We now briefly summarize the two descriptions of semisimple Frobenius manifolds, stressing their parallelism.

WDVV picture

Flat coordinates $(x^0, \ldots, x^{n-1})$, up to affine transformations, can be partially normalized in the presence of E.

Metric with constant coefficients $\sum g_{ab} dx^a dx^b$.

Potential $\Phi(x)$ satisfying the WDVV-equations (1.6), defined up to adding a quadratic polynomial in (x^a).

Flat identity $e = \partial_0$, additional equation $\partial_0 \Phi_{ab} = g_{ab}$.

Euler field $E = \sum E^a(x) \partial_a$, where E^a are of degree ≤ 1. Additional equation $E\Phi = (D + d_0)\Phi$ plus quadratic terms.

Darboux–Egoroff picture

Canonical coordinates $(u^1, \ldots, u^n)$, up to renumbering and constant shifts. Shifts can be fixed in the presence of E.

Diagonal potential metric $g = \sum_i e_i \eta (du^i)^2$, $e_i = \partial/\partial u^i$.

The metric potential $\eta(u)$ satisfying the Darboux–Egoroff equations (3.14), (3.15), and defined up to adding a constant.

Flat identity $e = \sum_i e_i$, additional equation $e\eta = $ const.

Euler field $E = d_0 \sum_i u^i e_i$. Additional equation $E\eta = (D - d_0)\eta + $ const.

Passage from WDVV to Darboux–Egoroff

In the presence of an Euler field and a flat identity:

$(u^1, \ldots, u^n) = $ the spectrum of $E\circ$ acting upon $\mathcal{T}_M$.

Metric potential $\eta = \sum_a x^a g(\partial_a, e)$.

3.9. A problem. It would be important to generalize the notion of semisimplicity to supermanifolds. Here are some scattered observations suggesting that there might be several different versions of it.

a) The main justification for considering Frobenius supermanifolds is the fact that quantum cohomology (theory of Gromov–Witten invariants) provides, for any projective algebraic or symplectic V, such a structure on an open (or formal) subspace of the conventional cohomology $H^*(V, \mathbf{C})$ considered as a linear superspace. Not many manifolds have pure even-dimensional cohomology, so we need odd coordinates.

b) If we look at Definition 1.2 from the vantage point of, say, supergravity, we will be tempted to replace the metric (g_{ab}) by a more refined structure. The standard nucleus of such a structure consists of a pair of pure odd integrable distributions $\mathcal{T}_l, \mathcal{T}_r \subset \mathcal{T}_M$ such that the supercommutator induces a maximally non-degenerate map $\mathcal{T}_l \otimes \mathcal{T}_r \to \mathcal{T}_M/(\mathcal{T}_l \oplus \mathcal{T}_r)$. There are two drawbacks to this. First, such a structure seems to be nowhere in sight in quantum cohomology. Second, in its natural habitat it is complemented by new constraints depending on dimension, so that there is no dimension–independent generalization of Riemannian geometry along these lines.

If one decides against this option, one should keep in mind alternative geometries peculiar to supergeometry, for instance, (a curved version of) Π–symmetry, where Π is the parity switch. For example, in the picture of Calabi–Yau Mirror Symmetry the cohomology spaces of mirror threefolds V, V' are, roughly speaking, connected by $H(V) = \Pi H(V')$.

Proceeding in this direction, we will have to rethink the ways to construct generating functions from Gromov–Witten invariants.

c) Finally, an extension of the notion of semisimplicity is suggested by the dominant role of the Euler field E, or rather of the Lie algebra spanned by $E^{\circ n}$. One can imagine a structure, consisting of a supermanifold M, a representation of the Neveu–Schwarz (or Ramond) Lie superalgebra in T_M, and a superversion of the equations $e\eta = \text{const}$, $E\eta = (D - d_0)\eta + \text{const}$.

One version of the definition of semisimple Frobenius supermanifolds was suggested in [MM]. It generalizes the reduction of the theory to that of the special solutions of Schlesinger's equations which will be treated in the next chapter. The superization of Schlesinger's equations is a relatively straightforward task, since one can rely upon the theory of SUSY–structures on algebraic curves developed earlier. The fact that it matches the Frobenius manifolds input so well results from a series of miraculous coincidences.

§4. Examples

4.1. Dimension one. Let M be a connected simply connected one–dimensional manifold; for definiteness, complex analytic.

The structure of a pre–Frobenius manifold on M is given by an arbitrary pair (∂, φ) where ∂ is a vector field without zeroes and φ is a function:

$$T_M^f := \mathbf{C}\partial, \ g(\partial, \partial) = 1, \ \partial \circ \partial = \varphi\partial.$$

Two pairs (∂, φ), (∂', φ') define the same structure iff they coincide or differ by a common sign.

Such a structure is automatically associative and potential, hence Frobenius. Let M_0 be the complement to the zeroes of φ. On M_0 there is an identity $e = \varphi^{-1}\partial$, which is flat iff $\partial\varphi = 0$. If $\partial = d/dx$, co–identity is $\varepsilon = \varphi(x)dx$.

M_0 is also the domain of semisimplicity. Solving the equation $e = d/du$ for u, we get $u = \int \varphi(x)dx = \int \varepsilon$.

A definite choice of u is equivalent to the choice of the would–be Euler field $E = u d/du$ with $d_0 = 1$. A metric potential is $\eta = u$, hence $E\eta = u$ so that (3.19) is satisfied with $D = 2$. Even if e is not flat, we have $[\partial, E] = \partial$, so that E is actually an Euler field.

This rather dull picture will give rise to quite non–trivial problems in the context of *formal* Frobenius manifolds when we introduce and calculate the operation of tensor product on them: see Chapter III, §6.

4.2. Dimension two. We will give a local classification of two–dimensional Frobenius structures with flat identity and a semisimple Euler field with $d_0 = 1$. The multiplication ∘ in this situation is automatically associative, so that the WDVV–equations are empty, and it remains to find all potentials satisfying the equations (2.2) and (2.7).

The final answer depends on the spectrum of E.

First, let $(d_0 = 1, d_1)$ be the spectrum of $-\text{ad}\, E$ on $\mathcal{T}_M^f$, (∂_0, ∂_1) the respective flat eigenvectors, and $e = \partial_0$. The classification starts branching depending on whether $d_1 \neq 0$ or $d_1 = 0$: this is our first *critical value* of d_1. We choose flat coordinates (x^0, x^1) such that (cf. 2.4)

(4.1a) $$d_1 \neq 0: \quad E = x^0 \partial_0 + d_1 x^1 \partial_1$$

(x^1 defined up to multiplication by a constant),

(4.1b) $$d_1 = 0: \quad E = x^0 \partial_0 + 2 \partial_1$$

(x^1 defined up to addition of a constant), or else

(4.1c) $$d_1 = 0: \quad E = x^0 \partial_0.$$

From (2.17) one sees that a compatible non–vanishing flat metric can exist only if $D \in \{2, 1 + d_1, 2d_1\}$, and for $d_1 \neq 1$ a non–degenerate flat metric exists only if $D = 1 + d_1$, so that $D = 1 + d_1$ always.

If $d_1 \neq 1$, we have $g_{00} = g_{11} = 0$, $g_{01} = \gamma \neq 0$; we can make $\gamma = 1$ by rescaling x_1. If $d_1 = 1$ (this is the second critical value of d_1), then (g_{ab}) can be an arbitrary symmetric non–degenerate matrix.

From (2.3) we obtain

(4.2) $$\Phi(x^0, x^1) = \frac{1}{2} x^0 (g_{11}(x^1)^2 + g_{01} x^0 x^1 + \frac{1}{3} g_{00}(x^0)^2) + \Psi(x^1),$$

and from (2.7)

(4.3) $$E\Phi = (d_1 + 2)\Phi + \text{a quadratic polynomial}.$$

In the case (4.1c) this leads to

$$x^0 \partial_0 [\frac{\gamma}{2}(x^0)^2 x^1 + \Psi(x^1)] = \gamma(x^0)^2 x^1 + 2\Psi(x^1) + \text{a quadratic polynomial}$$

so that we can take

(4.4) $$d_1 = 0, E = x^0 \partial_0: \quad \Phi = \frac{\gamma}{2}(x^0)^2 x^1.$$

The case (4.1b) leads to the equation

$$\partial_1 \Psi(x^1) = \Psi(x^1) + \text{a quadratic polynomial in } x^1$$

so that we can take, after rescaling x^1,

(4.5) $$d_1 = 0, E = x^0 \partial_0 + 2\partial_1: \quad \Phi = \frac{\gamma}{2}(x^0)^2 x^1 + e^{x^1}.$$

In the case (4.1a) with $d_1 = 1$, Φ can be reduced to a cubic form with constant coefficients:

(4.6)
$$d_1 = 0, E = x^0 \partial_0 + x^1 \partial_1: \quad \Phi = \frac{1}{2} x^0 (g_{11}(x^1)^2 + g_{01} x^0 x^1 + \frac{1}{3} g_{00}(x^0)^2) + c(x^1)^3.$$

Finally, the case (4.1a) produces two more critical values $d_1 = \pm 2$:

(4.7) $\qquad d_1 = -2,\ E = x^0 \partial_0 - 2x^1 \partial_1: \quad \Phi = \frac{1}{2}(x^0)^2 x^1 + c \log x^1,$

(4.8) $\qquad d_1 = 2,\ E = x^0 \partial_0 + 2x^1 \partial_1: \quad \Phi = \frac{1}{2}(x^0)^2 x^1 + c(x^1)^2 \log x^1,$

(4.9) $\qquad d_1 \neq 0, 1, \pm 2,\ E = x^0 \partial_0 + d_1 x^1 \partial_1: \quad \Phi = \frac{1}{2}(x^0)^2 x^1 + c(x^1)^{(2+d_1)/d_1}.$

4.3. Dimension three: A promise. This is the first dimension where the Associativity Equations become non–empty even in the presence of the flat identity. The beautiful theory of three–dimensional semisimple Frobenius manifolds essentially reduces their study to that of a subfamily of the sixth Painlevé equations. We will address this connection in Chapter II, §5.

4.4. Quantum cohomology: Brief encounter. Let V be a smooth projective algebraic manifold over $\mathbf{C}$ (another version of the theory exists for compact symplectic manifolds).

Denote by H the cohomology space $H^*(V, \mathbf{C})$ considered as *a complex analytic linear supermanifold*. We endow H with its natural flat structure $\mathcal{T}_H^f$, Poincaré form g, and two vector fields e, E which can be described as follows. First, H *as a linear space* can be identified with global flat vector fields. We denote by e the vector field corresponding to the identity in the cohomology ring, that is, the dual fundamental class of V. Second, $-\operatorname{ad} E$ is the semisimple operator on $\mathcal{T}_H^f$, with eigenvalue $1 - p/2$ on $H^p(X, \mathbf{C})$: this determines the first summand in the decomposition (2.14). The second (flat) one is the anticanonical class of V.

Explicitly, let $H^*(V, \mathbf{C}) = \bigoplus \mathbf{C}\Delta_a$, $\Delta_a \in H^{|\Delta_a|}(V, \mathbf{C})$, Δ_0 the dual fundamental class. Then the coordinates (x^a) in this basis are global flat coordinates on H, and

(4.10) $\qquad e = \partial_0,\ E = \sum_a (1 - \frac{|\Delta_a|}{2}) x^a \partial_a + \sum_{b:\ |\Delta_b|=2} r^b \partial_b,$

where r^b are defined by

(4.11) $\qquad c_1(\mathcal{T}_V) = -K_V = \sum_{b:\ |\Delta_b|=2} r^b \Delta_b.$

Moreover, $g_{ab} = \int_V \Delta_a \wedge \Delta_b$ (we imagine cohomology classes as differential forms, and use wedge for the cup product).

The relations (2.5) (resp. (2.16)) are satisfied with $D = 2 - \dim_{\mathbf{C}} V$ (resp. $d_0 = 1$) so that the total spectrum of E is

(4.12) $\qquad d_0 = 1,\ d_a = 1 - \frac{|\Delta_a|}{2}$ of multiplicity $\dim H^{|\Delta_a|}$, $D = 2 - \dim_{\mathbf{C}} V$.

The remaining and most important structure is the potential Φ. The theory of Gromov–Witten invariants furnishes (at least for manifolds with $K_V \leq 0$) *a formal series* $\Phi(x)$ in flat coordinates satisfying all the axioms of Frobenius structure, with flat identity e and the Euler field E, described above. Moreover, Φ can actually be represented as a series in E–homogeneous monomials (2.20) (notice that they are exponential in codimension two coordinates), with non–negative integers m_a and n_b, of E–degree $d_0 + D = 3 - \dim_{\mathbf{C}} V$. Coefficients of this series are certain numerical invariants of the space of stable maps of pointed curves of genus 0 to V.

If Φ converges in a subdomain $M \subset H$, it induces a Frobenius manifold structure on M. Generally, its maximal analytic continuation to an unramified covering of a subdomain of H should be considered as *the* Frobenius manifold representing the quantum cohomology of V.

One approach to the study of Φ consists in the identification of M (physicists' A–model) with a Frobenius manifold constructed by other methods, e.g. from isomonodromic deformations or periods of the families of algebraic manifolds (physicists' B–model.) This is a part of the general Mirror Program. The very first step in such an identification is the comparison of spectra. The famous $h_V^{11} = h_{\tilde{V}}^{12}$ mirror symmetry relation for the Calabi–Yau threefolds expresses such an identification.

As an elementary exercise, let us guess which of the manifolds (4.5)–(4.9) can represent quantum cohomology. Only $\mathbf{P}^1$ has two–dimensional pure even cohomology space, and $-K_{\mathbf{P}^1}$ has degree two, so that E must be of the type (4.1b). In fact, the potential of (the quantum cohomology of) $\mathbf{P}^1$ is given by (4.5) with $\gamma = 1/2$ in the natural basis.

We conclude this brief discussion by describing explicitly the potential Φ for all projective spaces $\mathbf{P}^r$. Put $\Delta_a =$ the dual class of the codimension a hyperplane, $\gamma = \sum x^a \Delta_a$, $(\gamma^3) =$ the triple self–intersection index. Then

$$(4.13) \qquad \Phi^{\mathbf{P}^r}(x) = \frac{1}{6}(\gamma^3) + \sum_{d, n_a \geq 0} N(d; n_2, \ldots, n_r) \frac{(x^2)^{n_2} \ldots (x^r)^{n_r}}{n_2! \ldots n_r!} e^{dx^1},$$

where $N(d; n_2, \ldots, n_r)$ is the number of rational curves of degree d in $\mathbf{P}^r$ intersecting n_a hyperplanes of codimension a in general position. This number (suitably interpreted in certain boundary cases) can be non–zero only for $\sum_a n_a(a-1) = (r+1)d + r - 3$ which is equivalent to the grading equation (2.7). The Associativity Equations (1.6) follow from a rather sophisticated analysis of degenerations. They allow us to calculate recursively all $N(d; n_1, \ldots, n_r)$ starting with a single number $N(1; 0, \ldots, 0, 2) = 1$ (there is only one line passing through two different points).

In fact, the recursive relations obtained from (1.6) form such an overdetermined system that it is not obvious how to prove the existence of a solution to (1.6) and (2.7) formally (i.e., without using the geometric interpretation). For a roundabout proof, see Ch. II, 4.2, below. The cases $r = 1$ and $r = 2$ are exceptional: we have respectively

$$(4.14) \qquad \Phi^{\mathbf{P}^1}(x\Delta_0 + z\Delta_1) = \frac{1}{2}x^2 z + e^z - (1 + z + \frac{z^2}{2}),$$

$$(4.15) \qquad \Phi^{\mathbf{P}^2}(x\Delta_0 + y\Delta_1 + z\Delta_2) = \frac{1}{2}(xy^2 + x^2 z) + \sum_{d=1}^{\infty} N(d) \frac{z^{3d-1}}{(3d-1)!} e^{dy}.$$

Here the Associativity Equations are equivalent to an explicit recursive formula for $N(d)$ (see Introduction, (0.19)).

All these Frobenius structures are generically (or formally) semisimple. Notice that in the semisimple case the potential η of the Poincaré metric is simply the linear function $\eta: H^*(V, \mathbf{C}) \to \mathbf{C}: \eta(\gamma) = \int_V \gamma$. This is a restatement of (2.4).

For further information, see II.4, III.5, and Chapter VI.

4.5. Space of polynomials. The following beautiful example furnishes another series of semisimple Frobenius manifolds of arbitrary dimension. This construction, due to B. Dubrovin and K. Saito, admits various generalizations.

Consider the n-dimensional affine space $\mathbf{A}^n$ with coordinate functions $a_1, \ldots, a_n$. Identify $\mathbf{A}^n$ with the space of polynomials $p(z) = z^{n+1} + a_1 z^{n-1} + \cdots + a_n$. Denote by $\pi: \tilde{\mathbf{A}}^n \to \mathbf{A}^n$ the covering space of degree $n!$ whose fiber over a point $p(z)$ consists of total orderings of the roots of $p'(z)$. In other words, $\tilde{\mathbf{A}}^n$ supports functions $\rho_1, \ldots, \rho_n$ such that

$$\pi^*(p'(z)) = (n+1) \prod_{i=1}^{n} (z - \rho_i); \tag{4.16}$$

$$\pi^*(a_i) = (-1)^{i+1} \frac{n+1}{n-i} \sigma_{i+1}(\rho_1, \ldots, \rho_n), \quad i = 1, \ldots, n-1, \tag{4.17}$$

and $\sigma_1(\rho_1, \ldots, \rho_n) = \rho_1 + \cdots + \rho_n = 0$. We will omit π^* in the notation of lifted functions.

Let $M \subset \tilde{\mathbf{A}}^n$ be the open dense subspace on which

A. $\forall i$, $p''(\rho_i) \neq 0$, that is, $\rho_i \neq \rho_j$ for $i \neq j$.

B. $u^i := p(\rho_i)$ form local coordinates at any point.

4.5.1. Theorem. *M is a semisimple Frobenius manifold with the following structure data:*

a) Canonical coordinates (u^i), *identity* $e = \sum_i e_i$, $e_i = \partial/\partial u^i$, *Euler field* $E = \sum u^i e_i$.

b) Flat metric

$$g := \sum_{i=1}^{n} \frac{(du^i)^2}{p''(\rho_i)} \tag{4.18}$$

with metric potential

$$\eta = \frac{a_1}{n+1} = \frac{1}{n-1} \sum_{i<j} \rho_i \rho_j = -\frac{1}{2(n-1)} \sum \rho_i^2. \tag{4.19}$$

Furthermore, e, E, and flat coordinates $x^{(1)}, \ldots, x^{(n)}$ can be calculated through $(a_1, \ldots, a_n)$ (which are generically local coordinates as well):

$$e = \partial/\partial a_n, \text{ i.e., } ea_n = 1, ea_i = 0 \text{ for } i < n, \tag{4.20}$$

$$E = \frac{1}{n+1} \sum_{i=1}^{n} (i+1) a_i \frac{\partial}{\partial a_i}, \tag{4.21}$$

$x^{(i)}$ are the first Laurent coefficients of the inversion of $w = \sqrt[n+1]{p(z)} = z + O(1/z)$ near $z = \infty$:

(4.22) $$z = w + \frac{x^{(1)}}{w} + \frac{x^{(2)}}{w^2} + \cdots + \frac{x^{(n)}}{w^n} + O(w^{-n-1}).$$

Finally, the spectrum is $D = \frac{n+3}{n+1}$, $d^{(i)} = \frac{i+1}{n+1}$, $1 \leq i \leq n$; more precisely, $Ex^{(i)} = \frac{i+1}{n+1} x^{(i)}$.

Proof. From (4.18) we find

(4.23) $$\eta_j := \frac{1}{p''(\rho_j)} = \frac{1}{(n+1) \prod_{i:\, i \neq j}(\rho_i - \rho_j)}.$$

Furthermore

(4.24) $$\delta_{ij} = \frac{\partial u^i}{\partial u^j} = \frac{\partial(p(\rho_i))}{\partial u^j} = \sum_{k=1}^{n} \frac{\partial a_k}{\partial u^j} z^{n-k} \bigg|_{z=\rho_i}$$

because $p'(\rho_i) = 0$. Therefore the polynomial on the right hand side of (4.24) (depending only on j) must be equal to

(4.25) $$\prod_{i:\, i \neq j} \frac{z - \rho_i}{\rho_j - \rho_i}$$

because it has the same degree $n - 1$ and takes the same values at $\rho_1, \ldots, \rho_n$. Comparing (4.23) and (4.25) we see first of all that

(4.26) $$\frac{\partial a_1}{\partial u^j} = \text{coeff. of } z^{n-1} \text{ in } \prod_{i:\, i \neq j} \frac{z - \rho_i}{\rho_j - \rho_i} = \frac{1}{\prod_{i:\, i \neq j}(\rho_j - \rho_i)} = (n+1)\eta_j.$$

This means that $\eta = \frac{a_1}{n+1}$ is the metric potential of g (cf. Theorem 3.3 b)). Now sum (4.24) for all j. We obtain that $\sum_{k=1}^{n} e a_k z^{n-k}$ is a polynomial of degree $n - 1$ taking value 1 at $z = \rho_1, \ldots, \rho_n$. Hence it is identically 1, that is,

$$ea_n = 1;\ ea_{n-1} = \cdots = ea_1 = 0.$$

This proves (4.20).

Let us now calculate $E\eta$. Multiplying (4.24) by u^j and summing over all j we see that $\sum_k E a_k z^{n-k}$ is a polynomial of degree $n - 1$ taking the value u^j at $z = \rho_j$. We know a polynomial of degree n taking the same values: it is $p(z)$. Hence $p(z) - \sum_k E a_k z^{n-k}$ is divisible by $p'(z)$ vanishing at all ρ_j. Comparing the top two coefficients we obtain

$$p(z) - \sum_k E a_k z^{n-k} = \frac{z}{n+1} p'(z),$$

that is,

$$a_k - E a_k = \frac{n-k}{n+1} a_k,$$

which proves (4.21). In particular, $E a_1 = \frac{2}{n+1} a_1$, so that $D = \frac{n+3}{n+1}$, because $d_0 = 1$.

We now turn to checking flatness of g. In fact, we can do rather more starting with a neat description of the multiplication $\circ$.

Let $p(z)$ be a point of M or its image in $\mathbf{A}^n$. Using $d\pi$ we can identify the tangent spaces at both points to the Milnor ring $\mathbf{C}[z] \bmod p'(z)$.

4.5.2. Lemma. *$d\pi$ identifies the $\circ$-multiplication with multiplication in the Milnor ring.*

Proof. Explicitly,

(4.27) $$d\pi(e_j|_p) = \frac{\partial p}{\partial u^j} \bmod p'(z).$$

In view of the previous calculations (see (4.24) and (4.25))

(4.28) $$\frac{\partial p}{\partial u^j} = e_j p = \prod_{i:\, i \neq j} \frac{z - \rho_i}{\rho_j - \rho_i} \bmod p'(z).$$

The polynomials on the right hand side of (4.28) are the basic idempotents in the Milnor ring, exactly as e_j in $\mathcal{T}_M$.

4.5.3. Lemma. *The metric (4.18) induces on the Milnor ring the scalar product*

(4.29) $$g(a(z)|_p, b(z)|_p) = -\mathrm{res}_{z=\infty} \frac{a(z)b(z)}{p'(z)} dz.$$

Proof. The right hand side of (4.29) equals

(4.30) $$\sum_{j=1}^n \mathrm{res}_{z=\rho_j} \frac{a(z)b(z)}{p'(z)} dz = \sum_{j=1}^n \frac{a(\rho_j)b(\rho_j)}{p''(\rho_j)}.$$

Choosing $a(z) = d\pi(e_i|_p)$, $b(z) = d\pi(e_j|_p)$, we see from (4.28) and (4.26) that the value of the right hand side of (4.30) is

$$\frac{\delta_{ij}}{p''(\rho_j)} = g(e_i|_p, e_j|_p).$$

4.5.4. End of the proof of Theorem 4.5.1. We can now prove simultaneously that g is flat and $x^{(a)}$ are flat coordinates by showing that $g(\partial_a, \partial_b)$ are constant, for $\partial_a := \partial/\partial x^{(a)}$.

In fact, from (4.22), considering z as a function of w and $x^{(a)}$: $p(z(w,x)) = w^{n+1}$, we get

(4.31) $$\frac{\partial p}{\partial x^{(a)}}(z(w,x)) = -p'(z(w,x))(w^{-a} + O(w^{-n-1})).$$

Substituting this into (4.29) we find

$$g(\partial_a|_p, \partial_b|_p) = -\mathrm{res}_{z=\infty}(\partial_a p\, \partial_b p) \frac{dz}{p'(z)}$$

$$= -\mathrm{res}_{z=\infty} p'(z) dz (w^{-a-b} + O(w^{-n-2})).$$

Replacing the local parameter z at infinity by w and taking into account that $p'(z)dz = (n+1)w^n dw$ we get

(4.32) $$g(\partial_a, \partial_b) = (n+1)\delta_{a+b,n+1}.$$

Finally, $p(z)$ becomes homogeneous of degree 1, if we assign to z the E–degree $\frac{1}{n+1}$. This implies that $x^{(i)}$ is of degree $\frac{i+1}{n+1}$.

4.5.5. Corollary. *The potential Φ is a polynomial of E-degree $D + d_0 = 2 + \frac{2}{n+1}$ and the usual degree $\leq n + 2$ in flat coordinates.*

Since Φ is analytic in $x^{(i)}$ and the spectrum of $-\operatorname{ad} E$ is strictly positive, the Taylor series can contain only finitely many terms of E–degree $D+d_0$. The maximal usual degree is furnished by $(x^1)^{(n+2)}$.

Notice that quantum cohomology cannot have spectrum of this type. However, using the theory of tensor product of Frobenius manifolds developed in Chapter III, one can produce manifolds with spectra of algebraic geometric type using tensor powers of this model and its generalizations treated by K. Saito and B. Dubrovin. See more details in Chapter III, §§7 and 8.

B. Dubrovin in [D2] generalized example 4.5 to the large class of Hurwitz moduli spaces, parametrizing ramified coverings of projective line. M. Rosellen's paper [Ros] contains a detailed study of this class of Frobenius manifolds.

§5. Weak Frobenius manifolds

5.1. Definition. *An F-manifold is a pair $(M, \circ)$, where M is a (super)manifold and $\circ$ is an associative supercommutative $\mathcal{O}_M$-bilinear multiplication $\mathcal{T}_M \times \mathcal{T}_M \to \mathcal{T}_M$ satisfying the following identity: for any (local) vector fields X, Y, Z, W we have*

(5.1)
$$\begin{aligned}&[X \circ Y, Z \circ W] - [X \circ Y, Z] \circ W - (-1)^{(X+Y)Z} Z \circ [X \circ Y, W] \\ &-X \circ [Y, Z \circ W] + X \circ [Y, Z] \circ W + (-1)^{YZ} X \circ Z \circ [Y, W] \\ &-(-1)^{XY} Y \circ [X, Z \circ W] + (-1)^{XY} Y \circ [X, Z] \circ W + (-1)^{X(Y+Z)} Y \circ Z \circ [X, W] = 0.\end{aligned}$$

Here and below we write, say, $(-1)^{(X+Y)Z}$ as shorthand for $(-1)^{(\widetilde{X}+\widetilde{Y})\widetilde{Z}}$.

5.1.1. Remarks. (i) The left hand side of (5.1) is $\mathcal{O}_M$–polylinear in X, Y, Z, and W. In other words, it is a tensor. This can be checked by a completely straightforward, although lengthy, calculation.

(ii) Introduce the expression measuring the deviation of the structure $(\mathcal{T}_M, \circ, [,])$ from that of Poisson algebra on $(\mathcal{T}_M, \circ)$:

(5.2) $$P_X(Z, W) := [X, Z \circ W] - [X, Z] \circ W - (-1)^{XZ} Z \circ [X, W].$$

Then (5.1) is equivalent to the following requirement:

(5.3) $$P_{X \circ Y}(Z, W) = X \circ P_Y(Z, W) + (-1)^{XY} Y \circ P_X(Z, W).$$

§5. WEAK FROBENIUS MANIFOLDS

5.2. Theorem. *a) Let (M, g, A) be a Frobenius manifold with multiplication $\circ$. Then $(M, \circ)$ is an F-manifold.*

b) Let $(M, \circ)$ be a pure even F-manifold, whose multiplication law is semisimple on an open dense subset. Assume that it admits an invariant flat metric g defining the cubic tensor A as in (1.2). Then (M, g, A) is a Frobenius manifold.

Proof. a) Since the left hand side of (5.1) is a tensor, it suffices to check that it vanishes on quadruples of flat fields $(X, Y, Z, W) = (\partial_a, \partial_b, \partial_c, \partial_d)$. Flat fields (super)commute so that only five summands of nine survive in (5.1). Denoting the structure constants $A_{ab}{}^c$ as in (1.4) and calculating the coefficient of ∂_f in (5.1), we can represent it as a sum of five summands, for which we introduce special notation in order to explain the pattern of cancellation:

$$\sum_e A_{ab}{}^e \partial_e A_{cd}{}^f - (-1)^{(a+b)(c+d)} \sum_e A_{cd}{}^e \partial_e A_{ab}{}^f = \alpha_1 + \beta_1,$$

$$(-1)^{(a+b)c} \sum_e \partial_c A_{ab}{}^e A_{ed}{}^f = \alpha_2, \quad (-1)^{(a+b+c)d} \sum_e \partial_d A_{ab}{}^e A_{ec}{}^f = \gamma_1,$$

$$-(-1)^{a(b+c+d)} \sum_e \partial_b A_{cd}{}^e A_{ea}{}^f = \gamma_2, \quad -(-1)^{(c+d)b} \sum_e \partial_a A_{cd}{}^e A_{eb}{}^f = \beta_2.$$

Use potentiality in order to interchange the subscripts e, c in α_1 (a sign emerges). After this we see that

$$\alpha_1 + \alpha_2 = (-1)^{(a+b)c} \partial_c \left(\sum_e A_{ab}{}^e A_{ed}{}^f \right).$$

Similarly, permuting a and e in β_1 we find

$$\beta_1 + \beta_2 = -(-1)^{(c+d)b} \partial_a \left(\sum_e A_{cd}{}^e A_{eb}{}^f \right).$$

Now rewrite γ_1 permuting a, d, and γ_2 permuting b, c. Calculating finally $\beta_1 + \beta_2 + \gamma_1 + \gamma_2$ we see that it cancels with $\alpha_1 + \alpha_2$ due to associativity relations (1.5).

b) Clearly, (M, g, A) is an associative pre–Frobenius manifold in the sense of Definition 1.3, so that it only remains to check its potentiality in the domain of semisimplicity. To this end we will use Theorem 3.3.

Let (e_i) be the idempotent local vector fields. Applying (5.3) to $X = Y = e_i$ we get $P_{e_i} = 2 e_i \circ P_{e_i}$ so that $P_{e_i} = 0$. Applying then (5.2) to $(X, Z, W) = (e_i, e_j, e_j)$, $i \neq j$, we see that $[e_i, e_j] = 0$. This is the first condition of Theorem 3.3. The second one expresses invariance and flatness of the metric in canonical coordinates, which we have already postulated.

5.2.1. Corollary (of the proof). *Semisimple F-manifolds are exactly those manifolds $(M, \circ)$ which everywhere locally admit a basis of pairwise commuting $\circ$-idempotent vector fields, or, which is the same, Dubrovin's canonical coordinates.*

In fact, we have already deduced from (5.1) that e_i pairwise commute. Conversely, if they commute, (5.1) holds for any quadruple of idempotents.

5.3. Definition. *A weak Frobenius manifold is an F-manifold $(M, \circ)$ such that in a neighborhood U of any point there exists a flat invariant metric g making $(U, g, \circ)$ a Frobenius manifold. We will call such metrics compatible (with the given F-structure).*

Thus, a weak Frobenius manifold is a Frobenius manifold without a fixed metric.

Any semisimple F-manifold is automatically weak Frobenius. This follows from the results of Chapter II, §3, which reduce the construction of compatible metrics to the solution to Schlesinger's equations. We do not know whether there exist non–semisimple F-manifolds which are not weak Frobenius.

5.4. Sheaves of compatible metrics and Euler fields. Let $(M, \circ)$ be a weak Frobenius manifold. Then compatible metrics on M form a sheaf $\mathcal{M}_M$. Assume now that M admits an identity e. Then there is an embedding $\mathcal{M}_M \hookrightarrow \Omega^1_M$ which sends each metric g to the respective coidentity ε_g defined by $i_X \varepsilon_g = g(e, X)$. In fact, knowing ε_g we can reconstruct g: $g(X, Y) = i_{X \circ Y}(\varepsilon_g)$. We will call ε_g compatible 1–forms.

We will now impose an additional restriction and denote by $\mathcal{F}_M$ the sheaf of those compatible metrics for which e is flat. We will call such metrics *admissible*. If $g \in \mathcal{F}_M$, then ε_g is closed and g–flat: see (2.4). It is important to understand the structure of this sheaf of sets. Again, the situation is rather transparent on the tame semisimple part of M. We will state and prove Dubrovin's theorem which provides a neat local description of admissible metrics *with fixed rotation coefficients* γ_{ij} considered as functions on the common definition domain of metrics.

This result should be compared with Theorem 3.4.3 of Chapter II below which depicts the set of metrics *with fixed v_{ij} at a point* where

$$(5.4) \qquad v_{ij} = \frac{1}{2}(u^j - u^i)\frac{\eta_{ij}}{\eta_j}, \; \gamma_{ij} = \frac{1}{2}\frac{\eta_{ij}}{\sqrt{\eta_i \eta_j}} = \frac{1}{u^j - u^i}\sqrt{\frac{\eta_j}{\eta_i}} \, v_{ij}$$

so that both statements refer to the closely related coordinates on the space of metrics.

5.4.1. Theorem. *a) Let $g = \sum_i \eta_i (du^i)^2$, $\widetilde{g} = \sum_i \widetilde{\eta}_i (du^i)^2$ be two $\circ$–invariant metrics in a simply connected domain of canonical coordinates $u^1, \ldots, u^n$ in M. Then there exist exactly 2^n vector fields ∂ in this domain such that*

$$(5.5) \qquad \widetilde{g}(X, Y) = g(\partial \circ X, \partial \circ Y)$$

for any X, Y. These fields are $\circ$–invertible and differ only by the signs of their e_i–components.

b) Assume that g is admissible. Then $\widetilde{g}$ defined by (5.5) is admissible and has the same rotation coefficients $\widetilde{\gamma}_{ij} = \gamma_{ij}$ iff ∂ is g–flat.

Proof. a) Put $\partial = \sum_i D_i e_i$. Then (5.5) is equivalent to

$$(5.6) \qquad D_i^2 = \frac{\widetilde{\eta}_i}{\eta_i}.$$

This proves the first statement.

b) Choose a solution (D_i) of (5.6). Let ∇_i denote the Levi–Civita covariant derivative in the direction e_i with respect to the metric g. Using (3.10), we see that

$$\nabla_i(\sum_j D_j e_j)$$

(5.7) $\quad = \sum_{j \neq i} \left(e_i D_j + \frac{1}{2} D_j \frac{\eta_{ij}}{\eta_j} - \frac{1}{2} D_i \frac{\eta_{ij}}{\eta_j} \right) e_j + \left(e_i D_i + \frac{1}{2} \sum_j D_j \frac{\eta_{ij}}{\eta_j} \right) e_i.$

Using (5.6) and (3.13), we can rewrite the first sum in (5.7) as

(5.8) $\quad \sum_{j \neq i} \left(\sqrt{\frac{\widetilde{\eta}_i}{\eta_j}} (\widetilde{\gamma}_{ij} - \gamma_{ij}) \right) e_j$

and the remaining terms as

(5.9) $\quad \left(\frac{1}{2} \frac{\widetilde{\eta}_{ii}}{\sqrt{\eta_i \widetilde{\eta}_i}} + \sum_{j \neq i} \sqrt{\frac{\widetilde{\eta}_j}{\eta_i}} \gamma_{ij} \right) e_i.$

If we replace γ_{ij} by $\widetilde{\gamma}_{ij}$ in (5.9), the resulting expression will vanish for admissible $\widetilde{g}$ because $(\sum_j e_j) \widetilde{\eta}_i = 0$ (see Proposition 3.5b). Subtracting this zero from (5.9) we finally find

(5.10) $\quad \nabla_i(\sum_j D_j e_j) = \sum_{j \neq i} \sqrt{\frac{\widetilde{\eta}_i}{\eta_j}} (\widetilde{\gamma}_{ij} - \gamma_{ij}) e_j - \sum_{j \neq i} \sqrt{\frac{\widetilde{\eta}_j}{\eta_i}} (\widetilde{\gamma}_{ij} - \gamma_{ij}) e_i.$

Hence admissibility of $\widetilde{g}$ and coincidence of the rotation coefficients imply the g-flatness of ∂, and vice versa. This proves the second statement of the theorem.

If one does not assume semisimplicity, a part of the preceding theorem still holds true.

5.4.2. Theorem. *Let g be an admissible metric and ∂ a g-flat even invertible vector field. Then $\widetilde{g}$ defined by (5.5) is admissible.*

Proof. Put $\widetilde{x}^a := \sum_b g^{ab} \partial_b \partial \Phi = \partial \Phi^a$, where $(\partial_a = \partial/\partial x^a)$ is a local basis of flat vector fields and Φ is a local Frobenius potential. Then we have

$$\frac{\partial \widetilde{x}^a}{\partial x^b} = \partial \Phi^a_b.$$

Hence the respective Jacobian is the matrix of the multiplication $\partial \circ$ in the basis (∂_a). Since the latter is invertible, $(\widetilde{x}^a)$ form a local coordinate system. For the dual basis of vector fields $\widetilde{\partial}_a$ we have

(5.11) $\quad \partial \circ \widetilde{\partial}_a = \sum_b \widetilde{\partial}_a(x^b) \circ \partial_b = \sum_{b,c} \widetilde{\partial}_a(x^b) \partial_b \partial_c \Phi g^{cd} \partial_d = \sum_{b,c} \widetilde{\partial}_a(x^b) \partial_b(\widetilde{x}^c) \partial_c = \partial_a.$

Hence $\widetilde{g}$ has the same coefficients in the basis $(\widetilde{\partial}_a)$ as g in the basis (∂_a) and is flat. Clearly, it is also $\circ$–invariant. It remains to show that the pre–Frobenius structure $(M, \circ, \widetilde{g})$ is potential. It is easy to see that any local function $\widetilde{\Phi}$ satisfying the equations

(5.12) $\quad \widetilde{\partial}_a \widetilde{\partial}_b \widetilde{\Phi} = \partial_a \partial_b \Phi$

for all a, b can serve as a local potential defining the same multiplication $\circ$. To prove its existence, we check the integrability condition:

$$\widetilde{\partial}_a \partial_b \partial_c \Phi = \sum_d \widetilde{\partial}_a(x^d) \partial_d \partial_b \partial_c \Phi = \sum_d \widetilde{\partial}_a(x^d) g(\partial_d, \partial_b \circ \partial_c)$$

(5.13) $\qquad = g(\widetilde{\partial}_a, \partial_b \circ \partial_c) = g(\partial^{-1} \circ \partial_a, \partial_b \circ \partial_c) = (-1)^{ab} \widetilde{\partial}_b \partial_a \partial_c \Phi.$

The same reasoning as in the end of the proof of Theorem 1.5 then shows the existence of $\widetilde{\Phi}$. The identity e remains flat because $\widetilde{g}$-flat fields form the sheaf $\partial^{-1} T_M^f$ and $e = \partial^{-1} \circ \partial$. This finishes the proof.

Dubrovin calls the passage from g to $\widetilde{g}$ *the Legendre-type transformation*. In Appendix B of [D2] he also constructs a different type of transformation which he calls *inversion*.

What Dubrovin calls *a twisted Frobenius manifold* in our language is a weak Frobenius manifold, endowed with local admissible metrics connected by the Legendre-type transformations on the overlaps of their definition domains.

We now turn to Euler fields. Again let $(M, \circ)$ be a weak Frobenius manifold. An even vector field E on M is called *a weak Euler field of (constant) weight d_0* if $\text{Lie}_E(\circ) = d_0 \circ$, that is, for all local vector fields X, Y we have

(5.14) $\qquad P_E(X, Y) = [E, X \circ Y] - [E, X] \circ Y - X \circ [E, Y] = d_0 X \circ Y.$

This is the same as (2.6). If $(M, \circ)$ admits an identity e, we get formally from (5.14) that $[e, E] = d_0 e$. Clearly, local weak Euler fields form a sheaf of vector spaces $\mathcal{E}_X$, and weight is a linear function on this sheaf. If $(M, \circ)$ comes from a Frobenius manifold with flat identity e, then any Euler vector field on the latter is a weak Euler field, and e itself is a (weak) Euler field of weight zero. The latter statement follows by combining 2.2.2 and (2.3).

5.4.3. Proposition. *The commutator of any two (local) weak Euler fields is a weak Euler field of weight zero.*

Proof. We start with the following general identity: for any local vector fields X, Y, Z, W we have

$$P_{[X,Y]}(Z, W)$$
$$= [X, P_Y(Z, W)] - (-1)^{XY} P_Y([X, Z], W) - (-1)^{X(Y+Z)} P_Y(Z, [X, W])$$
(5.15) $\quad -(-1)^{XY}[Y, P_X(Z, W)] + P_X([Y, Z], W) + (-1)^{YZ} P_X(Z, [Y, W]).$

In order to check this, replace the seven terms in (5.15) by their expressions from (5.2), and then rewrite the resulting three terms on the left hand side using the Jacobi identity. All the twenty four summands will cancel.

Now apply (5.15) to the two weak Euler fields $X = E_1, Y = E_2$. The right hand side will turn to zero. This proves our statement.

5.4.4. Example (Sh. Katz). The (formal) Frobenius manifold corresponding to the quantum cohomology of a projective algebraic manifold V admits at least two different Euler fields (besides e), if $h^{pq}(V) \neq 0$ for some $p \neq q$. To write them down explicitly, choose a basis (∂_a) of $H = H^*(V, \mathbf{C})$ considered as the space of flat vector fields, and let (x^a) be the dual flat coordinates vanishing at zero. Let $\partial_a \in H^{p_a, q_a}(V)$. Put $-K_V = \sum_{p_b + q_b = 2} r^b \partial_b$. Then

$$E_1 := \sum_a (1 - p_a) x^a \partial_a + \sum_b r^b \partial_b,$$

$$E_2 := \sum_a (1 - q_a) x^a \partial_a + \sum_b r^b \partial_b$$

are Euler.

Now let g be an admissible metric on $(M, \circ)$. Weak Euler fields which are conformal with respect to g form a subsheaf of linear spaces in $\mathcal{E}_M$ endowed with a linear function D, conformal weight: see (2.5). A direct calculation shows that the commutator of such fields is conformal of conformal weight zero. One can also say what happens to the weights (and the full spectrum) of E when one replaces g by another metric as in (5.5).

5.4.5. Proposition. *Let $(M, \circ, g)$ be a Frobenius manifold with flat identity e and an Euler field E, $[e, E] = d_0 e$, $\mathrm{Lie}_E(g) = Dg$, $(d_a) = $ the spectrum of $-\mathrm{ad}\, E$ on flat vector fields. Assume that ∂ is an invertible flat field such that $[\partial, E] = d\partial$. Then E is an Euler field on $(M, \widetilde{g}, \circ)$, $\mathrm{Lie}_E(\widetilde{g}) = (D + 2d_0 - 2d)\widetilde{g}$, and the spectrum of $-\mathrm{ad}\, E$ on $\partial^{-1} T_M^f$ is $(d_a + d_0 - d)$.*

We omit the straightforward proof.

5.5. Relation to the Poisson structure. Consider an abstract structure $(A, \circ, [,])$, where $\circ$, resp. $[,]$, induces on the $\mathbf{Z}_2$-graded additive group A the structure of the supercommutative, resp. Lie, ring. Assume that these operations satisfy the relation (5.1), or equivalently, (5.3). Then we will call $(A, \circ, [,])$, or simply A, an F-algebra. In particular, vector fields on a Frobenius manifold form a sheaf of F-algebras.

Every Poisson algebra is an F-algebra. Conversely, let A be an F-algebra, and

(5.16) $$B := \{X \in A \,|\, P_X \equiv 0\}.$$

5.5.1. Proposition. *a) B is closed with respect to $\circ$ and $[,]$ and hence forms a Poisson subalgebra. If A contains identity e, then $e \in B$.*

b) If A is the algebra of vector fields on a split semisimple Frobenius manifold, then B is spanned by the basic idempotent fields e_i over constants. In particular, the Lie bracket in B is trivial.

Proof. a) Assume that $P_X = P_Y = 0$. We have $P_{X \circ Y} = 0$ in view of (5.3). Putting $X = Y = e$ in (5.3), we get $P_e = 0$. Finally, $P_{[X,Y]} = 0$ follows from (5.15).

b) Writing X, Y, Z in the basis e_i with indeterminate coefficients, one easily checks that if $P_X(Y, Z) = 0$ for all Y, Z, then the coefficients of X are constant.

5.6. Theorem. *Let E be an Euler field on a Frobenius manifold with identity e such that $[e, E] = d_0 e$. Then for all $m, n \geq 0$*

(5.17) $$[E^{\circ n}, E^{\circ m}] = d_0(m - n)\, E^{\circ m+n-1}.$$

Proof. We will prove slightly more. Let X be an even vector field on an arbitrary F-manifold with identity e. Since in view of (5.3) the map $X \mapsto P_X$ is a $\circ$-derivation, we have $P_{X^{\circ n}} = n X^{\circ n-1} \circ P_X$. Moreover, from (5.2) we have

$$P_{X^{\circ n}}(e, e) = -[X^{\circ n}, e].$$

Hence

(5.18) $$[X^{\circ n}, e] = n X^{\circ n-1} \circ [X, e].$$

Let us *assume* now that X satisfies the following identities: for all $n \geq 1$

(5.19) $$[X^{\circ n}, X] = (1 - n)\, X^{\circ n} \circ [e, X].$$

Then we assert that for all $m, n \geq 0$

(5.20) $$[X^{\circ n}, X^{\circ m}] = (m - n)\, X^{\circ n+m-1} \circ [e, X].$$

In fact, the cases when m or n is ≤ 1 are covered by (5.18), (5.19). The general case can be treated by induction. We have

$$[X^{\circ n}, X^{\circ m}] = P_{X^{\circ n}}(X^{\circ m-1}, X) + [X^{\circ n}, X^{\circ m-1}] \circ X + [X^{\circ n}, X] \circ X^{\circ m-1}$$
$$= n X^{\circ n-1} \circ ([X, X^{\circ m}] - [X, X^{\circ m-1}] \circ X)$$
$$+ [X^{\circ n}, X^{\circ m-1}] \circ X + [X^{\circ n}, X] \circ X^{\circ m-1}$$
$$= (m - n)\, X^{\circ n+m-1} \circ [e, X].$$

It remains to notice that since $[e, E] = d_0 e$, E satisfies (5.19) in view of the general identity (2.12).

5.6.1. Remark. In the semisimple case the meaning of (5.19) is transparent: writing $X = \sum X_i e_i$, we must have $e_i X_j = 0$ for $i \neq j$.

CHAPTER II

Frobenius Manifolds and Isomonodromic Deformations

§1. The second structure connection

1.1. Preparation. Let (M, g, A) be a Frobenius (super)manifold, and ∇_0 the Levi–Civita connection (on $\mathcal{T}_M$) of the flat metric g. Recall that the (first) structure connection on M is actually a pencil of flat connections ∇_λ, determined by the formula $\nabla_{\lambda,X}(Y) = \nabla_{0,X}(Y) + \lambda X \circ Y$ (see I.(1.8)). If in addition M is endowed with an Euler field E with $d_0 = 1$, we can define the extended structure connection $\widehat{\nabla}$ on the sheaf $\widehat{\mathcal{T}} = \mathrm{pr}_M^*(\mathcal{T}_M)$ on $\widehat{M} = M \times (\mathbf{P}_\lambda^1 \setminus \{0, \infty\})$ such that for $X \in \mathcal{T}_M$, $Y \in \mathcal{T}_M^f$ we have

(1.1) $$\widehat{\nabla}_X(Y) = \lambda X \circ Y, \quad \widehat{\nabla}_{\partial/\partial \lambda}(Y) = E \circ Y - \frac{1}{\lambda}[E, Y]$$

(cf. I.2.5, in particular (2.24); we now omit a few extra hats in notation and commit the respective abuses of language).

In this section and chapter we will restrict ourselves to the case of semisimple complex Frobenius manifolds with an Euler field with $d_0 = 1$ admitting a global system of canonical coordinates (u^i). We will call *the second structure connection* $\check{\nabla}_\lambda$ the Levi–Civita connection of the flat metric

$$\check{g}_\lambda(X, Y) := g((E - \lambda)^{-1} \circ X, Y)$$

depending on a parameter λ and defined on the open subset $M_\lambda \subset M$ where $u^i \neq \lambda$ for all i. Put $\check{M} := \bigcup_\lambda (M_\lambda \times \{\lambda\}) \subset M \times \mathbf{P}_\lambda^1$ and denote by $\check{\mathcal{T}}$ the restriction of $\mathrm{pr}_M^*(\mathcal{T}_M)$ to $\check{M}$.

In this section we will construct a flat extension $\check{\nabla}$ of $\check{\nabla}_\lambda$ to $\check{\mathcal{T}}$ which will also be referred to as the second structure connection. Both extensions $\widehat{\nabla}$ and $\check{\nabla}$ will be further studied as isomonodromic deformations of their restrictions to the λ-direction parametrized by M.

More precisely, assume that $\mathcal{T}_M^f$ is a trivial local system (for instance, because M is simply connected). Put $T := \Gamma(M, \mathcal{T}_M^f)$. Then $\widehat{\nabla}$ (resp. $\check{\nabla}$) induces an integrable family of connections with singularities on the trivial bundle on $\mathbf{P}_\lambda^1$ with the fiber T. The first connection $\widehat{\nabla}$ is singular only at $\lambda = 0$ and $\lambda = \infty$, but whereas 0 is a regular (Fuchsian) singularity, ∞ is an irregular one, so that $\widehat{\nabla}$ cannot be an algebraic geometric Gauss–Manin connection, and its monodromy involves the Stokes phenomenon. To the contrary, the second connection $\check{\nabla}$ generally has only regular singularities at infinity and at $\lambda = u^i$ whose positions thus depend on the parameters. It is determined by the conventional monodromy representation and has a chance to define a variation of Hodge structure. For more details, see the next section.

It turns out that both deformations share a common moduli space and deserve to be studied together. In fact, fiberwise they are more or less formal Laplace transforms of each other. More to the point, they form a dual pair in the sense of [Har].

In this chapter we deal only with the geometry of the second structure connection. For a deep and important study of the first connection, especially of its monodromy, see [DZh3] and [Gu].

In our calculations the key role will be played by the $\mathcal{O}_M$-linear skew symmetric operator $\mathcal{V} : \mathcal{T}_M \to \mathcal{T}_M$ which is the unique extension of the operator defined in I.2.2 on flat vector fields by the formula

$$(1.2) \qquad \mathcal{V}(X) = [X, E] - \frac{D}{2}X \text{ for } X \in \mathcal{T}_M^f.$$

1.1.1. Proposition. *a) We have for arbitrary $X \in \mathcal{T}_M$:*

$$(1.3) \qquad \mathcal{V}(X) = \nabla_{0,X}(E) - \frac{D}{2}X.$$

b) Let $e_j = \partial/\partial u^j$, $f_j = e_j/\sqrt{\eta_j}$. Then

$$(1.4) \qquad \mathcal{V}(f_i) = \sum_{j \neq i}(u^j - u^i)\gamma_{ij}f_j.$$

Proof. The fact that $-\operatorname{ad} E - \frac{D}{2}\operatorname{Id}$ is skew symmetric with respect to g was checked in I.2.2. Formula (1.3) defines an $\mathcal{O}_M$-linear endomorphism of $\mathcal{T}_M$ which coincides with (1.2) on the flat fields, as a calculation in flat coordinates shows.

To check (1.4), we use (1.3) and Ch. I, (3.10):

$$(1.5)$$

$$\mathcal{V}(f_i) = \nabla_{0,f_i}(E) - \frac{D}{2}f_i = \nabla_{0,e_i/\sqrt{\eta_i}}\left(\sum_j u^j e_j\right) - \frac{D}{2}\frac{e_i}{\sqrt{\eta_i}}$$

$$= \frac{1}{\sqrt{\eta_i}}\left[e_i + u^i\nabla_{0,e_i}(e_i) + \sum_{j \neq i}u^j\nabla_{0,e_i}(e_j)\right] - \frac{D}{2}\frac{e_i}{\sqrt{\eta_i}}$$

$$= \frac{1}{\sqrt{\eta_i}}\left[e_i + u^i\left(\frac{\eta_{ii}}{2\eta_i}e_i - \sum_{j \neq i}\frac{\eta_{ij}}{2\eta_j}e_j\right) + \sum_{j \neq i}u^j\left(\frac{\eta_{ij}}{2\eta_i}e_i + \frac{\eta_{ij}}{2\eta_j}e_j\right)\right] - \frac{D}{2}\frac{e_i}{\sqrt{\eta_i}}.$$

For $j \neq i$, the coefficient of f_j in the right hand side of (1.5) is $(u^j - u^i)\gamma_{ij}$. For $j = i$ it vanishes, because

$$1 + \sum_j u^j \frac{\eta_{ij}}{2\eta_i} = 1 + \frac{E\eta_i}{2\eta_i} = \frac{D}{2}$$

(see Ch. I, Theorem 3.6b)).

We can now state the main result of this section. In addition to (1.2), define the operator $\mathcal{U} : \mathcal{T}_M \to \mathcal{T}_M$:

(1.6) $$\mathcal{U}(X) := E \circ X,$$

so that $\mathcal{U}(f_i) = u^i f_i$.

1.2. Theorem. *For $X, Y \in \mathrm{pr}_M^{-1}(\mathcal{T}_M) \subset \mathcal{T}_{\tilde{M}}$ (meromorphic vector fields on $\mathcal{T}_{M \times \mathbf{P}_\lambda^1}$ independent on λ) put*

(1.7) $$\check{\nabla}_X(Y) = \nabla_{0,X}(Y) - \left(\mathcal{V} + \frac{1}{2}\mathrm{Id}\right)(\mathcal{U} - \lambda)^{-1}(X \circ Y),$$

(1.8) $$\check{\nabla}_{\partial/\partial\lambda}(Y) = \left(\mathcal{V} + \frac{1}{2}\mathrm{Id}\right)(\mathcal{U} - \lambda)^{-1}(Y).$$

Then $\check{\nabla}$ is a flat connection on $\check{\mathcal{T}}$ whose restriction on $M \times \{\lambda\}$ defined by (1.7) is the Levi–Civita connection for $\check{g}_\lambda$.

Remark. Rewriting (1.1) in the same notation, we get

(1.9) $$\widehat{\nabla}_X(Y) = \nabla_{0,X}(Y) + \lambda X \circ Y,$$

(1.10) $$\widehat{\nabla}_{\partial/\partial\lambda}(Y) = \left[\mathcal{U} + \frac{1}{\lambda}\left(\mathcal{V} + \frac{D}{2}\mathrm{Id}\right)\right](Y).$$

Proof. We will first apply Ch. I, (3.10), in order to calculate the Levi–Civita connection for $\check{g}_\lambda$ in coordinates $\check{u}^i = \log(u^i - \lambda)$. As in I.3.7 we have

$$\check{e}_i = \frac{\partial}{\partial \check{u}^i} = (u^i - \lambda)e_i, \quad \check{\eta}_i = (u^i - \lambda)\eta_i, \quad \check{\eta}_{ij} = (u^i - \lambda)(u^j - \lambda)\eta_{ij} + \delta_{ij}(u^i - \lambda)\eta_i,$$

$$\check{\gamma}_{ij} = \gamma_{ij}(u^i - \lambda)^{1/2}(u^j - \lambda)^{1/2}.$$

Then for $i \neq j$

$$\check{\nabla}_{\check{e}_i}(\check{e}_j) = \frac{1}{2}\frac{\check{\eta}_{ij}}{\check{\eta}_i}\check{e}_i + \frac{1}{2}\frac{\check{\eta}_{ij}}{\check{\eta}_j}\check{e}_j = \frac{1}{2}(u^i - \lambda)(u^j - \lambda)\left(\frac{\eta_{ij}}{\eta_i}e_i + \frac{\eta_{ij}}{\eta_j}e_j\right)$$

so that

(1.11) $$\check{\nabla}_{e_i}(e_j) = \frac{1}{2}\frac{\eta_{ij}}{\eta_i}e_i + \frac{1}{2}\frac{\eta_{ij}}{\eta_j}e_j = \nabla_{0,e_i}(e_j).$$

Similarly,

$$\check{\nabla}_{\check{e}_i}(\check{e}_i) = \frac{1}{2}\frac{\check{\eta}_{ii}}{\check{\eta}_i}\check{e}_i - \frac{1}{2}\sum_{j \neq i}\frac{\check{\eta}_{ij}}{\check{\eta}_j}\check{e}_j$$

$$= \frac{1}{2}(u^i - \lambda)^2\left[\frac{\eta_{ii}}{\eta_i} + \frac{1}{u^i - \lambda}\right]e_i - \frac{1}{2}\sum_{j \neq i}(u^i - \lambda)(u^j - \lambda)\frac{\eta_{ij}}{\eta_j}e_j$$

so that

(1.12) $$\check{\nabla}_{e_i}(e_i) = \frac{1}{2}\left[\frac{\eta_{ii}}{\eta_i} - \frac{1}{u^i - \lambda}\right]e_i - \frac{1}{2}\sum_{j \neq i}\frac{u^j - \lambda}{u^i - \lambda}\frac{\eta_{ij}}{\eta_j}e_j.$$

Subtracting from this (3.10) (Ch. I), we get

$$(1.13) \qquad (\check{\nabla}_{e_i} - \nabla_{0,e_i})(e_i) = -\frac{1}{2}\frac{1}{u^i-\lambda}e_i - \frac{1}{2}\sum_{j:\,j\neq i}\frac{u^j-u^i}{u^i-\lambda}\frac{\eta_{ij}}{\eta_j}e_j$$

and

$$(1.14) \qquad (\check{\nabla}_{e_i} - \nabla_{0,e_i})(f_i) = -\frac{1}{2}\frac{1}{u^i-\lambda}f_i - \sum_{j\neq i}\frac{u^j-u^i}{u^i-\lambda}\gamma_{ij}f_j.$$

In view of (1.4), we can write (1.11) and (1.12) together as

$$(1.15) \qquad (\check{\nabla}_{e_i} - \nabla_{0,e_i})(f_j) = -\left(\mathcal{V} + \frac{1}{2}\mathrm{Id}\right)(\mathcal{U}-\lambda)^{-1}(e_i \circ f_j)$$

because $e_i \circ f_j = \delta_{ij}f_j$. This family of formulas is equivalent to (1.7) so that (1.7) is the Levi–Civita connection for $\check{g}_\lambda$. In particular, it is flat for each fixed λ.

Since $[X, \partial/\partial\lambda] = 0$ for $X \in \mathrm{pr}_M^{-1}(\mathcal{T}_M)$, it remains to show that the covariant derivatives (1.7) and (1.8) commute on $\check{M}$, i.e. that for all i, j

$$(1.16) \qquad \check{\nabla}_{e_i}\check{\nabla}_{\partial/\partial\lambda}(e_j) = \check{\nabla}_{\partial/\partial\lambda}\check{\nabla}_{e_i}(e_j).$$

First of all, from (1.8) and (1.14) we find

$$(1.17) \qquad \check{\nabla}_{\partial/\partial\lambda}(e_j) = \frac{1}{2}\frac{1}{u^j-\lambda}e_j + \frac{1}{2}\sum_{k\neq j}\frac{u^k-u^j}{u^j-\lambda}\frac{\eta_{jk}}{\eta_k}e_k.$$

Together with (1.11) and (1.12) this gives for $i \neq j$:

$$(1.18) \quad \check{\nabla}_{\partial/\partial\lambda}\check{\nabla}_{e_i}(e_j) = \frac{1}{2}\frac{\eta_{ij}}{\eta_i}\left[\frac{1}{2}\frac{1}{u^i-\lambda}e_i + \frac{1}{2}\sum_{k\neq i}\frac{u^k-u^i}{u^i-\lambda}\frac{\eta_{ik}}{\eta_k}e_k\right] + (i \leftrightarrow j),$$

$$\check{\nabla}_{e_i}\check{\nabla}_{\partial/\partial\lambda}(e_j) = \frac{1}{2}\frac{1}{u^j-\lambda}\left(\frac{1}{2}\frac{\eta_{ij}}{\eta_i}e_i + \frac{1}{2}\frac{\eta_{ij}}{\eta_j}e_j\right)$$

$$+\frac{1}{2}\sum_{k\neq j}e_i\left(\frac{u^k-u^j}{u^j-\lambda}\frac{\eta_{jk}}{\eta_k}\right)e_k + \frac{1}{2}\sum_{k\neq j,i}\frac{u^k-u^j}{u^j-\lambda}\frac{\eta_{jk}}{\eta_k}\left(\frac{1}{2}\frac{\eta_{ik}}{\eta_i}e_i + \frac{1}{2}\frac{\eta_{ik}}{\eta_k}e_k\right)$$

$$(1.19) \qquad +\frac{1}{2}\frac{u^i-u^j}{u^j-\lambda}\frac{\eta_{ij}}{\eta_i}\left[\frac{1}{2}\left(\frac{\eta_{ii}}{\eta_i} - \frac{1}{u^i-\lambda}\right)e_i - \frac{1}{2}\sum_{k\neq i}\frac{u^k-\lambda}{u^i-\lambda}\frac{\eta_{ik}}{\eta_k}e_k\right].$$

The coincidence of coefficients of e_k in (1.18) and (1.19) for $i \neq j \neq k \neq i$ can be checked with the help of the following identity, which is equivalent to the Darboux–Egoroff equation Ch. I, (3.14):

$$\eta_{ijk} = \frac{1}{2}\left(\frac{\eta_{ik}\eta_{jk}}{\eta_k} + \frac{\eta_{ij}\eta_{ik}}{\eta_i} + \frac{\eta_{ij}\eta_{jk}}{\eta_j}\right).$$

The coincidence of the coefficients of e_i requires a little more work, and we will give some details, again for the case $i \neq j$.

§1. THE SECOND STRUCTURE CONNECTION

In (1.18) the coefficient of e_i is

(1.20)
$$\frac{1}{4}\frac{1}{u^i-\lambda}\frac{\eta_{ij}}{\eta_i} + \frac{1}{4}\frac{u^i-u^j}{u^j-\lambda}\frac{\eta_{ij}^2}{\eta_i\eta_j},$$

whereas in (1.19) we get

$$\frac{1}{4}\frac{1}{u^j-\lambda}\frac{\eta_{ij}}{\eta_i} + \frac{1}{2}e_i\left(\frac{u^i-u^j}{u^j-\lambda}\frac{\eta_{ij}}{\eta_i}\right)$$

(1.21)
$$+\frac{1}{4}\sum_{k\neq i,j}\frac{u^k-u^j}{u^j-\lambda}\frac{\eta_{ik}\eta_{jk}}{\eta_i\eta_k} + \frac{1}{2}\left(\frac{\eta_{ii}}{\eta_i}-\frac{1}{u^i-\lambda}\right)\frac{u^i-u^j}{u^i-\lambda}\frac{\eta_{ij}}{\eta_i}.$$

To identify (1.20) and (1.21) we have to get rid of the sum $\sum_k$ in (1.21). This can be done with the help of Ch. I, (3.14), (3.21), and (3.22):

$$\frac{1}{4}\sum_{k\neq i,j}\frac{u^k-u^j}{u^j-\lambda}\frac{\eta_{ik}\eta_{jk}}{\eta_i\eta_k} = \frac{1}{u^j-\lambda}\frac{\eta_j^{1/2}}{\eta_i^{1/2}}\left[\sum_{k\neq i,j}u^k\gamma_{ik}\gamma_{kj} - u^j\sum_{k\neq i,j}\gamma_{ik}\gamma_{kj}\right]$$

$$= \frac{1}{u^j-\lambda}\frac{\eta_j^{1/2}}{\eta_i^{1/2}}\left[-\gamma_{ij} - u^i e_i\gamma_{ij} - u^j e_j\gamma_{ij} + u^j(e_i+e_j)\gamma_{ij}\right]$$

$$= \frac{1}{2}\frac{1}{u^j-\lambda}\left[-\frac{\eta_{ij}}{\eta_i} + (u^j-u^i)\left(\frac{\eta_{iij}}{\eta_i} - \frac{1}{2}\frac{\eta_{ij}\eta_{ii}}{\eta_i^2} - \frac{1}{2}\frac{\eta_{ij}^2}{\eta_i\eta_j}\right)\right].$$

The remaining part of the calculation is straightforward, and we leave it to the reader, as well as the case $i=j$ which is treated similarly.

1.3. Formal Laplace transform. Assume now that $\mathcal{T}_M^f$ is a trivial local system. This means that if we put $T:=\Gamma(M,\mathcal{T}_M^f)$, there is a natural isomorphism $\mathcal{O}_M\otimes T \to \mathcal{T}_M$.

Formulas (1.8) (resp. (1.10)) define two families of connections with singularities on the trivial vector bundle on $\mathbf{P}_\lambda^1$ with fiber T, parametrized by M. Namely, denote by ∂_λ the covariant derivative along $\partial/\partial\lambda$ on this bundle for which the constant sections are horizontal. Then the two connections are

(1.22)
$$\check{\nabla}_{\partial/\partial\lambda} = \partial_\lambda + \left(\mathcal{V}+\frac{1}{2}\mathrm{Id}\right)(\mathcal{U}-\lambda)^{-1},$$

(1.23)
$$\widehat{\nabla}_{\partial/\partial\lambda} = \partial_\lambda + \mathcal{U} + \frac{1}{\lambda}\left(\mathcal{V}+\frac{D}{2}\mathrm{Id}\right).$$

Let M,N be two $\mathbf{C}[\lambda,\partial_\lambda]$-modules. *A formal Laplace transform* $M\to N$: $Y\mapsto Y^t$ is a $\mathbf{C}$-linear map for which

(1.24)
$$(-\lambda Y)^t = \partial_\lambda(Y^t),\ (\partial_\lambda Y)^t = \lambda Y^t.$$

The archetypal Laplace transform is the Laplace integral

(1.25)
$$Y^t(\mu) = \int e^{-\lambda\mu}Y(\lambda)d\lambda$$

taken along a contour (not necessarily closed) in $\mathbf{P}^1(\mathbf{C})$. In an analytical setting we have to secure the convergence of (1.25), the possibility to derivate under the integral sign and the identity

$$\int \partial_\lambda(e^{-\lambda\mu}Y(\lambda))d\lambda = 0.$$

However, (1.25) may admit other interpretations, for instance, in terms of asymptotic series.

Now let M (resp. N) be two $\mathbf{C}[\lambda,\partial_\lambda]$-modules of local (or formal, or distribution) sections of $\mathbf{P}^1_\lambda \times T$ so that the operators $\check{\nabla} \cdot (\mathcal{U} - \lambda)$ (resp. $\lambda\widehat{\nabla}$) make sense in M (resp. N) (cf. (1.22), resp. (1.23)), and assume that we are given a formal Laplace transform $M \to N$.

1.3.1. Proposition. *We have:*

$$[\check{\nabla}_{\partial/\partial\lambda}((\mathcal{U} - \lambda)Y)]^t = (\lambda\widehat{\nabla}_{\partial/\partial\lambda} + \frac{1-D}{2})Y^t = \lambda^{\frac{D+1}{2}}\widehat{\nabla}_{\partial/\partial\lambda}(\lambda^{\frac{1-D}{2}}Y^t).$$

In particular, $\lambda^{\frac{1-D}{2}}Y^t$ is $\widehat{\nabla}$-horizontal, if $(\mathcal{U} - \lambda)Y$ is $\check{\nabla}$-horizontal.

Proof. Using (1.22)–(1.24), we find:

$$[\check{\nabla}_{\partial/\partial\lambda}((\mathcal{U} - \lambda)Y)]^t = \left[(\partial_\lambda \cdot (\mathcal{U} - \lambda) + \mathcal{V} + \frac{1}{2}\mathrm{Id})Y\right]^t$$

$$= \left[\lambda(\mathcal{U} + \partial_\lambda) + \mathcal{V} + \frac{1}{2}\mathrm{Id}\right]Y^t$$

$$= \left[\lambda\widehat{\nabla}_{\partial/\partial\lambda} + \frac{1-D}{2}\mathrm{Id}\right]Y^t = \lambda^{\frac{D+1}{2}}\widehat{\nabla}_{\partial/\partial\lambda}(\lambda^{\frac{1-D}{2}}Y^t).$$

This proves our assertion.

For a more detailed discussion of the formal Laplace transform, see [Sa], 1.6.

In [MM] a further deformation of $\check{\nabla}$ was constructed. It will be used later in §3. Namely, for any constant s put

(1.26) $$\check{\nabla}^{(s)}_X(Y) = \check{\nabla}_X(Y) - s(\mathcal{U} - \lambda)^{-1}(X \circ Y),$$

(1.27) $$\check{\nabla}^{(s)}_{\partial/\partial\lambda}(Y) = \check{\nabla}_{\partial/\partial\lambda}(Y) + s(\mathcal{U} - \lambda)^{-1}(Y).$$

1.4. Theorem. *$\check{\nabla}^{(s)}_X$ is a flat connection on $\check{T}$.*

This can be checked by a direct calculation similar to that in the proof of Theorem 1.2, but the following calculation of the formal Laplace transform is more illuminating:

$$[\check{\nabla}^{(s)}_{\partial/\partial\lambda}((\mathcal{U} - \lambda)Y)]^t = (\lambda\widehat{\nabla}_{\partial/\partial\lambda} + \frac{1-D+2s}{2})Y^t = \lambda^{\frac{D+1-2s}{2}}\widehat{\nabla}_{\partial/\partial\lambda}(\lambda^{\frac{1-D+2s}{2}}Y^t).$$

In particular, $\lambda^{\frac{1-D+2s}{2}}Y^t$ is $\widehat{\nabla}$-horizontal, if $(\mathcal{U} - \lambda)Y$ is $\check{\nabla}^{(s)}$-horizontal.

Now we will more systematically review the deformation picture.

§2. Isomonodromic deformations

2.1. Singularities of meromorphic connections. Let N be a complex manifold, $D \subset N$ a closed complex submanifold of codimension one, and $\mathcal{F}$ a locally free sheaf of finite rank on N. A meromorphic connection with singularities on D is given by a covariant differential $\nabla : \mathcal{F} \to \mathcal{F} \otimes \Omega^1_N((r+1)D)$ for some $r \geq 0$. It is called flat (or integrable) if it is flat outside D. We start with a list of elementary notions and constructions that will be needed later. They depend only on the local behavior of $\mathcal{F}$ and ∇ in a neighborhood of D, so we will assume D is irreducible.

i) Order of singularity. We will say that ∇ as above is of order $\leq r+1$ on D if $\nabla_X(\mathcal{F}) \subset \mathcal{F}(rD)$ for any vector field X tangent to D (i.e. satisfying $XJ_D \subset J_D$ where J_D is the ideal of D), and $\nabla_X(\mathcal{F}) \subset \mathcal{F}((r+1)D)$ in general. Locally, if $(t^0, t^1, \ldots, t^n)$ is a coordinate system on N such that $t^0 = 0$ is the equation of D, the connection matrix of ∇ in a basis of $\mathcal{F}$ can be written as

$$(2.1) \qquad G_0 \frac{dt^0}{(t^0)^{r+1}} + \sum_{i=1}^n G_i \frac{dt^i}{(t^0)^r},$$

where $G_i = G_i(t^0, t^1, \ldots, t^n)$ are holomorphic matrix functions.

Note that $G_0(0, t^1, \ldots, t^n) \in H^0(D, \operatorname{End}\mathcal{F})$ is well defined, i.e. it does not depend on the choice of local coordinates. It is called the residue of ∇ at D and will be denoted by $\operatorname{res}_D(\nabla)$.

ii) Restriction to a transversal submanifold. Let $i : N' \to N$ be a closed embedding of a submanifold transversal to D, $D' = N' \cap D$, and $\mathcal{F}' = i^*(\mathcal{F})$. Then the induced connection $\nabla' = i^*(\nabla)$ on $\mathcal{F}'$ is flat and of order $\leq r+1$ on D' if ∇ has these properties.

iii) Residual connection. Assume now that ∇ is of order ≤ 1 on D. For any given local trivialization f of $\mathcal{O}(D)$, one can define a connection without singularities $\nabla^{D,f}$ on $j^*(\mathcal{F})$ where j is the embedding of D in N. Namely, to define $\nabla^{D,f}_{X'}(s')$ where $s' \in j^*(\mathcal{F})$, $X' \in T_D$, we locally extend s' to a section s of $\mathcal{F}$, X' to a vector field X on N, $\operatorname{res}_D(\nabla)$ to a section $\operatorname{res}(\nabla)$ of $\mathcal{F} \otimes \mathcal{F}^*$ on N, calculate $(\nabla_X - \frac{Xf}{f}\operatorname{res}(\nabla))(s)$ and restrict it to D. One checks that the result does not depend on the choices made. In the notation of (2.1), the matrix of the residual connection can be written as ($r = 0$):

$$(2.2) \qquad \sum_{i=1}^n G_i(0, t^1, \ldots, t^n) dt^i.$$

If ∇ is flat, $\nabla^{D,f}$ is flat for any local trivialization f.

iv) Principal part of order $r+1$. Similarly to (2.2), we can consider the matrix function on D

$$(2.3) \qquad G_0(0, t^1, \ldots, t^n),$$

which we will call the principal part of order $r+1$ of ∇. In more invariant terms, it is the $\mathcal{O}_D$-linear map $j^*(\mathcal{F}) \to j^*(\mathcal{F})$ induced by $\mathcal{F} \to j^*(\mathcal{F}) : s \mapsto (t^0)^{r+1} \nabla_{\partial/\partial t^0}(s)|_D$. For $r \geq 1$ it depends on the choice of local coordinates, and is multiplied by an invertible local function on D when this choice is changed. Hence its spectrum is well defined globally on D for $r = 0$, and the simplicity of the spectrum makes sense for any r.

v) Tameness and resonance. Two general position conditions are important in the study of meromorphic singularities of order $\leq r + 1$.

If $r \geq 1$ (irregular case), the singularity is called *tame*, if the spectrum of its principal part at any point of D is simple.

If $r = 0$ (regular case), the singularity is called *non-resonant*, if it is tame and, moreover, the difference of any two eigenvalues never takes an integer value on D.

2.1.1. Example: The structure connections. As in 1.3, we will assume that $\mathcal{T}_M^f$ is trivial, and its fibers are identified with the space T of global flat vector fields.

Put $N = M \times \mathbf{P}_\lambda^1$, $\mathcal{F} = \mathcal{O}_N \otimes T$. We can apply the previous considerations to $\widehat{\nabla}$ and $\widecheck{\nabla}$.

Analysis of $\widehat{\nabla}$. Clearly, $\widehat{\nabla}$ has singularity of order 1 at $\lambda = 0$ (i.e. on $D_0 = M \times \{0\}$) and of order 2 at $\lambda = \infty$ (i.e. on $D_\infty = M \times \{\infty\}$): cf. (1.9) and (1.10). Restricting $\widehat{\nabla}$ to $\{y\} \times \mathbf{P}_\lambda^1$ for various $y \in M$ we get a family of meromorphic connections on $\mathbf{P}_\lambda^1$ parametrized by M.

The residual connection is defined on $D_0 = M$ and it coincides with the Levi–Civita connection of g. The principal part of order 1 on D_0 is $\mathcal{V} + \dfrac{D}{2}\,\mathrm{Id}$. The eigenvalues of this operator do not depend on $y \in D_0$: in Ch. I, 2.4, they were denoted (d_a). Their description for the case of quantum cohomology (see Ch. I, (4.12)) shows that in this case the principal part *is always resonant*.

The principal part of order 2 on $D_\infty = M$ is (proportional to) $\mathcal{U}$ (cf. (1.10), use the local equation $\mu = \lambda^{-1} = 0$ for D_∞). Its eigenvalues now depend on $y \in M$: they are just the canonical coordinates $u^i(y)$. We will call the point y *tame* if $u^i(y) \neq u^j(y)$ for $i \neq j$. We will call M tame if all its points are tame. Every M contains the maximum tame subset which is open and dense.

Analysis of $\widecheck{\nabla}$. According to (1.7), (1.8), $\widecheck{\nabla}$ has singularities of order 1 at the divisors $\lambda = u^i$ and $\lambda = \infty$. These divisors do not intersect pairwise iff M is tame. The principal part of order 1 at $\lambda = u^i$ is $-(\mathcal{V} + \dfrac{1}{2}\,\mathrm{Id}) \cdot (e_i \circ)$.

The residual connection of $\widecheck{\nabla}$ on $\lambda = \infty$ is again the Levi–Civita connection ∇_0 of g. In fact, using (1.15) we find

$$\widecheck{\nabla} = d\lambda \widecheck{\nabla}_{\partial/\partial\lambda} + \sum_i du^i \widecheck{\nabla}_{e_i}$$

$$= d\lambda \widecheck{\nabla}_{\partial/\partial\lambda} + \sum_i du^i [\nabla_{0.e_i} - (\mathcal{V} + \frac{1}{2}\,\mathrm{Id})(\mathcal{U} - \lambda)^{-1}(e_i \circ)].$$

Replacing λ by the local parameter $\mu = \lambda^{-1}$ at infinity, we have

$$\widecheck{\nabla} = d\mu \widecheck{\nabla}_{\partial/\partial\mu} + \sum_i du^i [\nabla_{0.e_i} - \mu(\mathcal{V} + \frac{1}{2}\,\mathrm{Id})(\mu\mathcal{U} - \mathrm{Id})^{-1}(e_i \circ)]$$

so that the expression (2.2) (with $(\mu, u^1, \ldots, u^m)$ in lieu of $(t^0, t^1, \ldots, t^n)$) becomes $\sum_i du^i \nabla_{0, e_i} = \nabla_0$.

§2. ISOMONODROMIC DEFORMATIONS

2.2. Versal deformation. We will now review the basic results on the deformation of meromorphic connections on $\mathbf{P}^1_\lambda$, restricting ourselves to the case of singularities of order ≤ 2. This suffices for applications to both structure connections; on the other hand, this is precisely the case treated in full detail by B. Malgrange in [Mal4], Theorem 3.1. It says that the positions of finite poles and the spectra of the principal parts of order 2 form coordinates on the coarse moduli space with tame singularities. To be more precise, one has to rigidify the data slightly.

Let ∇^0 be a meromorphic connection on a locally free sheaf $\mathcal{F}^0$ on $\mathbf{P}^1_\lambda$ of rank p, with $m+1 \geq 2$ tame singularities (including $\lambda = \infty$) of order ≤ 2. Call *the rigidity* for ∇^0 the following data:

a) A numbering of singular points: $a_0^1, \ldots, a_0^m$, $a^{m+1} = \infty$.

b) The subset $I \subset \{1, \ldots, m+1\}$ such that a_0^j is of order 2 exactly when $j \in I$.

c) For each $j \in I$, a numbering $(b_0^{j1}, \ldots, b_0^{jp})$ of the eigenvalues of the principal part at a_0^j.

Construct the space $B = B(m, p, S)$ as the universal covering of

$$(\mathbf{C}^m \setminus \text{diagonals}) \times \prod_{j \in I}(\mathbf{C}^p \setminus \text{diagonals})$$

with the base point $(a_0^i; b_0^{jk})$; let $b_0 \in B$ be its lift. We denote by a^i, b^{jk} the coordinate functions lifted to B. Let $i : \mathbf{P}^1_\lambda \to B \times \mathbf{P}^1_\lambda$ be the embedding $\lambda \mapsto (b_0, \lambda)$, and D_j the divisor $\lambda = a^j$ in $B \times \mathbf{P}^1_\lambda$.

2.2.1. Theorem ([Mal4], Th. 3.1). *For a given $(\nabla^0, \mathcal{F}^0)$ with rigidity, there exists a locally free sheaf $\mathcal{F}$ of rank p on $\mathbf{P}^1_\lambda \times B$, a flat meromorphic connection ∇ on it, and an isomorphism $i^0 : i^*(\mathcal{F}, \nabla) \to (\mathcal{F}^0, \nabla^0)$ with the following properties:*

D_j, $j = 1, \ldots, m+1$, *are all the poles of ∇, of order 1 (resp. 2) if $j \neq I$ (resp. $j \in I$). If $j \in I$, then $(b^{j1}, \ldots, b^{jp})$ (as functions on D_j) form the spectrum of the principal part of order 2 of ∇ at D_j.*

It follows that the restrictions of ∇ to the fibers $\{b\} \times \mathbf{P}^1_\lambda$ are endowed with the induced rigidity, and i^0 is compatible with it.

The data $(\mathcal{F}, \nabla, i^0)$ are unique up to unique isomorphism.

2.2.2. Comments on the proof. a) The case when all singularities are of order 1 is easier. It is treated separately in [Mal3], Th. 2.1; for the thorough study of this case and the treatment of the Gauss–Manin connections see [Del]. Since the second structure connection satisfies this condition, we sketch Malgrange's argument in this case.

Choose base points $a \in U := \mathbf{P}^1_\lambda \setminus \bigcup_{j=1}^{m+1}\{a_0^j\}$ and $(b_0, a) \in B \times \mathbf{P}^1_\lambda$. Notice that (b_0, a) belongs to $V := B \times \mathbf{P}^1_\lambda \setminus \bigcup_{j=1}^m D_j$.

The restriction of $(\mathcal{F}^0, \nabla^0)$ to U is determined uniquely up to unique isomorphism by the monodromy action of $\pi_1(U, a)$ on the space F, the geometric fiber $\mathcal{F}^0(a)$ at a, which can be arbitrary. Similarly, there is a bijection between flat connections $(\mathcal{F}, \nabla)$ on V with fixed identification $\mathcal{F}^0(a) \to \mathcal{F}(a) = F$ and actions of $\pi_1(V, (a, b))$ on F. Hence to construct an extension $(\mathcal{F}, \nabla)$ to V together with an isomorphism of its restriction to U with $(\mathcal{F}^0, \nabla^0)$, it suffices to check that i induces an isomorphism $\pi_1(U, a) \to \pi_1(V, (a, b))$, which follows from the homotopy exact sequence and the fact that B is contractible.

This argument explains the term "isomonodromic deformation".

Next, we must extend $(\mathcal{F}, \nabla)$ to $B \times \mathbf{P}^1_\lambda$. It suffices to do this separately in a tubular neighborhood of each D_j disjoint from other D_k. The coordinate change $\lambda \mapsto \lambda - a^j$ (or $\lambda \mapsto \lambda^{-1}$) allows us to assume that the equation of D_j is $\lambda = 0$. Take a neighborhood W of 0 in which $\mathcal{F}^0$ can be trivialized, describe ∇^0 by its connection matrix, lift $(\mathcal{F}^0, \nabla^0)$ to $B \times W$, and restrict to a tubular neighborhood of D_j. On the complement to D_j, this lifting can be canonically identified with $(\mathcal{F}, \nabla)$ through their horizontal sections. Clearly, it is of order ≤ 1 at D_j.

It remains to establish that any two extensions are canonically isomorphic. Outside singularities, an isomorphism exists and is unique. An additional argument which we omit shows that it extends holomorphically to $B \times \mathbf{P}^1_\lambda$.

b) When ∇ admits singularity of order 2, this argument must be completed. The extension of $(\mathcal{F}^0, \nabla^0)$ first to V and then to the singular divisors of order ≤ 1 can be done exactly as before. But both the existence and the uniqueness of the extension to the irregular singularities requires an additional local analysis in order to show that the simple spectrum of the principal polar part determines the singularity. When formulated in terms of the asymptotic behavior of horizontal sections, this analysis introduces the Stokes data as a version of irregular monodromy, which also proves to be deformation invariant.

2.3. The theta divisor and Schlesinger's equations.

In this subsection we will assume that $\mathcal{F}^0 = T \otimes \mathcal{O}_{\mathbf{P}^1_\lambda}$, where T is a finite-dimensional vector space which can be identified with the space of global sections of $\mathcal{F}^0$. This is the case of the two structure connections, when the local system $\mathcal{T}^f_M$ is trivial.

Then there exists a divisor Θ, eventually empty, such that the restriction of $\mathcal{F}$ to all fibers $\{b\} \times \mathbf{P}^1_\lambda$, $b \notin \Theta$, is free. This can be proved using the fact that a locally free sheaf $\mathcal{E}$ on $\mathbf{P}^1$ is free iff $H^0(\mathbf{P}^1, \mathcal{E}(-1)) = H^1(\mathbf{P}^1, \mathcal{E}(-1)) = 0$, and that the cohomology of fibers is semicontinuous. For an analytic treatment, see [Mal4], §§4 and 5.

Moreover, assume that $\lambda = \infty$ is a singularity of order 1 (to achieve this for the first structure connection, we must replace λ by λ^{-1}). Then we can identify the inverse image of $\mathcal{F}$ on $B \setminus \Theta \times \mathbf{P}^1_\lambda$ with $T \otimes \mathcal{O}_{B \setminus \Theta \times \mathbf{P}^1_\lambda}$ compatibly with the respective trivialization of $\mathcal{F}^0$. To this end trivialize $\mathcal{F}$ along $\lambda = \infty$ using the residual connection (see 2.1 iii)) and then take the constant extension of each residually horizontal section along $\mathbf{P}^1_\lambda$. (If there are no poles of order 1, one can extend this argument using a different version of the residual connection; see [Mal4], p.430, Remarque 1.4.)

Using this trivialization, we can define a meromorphic integrable connection ∂ on $\mathcal{F}$ with the space of horizontal sections T on $B \setminus \Theta \times \mathbf{P}^1_\lambda$. As sections of $\mathcal{F}$, they develop a singularity at Θ. Therefore, the respective connection form $\nabla - \partial$ is a meromorphic matrix one–form with eventual pole at Θ.

The following classical result clarifies the structure of this form in the case *when all poles of ∇ are of order 1*.

2.3.1. Theorem. *a) Let $(a^1, \ldots, a^m)$ be the functions on B describing the λ-coordinates of finite poles of ∇ (with given rigidity). Then*

$$(2.4) \qquad \nabla = \partial + \sum_{i=1}^m A_i(a^1, \ldots, a^m) \frac{d(\lambda - a^i)}{\lambda - a^i},$$

where A_i are meromorphic functions $B \to \text{End}(T)$ which can be considered as multivalued meromorphic functions of a_i.

b) The connection (2.4) is flat iff A_i satisfy the Schlesinger equations

$$\forall j, \quad dA_j = \sum_{i \neq j} [A_i, A_j] \frac{d(a^i - a^j)}{a^i - a^j}. \tag{2.5}$$

c) Fix a tame point $a_0 = (a_0^1, \ldots, a_0^m)$. Then arbitrary initial conditions $A_i^0 = A_i(a_0)$ define a solution of (2.5) holomorphic on $B \setminus \Theta$, with eventual pole at Θ of order 1.

d) For any such solution ∇ of (2.5), define the meromorphic 1-form on B:

$$\omega_\nabla := \sum_{i<j} \text{Tr}(A_i A_j) \frac{d(a^i - a^j)}{a^i - a^j}. \tag{2.6}$$

This form is closed, and for any local equation $t = 0$ of Θ the form $\omega_\nabla - \dfrac{dt}{t}$ is locally holomorphic.

2.3.2. Corollary. *For any solution ∇ to (2.5), there exists a holomorphic function τ_∇ on B such that $\omega_\nabla = d\log \tau_\nabla$. It is defined uniquely up to multiplication by a constant.*

In fact, B is simply connected.

Naturally enough, τ_∇ is called *the tau-function* of the respective solution. In the case of (semisimple) quantum cohomology, it turns out to be the main ingredient of the genus one potential: see [DZh2].

For a proof of Theorem 2.3.1, we refer to [Mal3]: a), b), and c) are proved on pp. 406–410, d) on pp. 420–425.

2.4. Hamiltonian structure of Schlesinger's equations. The equations (2.5) can be written in Hamiltonian form, with m times and m time-dependent Hamiltonians.

To be more precise, let X be a manifold with a Poisson structure given by the Poisson bracket $\{\,,\,\}$, S a manifold with a coordinate system $(t^1, \ldots, t^m)$, and $(\mathcal{H}_1, \ldots, \mathcal{H}_m)$ a family of functions on $X \times S$ called Hamiltonians. Extend the bracket to $X \times S$ fiberwise. Then we can define m flows on X such that the evolution of any function F is governed by the equations:

$$\frac{\partial F}{\partial t^j} = \{\mathcal{H}_j, F\}. \tag{2.7}$$

These flows commute iff

$$\forall j, k: \quad \{\mathcal{H}_j, \mathcal{H}_k\} = \frac{\partial \mathcal{H}_k}{\partial t^j} - \frac{\partial \mathcal{H}_j}{\partial t^k}. \tag{2.8}$$

To represent (2.5) in this form, we choose $X = (\text{End}\,T)^m$, $S = B$. The Poisson structure will be the product of m standard Poisson structures on the matrix spaces. If we choose a basis in T and identify $\text{End}\,T$ with the space of matrices $(A_{\alpha\beta})$, the bracket of two matrix elements is

$$\{A_{\alpha\beta}, A_{\gamma\delta}\} = \delta_{\beta\gamma} A_{\alpha\delta} - \delta_{\alpha\delta} A_{\gamma\beta}. \tag{2.9}$$

(I apologize for using the subscript δ in the Kronecker delta symbol.)

Finally, put:

$$(2.10) \qquad \mathcal{H}_j = -\sum_{i:\, i \neq j} \frac{\text{Tr}(A_i A_j)}{a^i - a^j}.$$

2.4.1. Theorem. *Schlesinger's equations (2.5) are equivalent to the equations*

$$(2.11) \qquad \forall\, i, j, \alpha, \beta: \qquad \frac{\partial A_{j\alpha\beta}}{\partial a^i} = \{\mathcal{H}_i, A_{j\alpha\beta}\}.$$

The flows (2.11) pairwise commute.

Proof. Rewrite (2.5) as

$$(2.12) \qquad \frac{\partial A_{j\alpha\beta}}{\partial a^j} = -\sum_{i:\, i \neq j} \frac{[A_i, A_j]_{\alpha\beta}}{a^i - a^j},$$

$$(2.13) \qquad \frac{\partial A_{j\alpha\beta}}{\partial a^i} = \frac{[A_i, A_j]_{\alpha\beta}}{a^i - a^j}, \quad i \neq j.$$

On the other hand, in view of (2.10),

$$(2.14) \qquad \{\mathcal{H}_j, A_{j\alpha\beta}\} = -\sum_{i:\, i \neq j} \frac{\{\text{Tr}(A_i A_j), A_{j\alpha\beta}\}}{a^i - a^j},$$

$$(2.15) \qquad \{\mathcal{H}_i, A_{j\alpha\beta}\} = \frac{\{\text{Tr}(A_i A_j), A_{j\alpha\beta}\}}{a^i - a^j}, \quad i \neq j.$$

(Notice that the matrix elements of A_j and A_k pairwise Poisson commute if $j \neq k$.) A straightforward calculation using (2.9) then shows that (2.12) (resp. (2.13)) coincides with (2.14) (resp. (2.15)).

The fact that flows (2.11) pairwise commute means that the trajectories of the flows starting at one point are all contained in a multisection of p which is equivalent to the flatness of ∇ and to (2.5).

§3. Semisimple Frobenius manifolds as special solutions to the Schlesinger equations

3.1. Special solutions. Slightly generalizing (2.5), we will call *a solution to Schlesinger's equations* any data $(M, (u^i), T, (A_i))$, where M is a complex manifold of dimension $m \geq 2$; $(u^1, \ldots, u^m)$ a system of holomorphic functions on M such that du^i freely generate Ω^1_M and, for any $i \neq j$, $x \in M$, we have $u^i(x) \neq u^j(x)$; T a finite-dimensional complex vector space; and $A_j : M \to \text{End}\, T$, $j = 1, \ldots, m$, a family of holomorphic matrix functions such that

$$(3.1) \qquad \forall j: \qquad dA_j = \sum_{i:\, i \neq j} [A_i, A_j] \frac{d(u^i - u^j)}{u^i - u^j}.$$

Let such a solution be given. Summing (3.1) over all j, we find $d(\sum_j A_j) = 0$. Hence $\sum_j A_j$ is a constant matrix function; denote its value by $\mathcal{W}$.

3.1.1. Definition. *A solution to Schlesinger's equations as above is called special, if* $\dim T = m = \dim M$; T *is endowed with a complex non-degenerate quadratic form* g; $\mathcal{W} = -\mathcal{V} - \frac{1}{2}\operatorname{Id}$, *where* $\mathcal{V} \in \operatorname{End} T$ *is a skew symmetric operator with respect to* g; *and finally*

$$(3.2) \qquad \forall j: \qquad A_j = -(\mathcal{V} + \frac{1}{2}\operatorname{Id})P_j,$$

where $P_j : M \to \operatorname{End} T$ *is a family of holomorphic matrix functions whose values at any point of* M *constitute a complete system of orthogonal projectors of rank one with respect to* g:

$$(3.3) \qquad P_i P_k = \delta_{ik} P_i, \quad \sum_{i=1}^{m} P_i = \operatorname{Id}_T, \quad g(\operatorname{Im} P_i, \operatorname{Im} P_j) = 0$$

if $i \neq j$. *Moreover, we require that the* A_j *do not vanish at any point of* M.

3.1.2. Comment. We commited a slight abuse of language: the notion of special solution involves a choice of additional data, the metric g. However, when it is chosen, the rest of the data is defined unambiguously if it exists at all.

In fact, assume that $A_j = \mathcal{W} P_j$ as above do not vanish anywhere. Then they have constant rank one. Hence at any point of M we have

$$\operatorname{Ker} A_j = \operatorname{Ker} \mathcal{W} P_j = \operatorname{Ker} P_j = \bigoplus_{i:\, i \neq j} \operatorname{Im} P_i,$$

so that

$$\operatorname{Im} P_i = \bigcap_{j:\, j \neq i} \bigoplus_{k:\, k \neq j} \operatorname{Im} P_k = \bigcap_{j:\, j \neq i} \operatorname{Ker} A_j.$$

This means that P_j can exist for given A_j only if the spaces $T_j = \bigcap_{i:\, i \neq j} \operatorname{Ker} A_i$ are one–dimensional and pairwise orthogonal at any point of M.

Conversely, assume that this condition is satisfied. Define P_j as the orthogonal projector onto T_j. Then $A_i P_j = 0$ for $i \neq j$ because $T_j = \operatorname{Im} P_j \subset \operatorname{Ker} A_i$. Hence

$$A_j = A_j(\sum_{i=1}^m P_i) = A_j P_j = (\sum_{i=1}^m A_i) P_j = \mathcal{W} P_j.$$

Notice that all A_j are conjugate to $\operatorname{diag}(-\frac{1}{2}, 0, \ldots, 0)$ and satisfy $A_j^2 + \frac{1}{2} A_j = 0$. These conditions, as well as $\sum_j A_j = -(\mathcal{V} + \frac{1}{2}\operatorname{Id})$, are compatible with the equations (3.1) and so must be checked at one point only.

3.1.3. Strictly special solutions. A special solution to Schlesinger's equations as above is called *strictly special* if the operators

$$A_j^{(t)} := A_j + t P_j$$

also satisfy Schlesinger's equations for any $t \in \mathbf{C}$.

3.1.4. Lemma. *If $\mathcal{W}$ is invertible, then any special solution with given $\mathcal{W}$ is strictly special.*

Proof. Inserting $A_j^{(t)}$ into (3.1) one sees that the solution is strictly special iff

$$\forall j: \quad dP_j = \sum_{i:\, i \neq j} (P_i \mathcal{W} P_j - P_j \mathcal{W} P_i) \frac{d(u^i - u^j)}{u^i - u^j}.$$

On the other hand, replacing A_k by $\mathcal{W} P_k$ in (3.1), one sees that after left multiplication by $\mathcal{W}$ this becomes a consequence of (3.1).

3.2. From Frobenius manifolds to special solutions. Given a semisimple Frobenius manifold with flat identity and an Euler field E with $d_0 = 1$, we can produce a special solution to Schlesinger's equations rephrasing the results of the previous two sections.

Namely, we first pass to a covering M of the subspace of tame points of the initial manifold such that $\mathcal{T}_M^f$ is trivial and a global splitting can be chosen, represented by the canonical coordinates (u^i). Then we put $T = \Gamma(M, \mathcal{T}_M^f)$ and $A_i = $ the coefficients of the second structure connection written as in (2.4).

Since this connection is flat, $(M, (u^i), T, (A_i))$ form a solution to (3.1).

Moreover, this solution is special. In fact, T comes equipped with the metric g. The operator A_i is the principal part of order 1 of $\check{\nabla}$ at $\lambda = u^i$ which is of the form (3.2), with $P_j = e_j \circ$.

Finally, this special solution comes with one more piece of data, the identity $e \in T$. We will axiomatize its properties in the following definition.

3.2.1. Definition. *Consider a special solution to Schlesinger's equations as in Definition 3.1.1. A vector $e \in T$ is called an identity of weight D for this solution, if*

a) $\mathcal{V}(e) = (1 - \dfrac{D}{2}) e.$

b) $e_j := P_j(e)$ *do not vanish at any point of M.*

For Frobenius manifolds with $d_0 = 1$, a) is satisfied by Ch. I, (2.16) and (1.2).

Merkulov's Theorem 1.4 shows that in this way we always obtain strictly special solutions, although the operator $\mathcal{W}$ need not be invertible. For example, from Ch. I, (4.12), it follows that for quantum cohomology of $\mathbf{P}^r$ (which is semisimple, cf. below) the spectrum of $\mathcal{W}$ is $\{a - \dfrac{r+1}{2} \mid a = 0, \ldots, r\}$. It contains 0 if r is odd.

3.3. From special solutions to Frobenius manifolds. Let $(M, (u^i), T, g, (A_i))$ be a strictly special solution, and $e \in T$ an identity of weight D for it.

3.3.1. Theorem. *These data come from the unique structure of the semisimple split Frobenius manifold on M, with flat identity and Euler field, as described in 3.2.*

Proof. Proceeding as in 3.2, but in the reverse direction, we are bound to make the following choices.

Put $e_j = P_j(e) \subset \mathcal{O}_M \otimes T$, $j = 1, \ldots, m$. Identify $\mathcal{O}_M \otimes T$ with $\mathcal{T}_M$ by setting $e_j = \partial/\partial u^j$. Transfer the metric g from T to $\mathcal{T}_M$. Define the multiplication on $\mathcal{T}_M$ for which $e_i \circ e_j = \delta_{ij} e_j$. Put $\eta_i := g(e_i, e_i)$.

Let $\mathcal{T}_M^f$ be the image of T under this identification. We will first check that it is an abelian Lie subalgebra of $\mathcal{T}_M$. It will then follow that g is flat, so that we get a structure of the semisimple pre–Frobenius manifold in the sense of Ch. I, Definition 3.2.

Choose $t \in \mathbf{C}$ in such a way that

$$\mathcal{W}^{(t)} := \sum_j A_j^{(t)} = \mathcal{W} + t\,\mathrm{Id} \in \mathrm{End}\, T$$

is invertible. The section $X = \sum_j f_j e_j$ of $\mathcal{O}_M \otimes T$ lands in $\mathcal{T}_M^f$ iff

$$\mathcal{W}^{(t)} X = \sum_j f_j \mathcal{W}^{(t)} P_j(e) = \left(\sum_j f_j A_j^{(t)}\right)(e) \in T.$$

Let ∇ be the connection on $\mathcal{T}_M$ for which $\mathcal{T}_M^f$ is horizontal. Applying it to $\left(\sum_j f_j A_j^{(t)}\right)(e)$ we see that the last condition is in turn equivalent to

$$(*) \qquad \forall k: \quad \sum_j \frac{\partial f_j}{\partial u^k} A_j^{(t)}(e) = -\sum_j f_j \frac{\partial A_j^{(t)}}{\partial u^k}(e).$$

We can similarly rewrite the condition $Y := \sum_j g_j e_j \in \mathcal{T}_M^f$.

The commutator of vector fields induces on $\mathcal{O}_M \otimes T$ the bracket

$$[X, Y] = \sum_{j,k} \left(f_j \frac{\partial g_k}{\partial u^j} - g_j \frac{\partial f_k}{\partial u^j}\right) e_k.$$

From (*) for Y and X we find:

$$\sum_{j,k} f_j \frac{\partial g_k}{\partial u^j} A_k^{(t)}(e) = -\sum_j f_j \sum_k g_k \frac{\partial A_k^{(t)}}{\partial u^j}(e),$$

$$\sum_{j,k} g_j \frac{\partial f_k}{\partial u^j} A_k^{(t)}(e) = -\sum_j g_j \sum_k f_k \frac{\partial A_k^{(t)}}{\partial u^j}(e).$$

The terms $j = k$ on the right hand sides are the same. For $j \neq k$, using the strict speciality of our solution, we find

$$f_j g_k \frac{\partial A_k^{(t)}}{\partial u^j} = f_j g_k \frac{[A_j^{(t)}, A_k^{(t)}]}{u^j - u^k}$$

so that the (j, k)-term of the first identity cancels with the (k, j)-term of the second one.

To establish that this pre–Frobenius manifold is Frobenius, it suffices to prove that $e_i \eta_j = e_j \eta_i$ for all i, j: see Ch. I, Theorem 3.3.

We have $\eta_j = g(e, e_j)$. Therefore

(3.4) $\quad g(e, A_j(e)) = -g(e, (\mathcal{V} + \frac{1}{2}\text{Id})P_j e) = g(\mathcal{V}e, e_j) - \frac{1}{2}g(e, e_j) = \frac{1-D}{2}\eta_j$

since $\mathcal{V}$ is skew symmetric, and e is an eigenvector of $\mathcal{V}$. Furthermore, let ∇ be the Levi–Civita connection of the flat metric g. Then derivating (3.4) we find for every i, j:

$$\frac{1-D}{2}\frac{\partial}{\partial u^i}\eta_j = g(\nabla_{e_i}(e), A_j(e)) + g(e, \nabla_{e_i}(A_j(e)))$$

(3.5) $$= g(e, \frac{\partial A_j}{\partial u^i}(e)),$$

because $e \in T$ so that $\nabla(e) = 0$. If $i \neq j$, we find from (3.1) that

(3.6) $$\frac{\partial A_j}{\partial u^i} = \frac{[A_i, A_j]}{u^i - u^j} = \frac{\partial A_i}{\partial u^j}.$$

This shows that if $D \neq 1$, $e_i\eta_j = e_j\eta_i$. To see that $D = 1$ is not exceptional, one can replace A_j by $A_j^{(t)}$ in this argument for any $t \neq 0$, so that $\frac{1-D}{2}$ in (3.4) will become $\frac{1-D}{2} + t$.

It remains to check that $E = \sum_i u^i e_i$ is the Euler field. According to Theorem 3.6 b) of Ch. I, we must prove that $E\eta_j = (D-2)\eta_j$ for all j. Insert (3.6) into (3.5) and sum over $i \neq j$. We obtain:

$$\frac{1-D}{2}E\eta_j = \frac{1-D}{2}\sum_{i:\, i\neq j} u^i \frac{\partial \eta_j}{\partial u^i} + \frac{1-D}{2} u^j \frac{\partial \eta_j}{\partial u^j}$$

(3.7) $$= \sum_{i:\, i\neq j} g\left(e, u^i \frac{[A_i, A_j]}{u^i - u^j}(e)\right) + u^j g(e, \frac{\partial A_j}{\partial u^j}(e)).$$

From (3.1) it follows that

(3.8) $$\frac{\partial A_j}{\partial u^j} = -\sum_{i:\, i\neq j} \frac{[A_i, A_j]}{u^i - u^j}.$$

On the other hand,

(3.9) $$u^i \frac{[A_i, A_j]}{u^i - u^j} = [A_i, A_j] + u^j \frac{[A_i, A_j]}{u^i - u^j}.$$

Inserting (3.8) and (3.9) into (3.7), we find

$$\frac{1-D}{2}E\eta_j = \sum_{i:\, i\neq j} g(e, [A_i, A_j](e)) + u^j \sum_{i:\, i\neq j} g\left(e, \frac{[A_i, A_j]}{u^i - u^j}(e)\right)$$

$$+ u^j g(e, \frac{\partial A_j}{\partial u^j}(e)) = g(e, [\sum_{i:\, i\neq j} A_i, A_j](e))$$

(3.10) $$= -g(e, [\mathcal{V} + \frac{1}{2}\text{Id}, (\mathcal{V} + \frac{1}{2}\text{Id})P_j](e)).$$

Using the skew symmetry of $\mathcal{V}$, we see that the last expression in (3.10) equals $\dfrac{1-D}{2}(D-2)\eta_j$. Hence $E\eta_j = (D-2)\eta_j$ if $D \neq 1$.

Again, replacing A_j by $A_j^{(t)}$ in this argument we see that the restriction $D \neq 1$ is irrelevant.

3.4. Special initial conditions. Theorem 2.3.1 c) shows that arbitrary initial conditions to Schlesinger's equations determine a global meromorphic solution on the universal covering $B(m)$ of $\mathbf{C}^m \setminus \{\text{diagonals}\}, m \geq 2$.

Fix a base point $b_0 \in B(m)$. Studying the special solutions, we may and will identify T with the tangent space at b_0, thus eliminating the gauge freedom. This tangent space is already coordinatized: we have e_i and e.

We will call a family of matrices $A_1^0, \ldots, A_m^0 \in \operatorname{End} T$ *special initial conditions* if we can find a diagonal metric g and a skew symmetric operator $\mathcal{V}$ such that $A_j^0 = -(\mathcal{V} + \tfrac{1}{2}\operatorname{Id}) P_j$, where P_j is the projector onto $\mathbf{C} e_j$.

We will describe explicitly the space $I(m)$ of the special initial conditions.

3.4.1. Notation. Let R be any equivalence relation on $\{1, \ldots, m\}$, and $|R|$ the number of its classes. Put $F(m) = (\operatorname{End} \mathbf{C}^m)^m$. Furthermore, denote by $F_R(m)$ the subset of families $(A_1, \ldots, A_m)$ in $F(m)$ such that R coincides with the minimal equivalence relation for which iRj if $\operatorname{Tr} A_i A_j \neq 0$, and put $I_R(m) = F_R(m) \cap I(m)$.

3.4.2. Construction. Denote by $\overline{I}(m) \subset \mathbf{C}^m \times \mathbf{C}^{m(m-1)/2}$ the locally closed subset defined by the equations:

(3.11) $$\sum_{i=1}^m \eta_i = 0, \quad \eta_i \neq 0 \quad \text{for all } i;$$

(3.12) $$v_{ij}\eta_j = -v_{ji}\eta_i \quad \text{for all } i,j;$$

(3.13) $$\sum_{i=1}^m v_{ij} := 1 - \frac{D}{2} \quad \text{does not depend on } j.$$

Each point of $\overline{I}(m)$ determines the diagonal metric $g(e_i, e_i) = \eta_i$ and the operator $\mathcal{V}: e_i \mapsto \sum_i v_{ij} e_j$ which is skew symmetric with respect to g and for which e is an eigenvector. Setting $A_i = -(\mathcal{V} + \dfrac{1}{2}\operatorname{Id}) P_i$ we get a point in $I(m)$.

This amounts to forgetting (η_i) which furnishes the surjective map $\overline{I}(m) \to I(m)$ because

$$A_i(e_j) = 0 \text{ for } i \neq j, \quad A_i(e_i) = -\frac{1}{2} e_i - \sum_{j=1}^m v_{ij} e_j.$$

3.4.3. Theorem. *a) The space $\overline{I}(m)$ can be realized as a Zariski open dense subset in $\mathbf{C}^{m+(m-1)(m-2)/2}$.*

b) *The inverse image in $\bar{I}(m)$ of any point in $I_R(m)$ is a manifold of dimension 1 for $|R| = 1$, and $|R| - 1$ for $|R| \geq 2$.*

Proof. Fixing η_i, we can solve (3.12) and (3.13) explicitly. Put $w_{ij} = v_{ij}\eta_j$ so that $w_{ij} = -w_{ji}$ and (3.13) becomes

$$(3.14) \qquad \forall j: \quad \sum_{i=1}^{m} w_{ij} = \eta_j\left(1 - \frac{D}{2}\right).$$

If we arbitrarily choose the values (w_{ij}) for all $1 \leq i < j \leq m - 1$, we can find w_{mj} from the first $m - 1$ equations (3.14), and then the last equations will hold automatically:

$$w_{mk} = \eta_k\left(1 - \frac{D}{2}\right) - \sum_{i=1}^{m-1} w_{ik},$$

$$\sum_{i=1}^{m} w_{im} = -\sum_{k=1}^{m} w_{mk} = -\sum_{k=1}^{m-1} \eta_k\left(1 - \frac{D}{2}\right) + \sum_{i,k=1}^{m-1} w_{ik} = \eta_m\left(1 - \frac{D}{2}\right)$$

because of (3.11).

It remains to determine the fiber of the projection onto $I(m)$.

We have for $i \neq j$: $\operatorname{Tr} A_i A_j = v_{ij}v_{ji}$. Hence in the generic case when all these traces do not vanish, we can reconstruct η_i compatible with given v_{ij} from (3.12) uniquely up to a common factor. Generally, for i, j in the same R–equivalence class, (3.12) allows us to determine the value η_i/η_j so that we have $|R|$ overall arbitrary factors constrained by (3.11).

3.4.4. Question. If we choose a special initial condition for the Schlesinger equation, does the solution remain special at every point?

Generically, the answer is positive. If this is the case, we obtain the action of the braid group Bd_m as the group of deck transformations on the space $I(m)$.

3.5. Analytic continuation of the potential. The picture described in this section gives a good grip on the analytic continuation of a germ of semisimple Frobenius manifold (M_0, m_0) in terms of its canonical coordinates. Namely, construct the universal covering M of the subset of the tame points of M_0, then fix at the point $b_0 = (u^i(m_0)) \in B(m)$ the initial conditions of M at m_0. This provides an open embedding $(M, m_0) \subset (B(m), b_0)$. Loosely speaking, in this way we find a maximal tame analytic continuation of the initial germ.

Now construct some global flat coordinates (x^a) on $B(m)$ corresponding to a given Frobenius structure. They map $B(m)$ to a subdomain in $\mathbf{C}^m$. This is the natural domain of the analytic continuation of the potential Φ of this Frobenius structure, which is the most important object for Quantum Cohomology. Unfortunately, its properties are not clear from this description.

§4. Quantum cohomology of projective spaces

In this section we will apply the developed formalism to the study of the quantum cohomology of projective spaces $\mathbf{P}^r$, $r \geq 2$, first introduced in Ch. I, 4.4. Our main goal is the calculation of the initial conditions of the relevant solutions to Schlesinger's equations.

§4. QUANTUM COHOMOLOGY OF PROJECTIVE SPACES

4.1. Notation. We start by recalling (and somewhat revising) the basic notation. Put $H = H^*(\mathbf{P}^r, \mathbf{C}) = \sum_{a=0}^{r} \mathbf{C}\Delta_a$, $\Delta_a =$ the dual class of $\mathbf{P}^{r-a} \subset \mathbf{P}^r$. Denote the dual coordinates on H by $x_0, \ldots, x_r$ (lowering indices for visual convenience), $\partial_a = \partial/\partial x_a$. The Poincaré form is $(g_{ab}) = (g^{ab}) = (\delta_{a+b,r})$. The term $\frac{1}{6}(\gamma^3)$ in Ch. I, (4.13), is the cubic self–intersection form, the classical part of the Frobenius potential

$$(4.1) \qquad \Phi_{\mathrm{cl}}(x) := \frac{1}{6} \sum_{a_1+a_2+a_3=r} x_{a_1} x_{a_2} x_{a_3}.$$

The remaining part of the potential is the sum of physicists' instanton corrections to the self–intersection form:

$$(4.2) \qquad \Phi_{\mathrm{inst}}(x) := \sum_{d=1}^{\infty} \Phi_d(x_2, \ldots, x_r) e^{dx_1},$$

where we will now write Φ_d as

$$(4.3) \qquad \Phi_d(x_2, \ldots, x_r) = \sum_{n=2}^{\infty} \sum_{\substack{a_1+\cdots+a_n= \\ r(d+1)+d-3+n}} I(d; a_1, \ldots, a_n) \frac{x_{a_1} \cdots x_{a_n}}{n!}.$$

This means that if we assign the weight $a-1$ to x_a, $a = 2, \ldots, n$, then Φ_d becomes the weighted homogeneous polynomial of weight $(r+1)d + r - 3$. Moreover, if we assign to e^{dx_1} the weight $-(r+1)$, then Φ_{cl} and Φ become weighted homogeneous formal series of weight $r - 3$. (Notice that e in the expressions e^{dx_1} and the like is $2.71828\ldots$, whereas in other contexts e means the identity vector field. This cannot lead to confusion.)

The starting point of our study in this section will be the following result.

4.2. Theorem. *a) For each $r \geq 2$, there exists a unique formal solution of the Associativity Equations I.(1.6) of the form*

$$(4.4) \qquad \Phi(x) = \Phi_{\mathrm{cl}}(x) + \Phi_{\mathrm{inst}}(x)$$

for which $I(1; r, r) = 1$.

b) This solution has a non–empty convergence domain in H on which it defines the structure of the semisimple Frobenius manifold $H_{\mathrm{quant}}(\mathbf{P}^r)$ with flat identity $e = \partial_0$ and Euler field

$$(4.5) \qquad E = \sum_{a=0}^{r} (1-a) x^a \partial_a + (r+1) \partial_1$$

with $d_0 = 1, D = 2 - r$.

c) The coefficient $I(d; a_1, \ldots, a_n)$ is the number of rational curves of degree d in $\mathbf{P}^r$ intersecting n projective subspaces of codimensions $a_1, \ldots, a_n \geq 2$ in general position.

Uniqueness of the formal solution can be established by showing that the Associativity Equations imply recursive relations for the coefficients of Φ which allow one to express all of them through $I(1; r, r)$. This is an elementary exercise for $r = 2$ (cf. Introduction, (0.19)). A more general result (stated in the language of Gromov–Witten invariants but of essentially combinatorial nature) is proved in [KM1], Theorem 3.1, and applied to the projective spaces in [KM1], Claim 5.2.2.

Existence is a subtler fact. The algebraic geometric (or symplectic) theory of the Gromov–Witten invariants provides the numbers $I(d; a_1, \ldots, a_n)$ satisfying the necessary relations, together with their numerical interpretation: see III.5 below. Another approach consists in calculating *ad hoc* the "special initial conditions" for the semisimple Frobenius manifold $H_{\text{quant}}(\mathbf{P}^r)$ in the sense of the previous section and identifying the appropriate special solution to the Schlesinger equations with this manifold. For $r = 2$, direct estimates of the coefficients showing convergence can be found in [D2], p. 185. Probably, they can be generalized to all r.

Our approach in this section consists in taking Theorem 4.2 for granted and investigating the passage to the Darboux–Egoroff picture as a concrete illustration of the general theory. The net outcome are formulas (4.18) and (4.19) for the special initial conditions.

Conversely, starting with them, we can construct the Frobenius structure on the space $B(r+1)$ as was explained in 3.5 above. Expressing the E–homogeneous flat coordinates $(x_0, \ldots, x_r)$ on this space satisfying (4.17) in terms of the canonical coordinates and then calculating the multiplication table of the flat vector fields, we can reconstruct the potential which now will be a germ of holomorphic functions of (x_a). Because of the unicity, it must have the Taylor series (4.4). So Theorem 4.2 a),b) can be proved essentially by reading this section in reverse order. Of course, the last statement is of a different nature.

4.3. Tensor of the third derivatives.
Most of our calculations in $(\mathcal{T}, \circ)$ will be restricted to the first infinitesimal neighborhood of the plane $x_2 = \cdots = x_r = 0$ in H. This just suffices for the calculation of the Schlesinger initial conditions. We denote by J the ideal $(x_2, \ldots, x_r)$.

Multiplication by the identity $e = \partial_0$ is described by the components $\Phi_{0a}{}^b = \delta_{ab}$ of the structure tensor. Of the remaining components, we will need only $\Phi_{1a}{}^b$, which allows us to calculate multiplication by ∂_1 and to proceed inductively. This is where the Associativity Equations are implicitly used.

Obviously, $\Phi_{10}{}^b = \delta_{1b}$.

4.3.1. Claim. *We have*

(4.6) $\quad$ for $1 \leq a \leq r-1:$ $\quad \Phi_{1a}{}^b = \delta_{a+1,b} + x_{r+1-a+b} e^{x_1} + O(J^2),$

(4.7) $\quad \Phi_{1r}{}^b = \delta_{b0} e^{x_1} + x_{b+1} e^{x_1} + O(J^2).$

(Here and below we agree that $x_c = 0$ for $c > r$.)

Proof. The term $\delta_{a+1,b}$ in (4.6) comes from Φ_{cl}. The remaining terms are provided by the summands of total degree ≤ 3 in $x_2, \ldots, x_r$ in

$$\partial_1 \Phi_{\text{inst}} = \sum_{d \geq 1} d e^{dx_1} \left(\sum I(d; a_1, a_2) \frac{x_{a_1} x_{a_2}}{2} + \sum I(d; a_1, a_2, a_3) \frac{x_{a_1} x_{a_2} x_{a_3}}{6} \right) + O(J^4).$$

For $n = 2$, the grading condition means that $d = 1$, $a_1 = a_2 = r$. For $n = 3$, it means that $d = 1$, $a_1 + a_2 + a_3 = 2r + 1$. We know that $I(1; r, r) = 1$. Similarly, $I(1; a_1, a_2, a_3) = 1$ in this range. This can be deduced formally from the Associativity Equations. A nice exercise is to check that this agrees also with the geometric description (for instance, only one line intersects two given generic lines and passes through a given point in three–space). So finally

$$\partial_1 \Phi_{\text{inst}} = \left(\frac{x_r^2}{2} + \frac{1}{6} \sum_{a_1+a_2+a_3=2r+1} x_{a_1} x_{a_2} x_{a_3} \right) e^{x_1} + O(J^4).$$

The term $\delta_{b0} e^{x_1}$ in (4.7) comes from $\dfrac{x_r^2}{2}$. Furthermore,

$$\Phi_{\text{inst};1ab} = x_{2r+1-a-b} e^{x_1} + O(J^2)$$

and

$$\Phi_{\text{inst};1a}{}^b = \Phi_{\text{inst};1,a,r-b} = x_{r+1-a+b} e^{x_1} + O(J^2).$$

4.4. Multiplication table. The main formula of this subsection is

(4.8) $$\partial_1^{\circ(r+1)} = e^{x_1} \left(\partial_0 + \sum_{b=1}^{r-1} (b+1) x_{b+1} \partial_b \right) + O(J^2).$$

We will prove it by consecutively calculating the powers $\partial_1^{\circ a}$. The intermediate results will also be used later. (Notice that $O(J^2)$ in (4.8) now means $O(\sum_i J^2 \partial_i)$.)

First, we find from (4.6) and (4.7) for $1 \leq a \leq r-1$:

(4.9) $$\partial_1 \circ \partial_a = \sum_{b=0}^{r} \Phi_{1a}{}^b \partial_b = \partial_{a+1} + e^{x_1} \sum_{b=0}^{a-1} x_{r+1-a+b} \partial_b + O(J^2),$$

(4.10) $$\partial_1 \circ \partial_r = \sum_{b=0}^{r} \Phi_{1r}{}^b \partial_b = e^{x_1} \left(\partial_0 + \sum_{b=1}^{r-1} x_{b+1} \partial_b \right) + O(J^2).$$

Then using (4.9) and induction, we obtain

(4.11) for $1 \leq a \leq r$: $$\partial_1^{\circ a} = \partial_a + e^{x_1} \sum_{b=0}^{a-2} (b+1) x_{r+2-a+b} \partial_b + O(J^2).$$

Multiplying this formula for $a = r$ by ∂_1 and using (4.10), we finally find (4.8).

From (4.11) it follows that $\partial_1^{\circ a}$ for $0 \leq a \leq r$ freely span the tangent sheaf.

4.5. Idempotents. Formula (4.8) allows us to calculate all e_i mod J^2, thus demonstrating semisimplicity. Namely, denote by q the $(r+1)$–th root of the right hand side of (4.8) congruent to $e^{\frac{x_1}{r+1}}$ mod J and put $\zeta = \exp\left(\dfrac{2\pi i}{r+1} \right)$. Then

(4.12) $$e_i = \frac{1}{r+1} \sum_{j=0}^{r} \zeta^{-ij} (\partial_1 \circ q^{-1})^{\circ j}$$

satisfy

$$e_i \circ e_j = \delta_{ij} e_i, \quad \sum_i e_i = \partial_0$$

for all $i = 0, \ldots, r$. A straightforward check shows this.

4.5.1. Proposition. *We have*

$$e_i = \frac{1}{r+1} \sum_{j=0}^{r} \zeta^{-ij} e^{-x_1 \frac{j}{r+1}} \left(e^{x_1} \sum_{b=0}^{j-2} \frac{(b+1-j)(r+1-j)}{r+1} x_{r+b+2-j} \partial_b \right.$$

(4.13)
$$\left. + \partial_j - \sum_{b=j+1}^{r} \frac{(b+1-j)j}{r+1} x_{b+1-j} \partial_b \right) + O(J^2).$$

Proof. We have

$$q^{-1} = e^{-\frac{x_1}{r+1}} \left(\partial_0 - \sum_{b=1}^{r-1} \frac{b+1}{r+1} x_{b+1} \partial_b \right) + O(J^2).$$

Together with (4.9) this gives

$$\partial_1 \circ q^{-1} = e^{-\frac{x_1}{r+1}} \left(\partial_1 - \sum_{b=1}^{r-1} \frac{b+1}{r+1} x_{b+1} \partial_{b+1} \right) + O(J^2).$$

Hence

$$(\partial_1 \circ q^{-1})^j = e^{-\frac{jx_1}{r+1}} \left(\partial_1^{\circ j} - j \partial_1^{\circ (j-1)} \circ \sum_{b=1}^{r-1} \frac{b+1}{r+1} x_{b+1} \partial_{b+1} \right) + O(J^2).$$

Inserting this into (4.12) and using (4.9)–(4.11) once again, we finally obtain (4.13).

4.6. Metric coefficients in canonical coordinates. The metric potential η is simply x_r (see Ch. I, (2.4)). Hence we can easily calculate $\eta_i = e_i x_r$. The answer is

(4.14) $$\eta_i = \frac{\zeta^i}{r+1} e^{-x_1 \frac{r}{r+1}} - \sum_{b=2}^{r} \frac{\zeta^{ib}}{(r+1)^2} b(r+1-b) e^{-x_1 \frac{r+1-b}{r+1}} x_b + O(J^2).$$

As an exercise, the reader can check that the same answer results from the (longer) calculation of $\eta_i = g(e_i, e_i)$.

4.7. Derivatives of the metric coefficients. We now see that the chosen precision just suffices to calculate the restriction of η_{ij}, γ_{ij} and the matrix elements of A_j to the plane $x_2 = \cdots = x_r = 0$, any point of which can be taken as the initial one.

4.7.1. Claim. *We have*

(4.15) $$\eta_{ki} = e_k \eta_i = -2 \frac{\zeta^{i-k}}{(\zeta^{i-k} - 1)^2} \frac{e^{-x_1}}{(r+1)^2} + O(J).$$

Notice that (4.15) is symmetric in i, k as it should be.

This is obtained by a straightforward calculation from (4.13) and (4.14). The numerical coefficient in (4.15) comes as a combination of $\sum_{j=1}^{r} j\zeta^j$ and $\sum_{j=1}^{r} j^2 \zeta^j$ which are then summed by standard tricks.

4.8. Canonical coordinates. We find u^i from the formula $E \circ e_i = u^i e_i$. To calculate $E \circ e_i$, use (4.5), (4.13) and (4.9)–(4.11). We omit the details. The result is:

4.8.1. Claim. *We have*

(4.16) $$u^i = x_0 + \zeta^i(r+1)e^{\frac{x_1}{r+1}} + \sum_{a=2}^{r} \zeta^{ai} e^{\frac{ax_1}{r+1}} x_a + O(J^2).$$

The reader can check that $e_i u^j = \delta_{ij} + O(j)$.

4.9. Schlesinger's initial conditions. Recall that the matrix residues A_i of Schlesinger's equations for Frobenius manifolds are

$$A_j(e_i) = 0 \text{ for } i \neq j,$$

(4.17) $$A_j(e_j) = -\frac{1}{2}e_j - \frac{1}{2}\sum_k (u^k - u^j)\frac{\eta_{jk}}{\eta_k}e_k$$

(cf (1.13)). Substituting here (4.14), (4.15), and (4.16), we finally get the main result of this section.

4.9.1. Theorem. *The point* $(x_0, x_1, 0, \ldots, 0)$ *has canonical coordinates* $u^i = x_0 + \zeta^i(r+1)e^{\frac{x_1}{r+1}}$.

The special initial conditions at this point (in the sense of 3.4) corresponding to $H_{\text{quant}}(\mathbf{P}^r)$ *are given by*

(4.18) $$v_{jk} = -\frac{\zeta^{j-k}}{1 - \zeta^{j-k}}$$

and

(4.19) $$\eta_i = \frac{\zeta^i}{r+1} e^{-x_1 \frac{r}{r+1}}.$$

As an exercise, the reader can check that

$$-\sum_{k:\, k \neq j} \frac{\zeta^{j-k}}{1 - \zeta^{j-k}} = 1 - \frac{D}{2} = \frac{r}{2}.$$

§5. Dimension three and Painlevé VI

The equations for the potential Φ or metric potential η generally form a system of PDE. However, in the three-dimensional semisimple case, in the presence of a flat identity and an Euler field, they can be effectively reduced to one non-linear ODE belonging to the family Painlevé VI. This section contains some details of this study.

5.1. Normalization. i) *Spectrum and normalized flat coordinates.* We start along the lines of Ch. I, 4.2, but with some additional assumptions; see [D2], pp. 127–129, for the general case.

Let M be a connected simply connected Frobenius manifold with flat identity and Euler field with $d_0 = 1$. The most important spectrum point is D.

We will assume that $D \neq 2$ which guarantees that the spectrum of $-\operatorname{ad} E$ on T_M^f is simple.

In fact, in the notation of Ch. I, 2.4, this spectrum must be of the form $(d_0, d_1, d_2) = (1, \frac{D}{2}, D-1)$, where the eigenvector for $d_0 = 1$ is $\partial_0 = e$, $g(e,e) = 0$;

the eigenvector for $d_2 = D - 1$ is uniquely normalized by the condition $g(e, \partial_2) = 1$; and the one for $\dfrac{D}{2}$ is uniquely up to sign normalized by $g(\partial_1, \partial_1) = 1$. Thus $(g_{ab}) = (g^{ab}) = (\delta_{a+b,2})$.

We can now consider three flat coordinates (x_0, x_1, x_2) such that $\partial_a = \partial/\partial x_a$ defined up to a shift (and sign change for x_1). Their final normalization will depend on the Euler field.

The spectrum of $\mathcal{V} = -\operatorname{ad} E - \dfrac{D}{2}\operatorname{Id}$ is $\left(1 - \dfrac{D}{2}, 0, \dfrac{D}{2} - 1\right)$.

ii) *Euler field and normalized potential.* If $D \neq 0, 1$, then all d_a do not vanish, and we can choose x_a so that

(5.1) $\qquad D \neq 0: \qquad E = x_0 \partial_0 + \dfrac{D}{2} x_1 \partial_1 + (D - 1) x_2 \partial_2.$

(Notice that the origin $(x_a) = (0)$ cannot be tame semisimple because E vanishes there.)

For $D = 0$ we obtain an extra parameter (cf. Ch. I, 2.4) which we denote $r + 1$ to conform with (4.5); x_1 remains defined only up to a sign change and shift:

(5.2) $\qquad D = 0: \qquad E = x_0 \partial_0 + (r + 1) \partial_1 - x_2 \partial_2.$

We will assume $r + 1 \neq 0$; then the sign can be normalized by $\operatorname{Re}(r + 1) > 0$.

The potential can be written in the form (Ch. I, (2.3)):

$$\Phi(x_0, x_1, x_2) = \dfrac{1}{2}(x_0 x_1^2 + x_0^2 x_2) + \varphi(x_1, x_2).$$

It is defined up to a quadratic polynomial in (x_a) and must satisfy $E\Phi = (D+1)\Phi + q$, where q is also a quadratic polynomial. We can try to make $q = 0$ by replacing Φ with $\Phi + p$ and solving $(E - 1 - D)p = q$. If $D \neq 0$ and $D \neq -1$, such p exists and is unique. If $D = -1$, we cannot kill a possible constant term c in q, which is a new parameter. If $D = 0$, we can unambiguously kill any quadratic polynomial in (x_1, x_2), but the term containing x_0 will remain. So our final normalization is:

(5.3) $\qquad D \neq -1, 2: \qquad E\varphi = (D + 1)\varphi,$

$\qquad\qquad\qquad D = -1: \qquad E\varphi = c.$

iii) *Associativity Equations.* A straightforward check shows that all the Associativity Equations follow from one of them, which can be written as

(5.4) $\qquad\qquad \varphi_{222} = \varphi_{112}^2 - \varphi_{111}\varphi_{122}.$

In [D2], p.128, equations (5.3) and (5.4) are reduced to an ODE for the function f which is defined in the following way.

If $D \neq 0, \pm 1, 2$, put $\delta = \dfrac{2}{D} - 2$. Then (5.3) means that locally φ can be written as $x_1^4 x_2^{-1} f(x_2 x_1^\delta)$.

If $D = -1$, we can similarly put $\varphi = 2c \log x_1 + f(x_2 x_1^{-4})$.

If $D = 0$, we have $\varphi = x_2^{-1} f(x_1 + (r + 1)\log x_2)$. We will copy Dubrovin's equation for f in this case:

(5.5) $\; f'''[(r+1)^3 + 2f' - (r+1)f''] - f''^2 - 6(r+1)^2 f'' + 11(r+1) f' - 6f = 0.$

The case $D = 0$ is the most interesting for us because it includes the quantum cohomology of $\mathbf{P}^2$. It is not easy to recognize a classical equation in (5.5). Below we will describe how Dubrovin uses the additional semisimplicity condition in order to reduce it to PVI.

5.2. Semisimplicity and tameness. At a tame semisimple point of M, the operator $E\circ$ has simple spectrum (canonical coordinates of this point). Conversely, if this is true, one can explicitly write the idempotents e_i as polynomials in E. This criterion is sufficiently practical for use in flat coordinates.

5.3. Analyticity. Consider now the case when Φ is analytic at the origin. (Recall that if $D = 0$, the origin can be any point along the x_1-axis, so its choice is the same as the choice of x_1.)

5.3.1. Proposition. *a) The origin can be tame semisimple only if $D = 0$. In this case the normalized analytic potential can be written in the form*

$$(5.6) \qquad \Phi(x_0, x_1, x_2) = \frac{1}{2}(x_0 x_1^2 + x_0^2 x_2) + \sum_{n=0}^{\infty} \frac{M(n)}{n!} e^{\frac{n+1}{r+1} x_1} x_2^n$$

so that $E\Phi = \Phi + (r+1)x_0 x_1$.

b) The Associativity Equations are equivalent to the following recursive relations for the coefficients $M(n)$:

$$
M(n+3) = \frac{1}{(r+1)^4} \sum_{\substack{k+l=n \\ k,l \geq 0}} \binom{n}{k} \left[M(k+1)M(l+1)(k+2)^2(l+2)^2 \right.
$$
$$(5.7) \qquad \left. - M(k)M(l+2)(k+1)^3(l+3) \right].$$

Hence any formal solution is uniquely defined by the choice of $M(0), M(1), M(2)$, which can be arbitrary.

c) The point (000) is tame semisimple iff the polynomial

$$u^3 - \frac{M(0)}{(r+1)^3} u^2 - \frac{8M(1)}{(r+1)^2} u - \frac{3M(2)}{r+1}$$

has no multiple roots.

For the quantum cohomology of $\mathbf{P}^2$, we have $r + 1 = 3$, $M(n) = 0$ unless $n = 3d - 1$, and $M(2) = 1$. If we put $N(d) := M(3d - 1)$, (5.7) becomes (0.19).

Proof. a) As we have already remarked, (000) cannot be tame semisimple with E of the form (5.1) since E vanishes at this point. One easily sees that for $D = 0$, (5.6) is normalized.

b) This is a restatement of (5.4).

c) We will use the criterion of 5.2. From (5.2) one sees that one can look at the spectrum of $\partial_1 \circ$ in lieu of $E\circ$. The multiplication table at the origin is

$$\partial_1 \circ \partial_0 = \partial_1,$$
$$\partial_1 \circ \partial_1 = \frac{4M(1)}{(r+1)^2} \partial_0 + \frac{M(0)}{(r+1)^3} \partial_1 + \partial_2,$$
$$\partial_1 \circ \partial_2 = \frac{3M(2)}{r+1} \partial_0 + \frac{4M(1)}{(r+1)^2} \partial_1.$$

Hence
$$\det(\partial_1 \circ -u\,\mathrm{Id}) = -u^3 + \frac{M(0)}{(r+1)^3}u^2 + \frac{8M(1)}{(r+1)^2}u + \frac{3M(2)}{r+1}.$$
This finishes the proof.

5.3.2. Exercises. a) Calculate formal (at the origin) potentials for $D \neq 0$.

b) Calculate the special Schlesinger initial conditions at the origin for the potential (5.6).

5.4. Introduction to the PVI equations. These equations form a family $\mathrm{PVI}_{\alpha,\beta,\gamma,\delta}$ depending on four parameters $\alpha, \beta, \gamma, \delta$, and are classically written as:

$$\frac{d^2X}{dt^2} = \frac{1}{2}\left(\frac{1}{X} + \frac{1}{X-1} + \frac{1}{X-t}\right)\left(\frac{dX}{dt}\right)^2 - \left(\frac{1}{t} + \frac{1}{t-1} + \frac{1}{X-t}\right)\frac{dX}{dt}$$

(5.8)
$$+ \frac{X(X-1)(X-t)}{t^2(t-1)^2}\left[\alpha + \beta\frac{t}{X^2} + \gamma\frac{t-1}{(X-1)^2} + \delta\frac{t(t-1)}{(X-t)^2}\right].$$

They were discovered around 1906 and have been approached from at least three different directions.

a. Study of non-linear ordinary differential equations of the second order whose solutions have no movable critical points.

Their classification program was initiated by Painlevé, but he inadvertently omitted (5.8) due to an error in calculations. It was B. Gambier [G] who completed Painlevé's list and found (5.8).

b. Study of the isomonodromic deformations of linear differential equations.

c. Theory of abelian integrals depending on parameters and taken over chains with boundary (not necessarily cycles).

These two approaches are due to R. Fuchs [F].

In the subsequent development of the theory, relationship with isomonodromic deformations proved to be most fruitful. Briefly speaking, (5.8) can be obtained by a change of variables from Schlesinger's equations with four singular points and the two–dimensional space T. This description can be used in order to connect PVI to the three–dimensional Frobenius manifolds. For recent research and a bibliography the reader may consult [JM], [O1], [H1], [H2].

In this section we take up the somewhat neglected approach via abelian integrals and algebraic geometry.

The main outcome of this approach is the representation of (5.8) as an equation on the multisection of an (arbitrary non–constant) pencil of elliptic curves with marked sections of order two. In particular, passing to the classical uniformization, we will find the following equivalent form of (5.8):

5.4.1. Theorem. *The equation (5.8) is equivalent to*

(5.9) $$\frac{d^2z}{d\tau^2} = \frac{1}{(2\pi i)^2}\sum_{j=0}^{3}\alpha_j \wp_z\left(z + \frac{T_j}{2}, \tau\right),$$

where

(5.10) $$(\alpha_0, \ldots, \alpha_3) := (\alpha, -\beta, \gamma, \frac{1}{2} - \delta),$$

$(T_0, \ldots, T_3) = (0, 1, \tau, 1 + \tau)$, and $\wp(z, \tau)$ is the Weierstrass function.

5.4.2. Theorem. *Any potential of the form (5.6) can be expressed through a solution to (5.9) with $(\alpha_0, \ldots, \alpha_3) = (\frac{1}{2}, 0, 0, 0)$, that is,*

$$\tag{5.11} \frac{d^2 z}{d\tau^2} = -\frac{1}{8\pi^2} \wp_z(z, \tau).$$

In particular, the solution corresponding to $\mathbf{P}^2$ passes through a point of order three on an elliptic curve with complex multiplication by the cubic root of unity.

Below we will give a more detailed version and a proof of both theorems.

The last result gives an exact meaning to the statement "mirror of $\mathbf{P}^2$ is a pencil of elliptic curves with marked sections of order two and an additional multisection". Namely, (5.10) can be considered as a "non–homogeneous Picard–Fuchs equation" associated to such a pencil (see 5.5.1 and 5.5.2 below). It replaces the conventional Picard–Fuchs equations for the periods of the Mirror dual Calabi–Yau threefold, and Claim 5.6.1 takes the place of the relevant Mirror Identity. It is conceivable that a similar picture will emerge for all homogeneous and toric Fano manifolds and for all Fano complete intersections in them. For alternative approaches to the mirror symmetry for Fano varieties, see [Va], [Giv4].

An intriguing question about the analytic nature of the particular solution corresponding to $\mathbf{P}^2$ remains open. There are theorems saying that solutions to (5.8) are generically "new" transcendents. There are also many examples of the particular solutions reducible to more classical functions, like hypergeometric ones.

5.5. Painlevé equations and elliptic pencils. We start with the following classical result.

5.5.1. Theorem (R. Fuchs, 1907). *The equation (5.8) can be written in the form*

$$t(1-t) \left[t(1-t) \frac{d^2}{dt^2} + (1 - 2t) \frac{d}{dt} - \frac{1}{4} \right] \int_\infty^{(X,Y)} \frac{dx}{\sqrt{x(x-1)(x-t)}}$$

$$\tag{5.12} = \alpha Y + \beta \frac{tY}{X^2} + \gamma \frac{(t-1)Y}{(X-1)^2} + \left(\delta - \frac{1}{2}\right) \frac{t(t-1)Y}{(X-t)^2},$$

where $Y^2 = X(X-1)(X-t)$.

Proof. First, let us clarify the meaning of (5.12). Consider the family of elliptic curves $E \to B$ parametrized by $t \in \mathbf{P}^1 \setminus \{0, 1, \infty\} := B$: the curve E_t is the projective closure of $Y^2 = X(X-1)(X-t)$. Points at infinity of $\{E_t\}$ form a section D_0 of this family which is the zero section for the standard group law on fibers. Choose in $E_t(\mathbf{C})$ a path from $D_0(t)$ to the point $(X(t), Y(t))$ of a local section. The operator

$$\tag{5.13} L_t := t(1-t) \frac{d^2}{dt^2} + (1 - 2t) \frac{d}{dt} - \frac{1}{4}$$

annihilates the periods $\int \frac{dx}{y}$ along closed paths in $E_t(\mathbf{C})$ because

(5.14) $$\left[t(1-t)\frac{\partial^2}{\partial t^2} + (1-2t)\frac{\partial}{\partial t} - \frac{1}{4}\right]\frac{d_{E/B}x}{y} = \frac{1}{2}d_{E/B}\frac{y}{(x-t)^2},$$

where we put $\frac{\partial}{\partial t}(x) = 0$ and $d_{E/B}t = 0$. Applying L_t to $\int_\infty^{(X,Y)} \frac{dx}{y}$ we get $\frac{1}{2}\frac{y}{(x-t)^2}\Big|_\infty^{(X,Y)}$ plus the contribution of the boundary sections which together with the right hand side of (5.12) amounts to (5.8).

5.5.2. μ–equations. The equation (5.12) is an instance of the general construction which was used in [Ma1] to prove the functional Mordell conjecture. We will briefly describe it now.

A μ-equation is a system of non-linear PDE in which independent variables are (local) coordinates on a manifold B and unknown functions are represented by a section s of a family of abelian varieties (or complex tori) $\pi: A \to B$. To write this system explicitly, assume B small enough so that $\pi_*(\Omega^1_{A/B})$ and $\mathcal{D}_B$ (sheaf of differential operators on B) are $\mathcal{O}_B$-free, and make the following choices:

a. An $\mathcal{O}_B$-basis of vertical 1-forms $\omega_1, \ldots, \omega_n \in \Gamma(B, \pi_*(\Omega^1_{A/B}))$.

b. A system of generators of the $\mathcal{D}_B$-module of the Picard–Fuchs equations

(5.15) $$\sum_{i=1}^n L_i^{(j)} \int_\gamma \omega_i = 0, \quad j = 1, \ldots, N,$$

where γ runs over families of closed paths in the fibers spanning $H_1(B_t)$.

c. A family of meromorphic functions $\Phi^{(j)}$, $j = 1, \ldots, N$, on A.

The respective μ-equation for a local (multi)section $s: B \to A$ then reads

(5.16) $$\sum_{i=1}^n L_i^{(j)} \int_0^s \omega_i = s^*(\Phi^{(j)}), \quad j = 1, \ldots, N,$$

where 0 denotes the zero section.

One drawback of (5.16) is its dependence on arbitrary choices. Clearly, this can be reduced by taking account of the transformation rules with respect to the changes of various generators. For elliptic pencils, the result takes a neat form.

Again let $E \to B$ be a non–constant one–dimensional family of elliptic curves. We temporarily keep the assumption that $\pi_*(\Omega^1_{E/B})$ and the tangent sheaf $\mathcal{T}_B$ are free. For any symbol $\sigma \in S^2(\mathcal{T}_B)$ of order two and any generator ω of $\pi_*(\Omega^1_{E/B})$ denote by $L_{\sigma,\omega}$ the Picard–Fuchs operator on B with the symbol σ annihilating all periods of ω.

5.5.3. Lemma. *For any local section s, the expression $L_{\sigma,\omega} \int_0^s \omega$ is $\mathcal{O}_B$-bilinear in σ and ω.*

Proof. Obviously,

$$L_{f\sigma,\omega} = fL_{\sigma,\omega}, \quad L_{\sigma,g\omega} = gL_{\sigma,\omega} \circ g^{-1},$$

where f, g are functions on B. The lemma follows.

Thus the expression

$$\mu(s) := \left(L_{\sigma,\omega} \int_0^s \omega\right) \otimes \sigma^{-1} \otimes \omega^{-1} \in S^2(\Omega_B^1) \otimes (\pi_*\Omega_{E/B}^1)^{-1} \tag{5.17}$$

depends only on s and is compatible with restrictions to open subsets of B. This means that the natural domain of the right hand sides for elliptic μ-equations is the set of meromorphic sections Φ of the sheaf $\pi^*\left[S^2(\Omega_B^1) \otimes (\pi_*\Omega_{E/B}^1)^{-1}\right]$.

Notice that the Kodaira–Spencer isomorphism (and eventually a choice of the theta–characteristic of B) allows us to identify Φ with a meromorphic section of $(\Omega_{E/B}^1)^3$ or $\pi^*(\Omega_B^1)^{3/2}$ as well.

We will now lift the Fuchs–Painlevé equation (5.12) to the classical covering space, which in particular will make transparent the nature of its right hand side.

5.5.4. Uniformization. Consider the family of elliptic curves parametrized by the upper half–plane H: $E_\tau := \mathbf{C}/(\mathbf{Z} + \mathbf{Z}\tau) \mapsto \tau \in H$. Recall that

$$\wp(z,\tau) := \frac{1}{z^2} + \sum{}'\left(\frac{1}{(z+m\tau+n)^2} - \frac{1}{(m\tau+n)^2}\right), \tag{5.18}$$

$$\wp_z(z,\tau) = -2\sum \frac{1}{(z+m\tau+n)^3}. \tag{5.19}$$

We have

$$\wp_z(z,\tau)^2 = 4(\wp(z,\tau) - e_1(\tau))(\wp(z,\tau) - e_2(\tau))(\wp(z,\tau) - e_3(\tau)), \tag{5.20}$$

where

$$e_i(\tau) = \wp\left(\frac{T_i}{2}, \tau\right), \quad (T_0,\ldots,T_3) = (0, 1, \tau, 1+\tau) \tag{5.21}$$

and $e_1 + e_2 + e_3 = 0$. Functions $\wp$ and $\wp_z$ are invariant with respect to the shifts $\mathbf{Z}^2 : (z,\tau) \mapsto (z+m\tau+n, \tau)$ and behave in the following way under the full modular group Γ:

$$\wp\left(\frac{z}{c\tau+d}, \frac{a\tau+b}{c\tau+d}\right) = (c\tau+d)^2 \wp(z,\tau), \tag{5.22}$$

$$\wp_z\left(\frac{z}{c\tau+d}, \frac{a\tau+b}{c\tau+d}\right) = (c\tau+d)^3 \wp_z(z,\tau). \tag{5.23}$$

Consider now the morphism of families $\varphi : \{E_\tau\} \to \{E_t\}$ induced by

$$(z,\tau) \mapsto \left(X = \frac{\wp(z,\tau) - e_1}{e_2 - e_1}, Y = \frac{\wp_z(z,\tau)}{2(e_2-e_1)^{3/2}}, t = \frac{e_3 - e_1}{e_2 - e_1}\right). \tag{5.24}$$

This is a Galois covering with the group $\Gamma(2) \ltimes \mathbf{Z}^2$. We have

$$\varphi^*\left(\frac{d_{E/B}X}{Y}\right) = 2(e_2 - e_1)^{1/2} d_{E/H}z. \tag{5.25}$$

In future formulas of this type we will omit φ^* and denote differentials over a base B by $d_\downarrow$. For instance, $d_\downarrow\left(\dfrac{z}{c\tau+d}\right) = \dfrac{d_\downarrow z}{c\tau+d}$, whereas $d\left(\dfrac{z}{c\tau+d}\right) = \dfrac{dz}{c\tau+d} - \dfrac{czd\tau}{(c\tau+d)^2}$.

It follows from (5.25) that if we denote by γ_1 (resp. γ_2) the image of $[0,1]$ (resp. $[0,1]\tau$) in $\{E_t\}$, then

$$(5.26) \qquad \int_{\gamma_1} \frac{d_\downarrow X}{Y} = 2(e_2 - e_1)^{1/2}, \quad \int_{\gamma_2} \frac{d_\downarrow X}{Y} = 2\tau(e_2 - e_1)^{1/2}$$

so that the operator L_t from (5.13) annihilates periods (5.26) as functions of τ.

We can now prove Theorem 5.4.1 in the following form:

5.5.5. Claim. *The lift of (5.12) to the (z,τ)-space $\mathbf{C} \times H$ is (5.9).*

Proof. Following the lead of 5.5.3, we will directly calculate the μ-equation for $\{E_\tau\}$, choosing $\omega = d_\downarrow z$ (instead of $d_\downarrow X/Y$) and $\sigma = \dfrac{d^2}{d\tau^2}$ (instead of $t^2(1-t)^2 \dfrac{d^2}{dt^2}$). Since periods of $d_\downarrow z$ are generated by 1 and τ, the relevant Picard–Fuchs operator is simply $\dfrac{d^2}{d\tau^2}$. From Lemma 5.5.3 and (5.26) it follows that

$$t(1-t)L_t \circ 2(e_2 - e_1)^{1/2} = Z(\tau)\frac{d^2}{d\tau^2}.$$

Using (5.24) and comparing symbols, we see that

$$Z(\tau) = 2\left(\frac{e_3 - e_1}{e_2 - e_1}\right)^2 \left(\frac{e_3 - e_2}{e_2 - e_1}\right)^2 \frac{(e_2 - e_1)^4}{9(e_1 e_2' - e_2 e_1')^2}(e_2 - e_1)^{1/2}$$

$$(5.27) \qquad = \frac{2}{9}\frac{\prod_{i>j}(e_i - e_j)^2}{(e_1 e_2' - e_2 e_1')^2}(e_2 - e_1)^{-3/2}.$$

Since $e_1 + e_2 + e_3 = 0$, we can replace $(e_1 e_2' - e_2 e_1')^2$ by $(e_i e_j' - e_j e_i')^2$ for any $i \neq j$. It follows that

$$C := \frac{\prod_{i>j}(e_i - e_j)^2}{(e_1 e_2' - e_2 e_1')^2}$$

is a modular function for the full modular group without zeroes and poles, hence a constant. A calculation with theta–functions, here omitted, shows that $C = -9\pi^2$, so that finally

$$(5.28) \qquad t(1-t)L_t \int_\infty^{(X(t),Y(t))} \frac{d_\downarrow x}{y} = -2\pi^2(e_2 - e_1)^{-3/2}\frac{d^2}{d\tau^2}\int_0^{z(\tau)} d_\downarrow z$$

for the respective sections. We can now consecutively compare the summands on the right hand side of (5.8) with those in (5.9). The first summand gives

$$\alpha Y = \frac{\alpha}{2}(e_2 - e_1)^{-3/2}\wp_z(z,\tau).$$

For the remaining ones we have to use the addition formulas

$$\wp_z\left(z + \frac{T_i}{2},\tau\right) = -\frac{(e_i - e_j)(e_i - e_k)}{(\wp(z,\tau) - e_i)^2}\wp_z(z,\tau), \quad \{i,j,k\} = \{1,2,3\},$$

so that, say, for $i = 3$ we get

$$(\delta - \frac{1}{2})\frac{t(t-1)Y}{(X-t)^2} = (\delta - \frac{1}{2})\frac{(e_3 - e_1)(e_3 - e_2)}{(e_2 - e_1)^2} \cdot \frac{\wp_z(z,\tau)}{2(e_2 - e_1)^{3/2}} \cdot \frac{(e_2 - e_1)^2}{(\wp(z,\tau) - e_3)^2}$$

$$= -\frac{1}{2}(\delta - \frac{1}{2})(e_2 - e_1)^{-3/2} \cdot \frac{-(e_3 - e_1)(e_3 - e_2)}{(\wp(z,\tau) - e_3)^2}\wp_z(z,\tau)$$

$$= -\frac{1}{2}(\delta - \frac{1}{2})(e_2 - e_1)^{-3/2}\wp_z(z + \frac{1+\tau}{2}, \tau).$$

The remaining two summands are treated similarly. This finishes the proof.

In [Ma5], Theorem 5.4.1 was used in order to give an algebraic geometric description of the Painlevé VI equations and of their Hamiltonian structure.

5.5.6. S_4-symmetry and the Landin transform. As an application of (5.9) we will construct some natural transformations of PVI.

a. *The classical S_4-symmetry.* Isomorphisms of elliptic pencils with marked sections of order two, (E, D_i), which do not conserve the labeling of D_i induce transformations of PVI permuting α_i. In the form (5.9), they act on the solutions as compositions of the transformations of two types: $(z, \tau) \mapsto \left(\frac{z}{c\tau + \tau}, \frac{a\tau + b}{c\tau + d}\right)$ indexed by cosets $\Gamma/\Gamma(2)$, and $(z, \tau) \mapsto (z + \frac{T_i}{2}, \tau)$ shifting the zero section.

b. *The Landin transform.* From (5.19) one easily deduces Landin's identity

$$\wp_z(z, \frac{\tau}{2}) = -2\left[\sum \frac{1}{(z + 2m\frac{\tau}{2} + n)^3} + \sum \frac{1}{(z + \frac{\tau}{2} + 2m\frac{\tau}{2} + n)^3}\right]$$

$$= \wp_z(z, \tau) + \wp_z(z + \frac{\tau}{2}, \tau).$$

Hence if $z(\tau)$ is a solution to PVI with parameters $(\alpha_0, \alpha_1, \alpha_0, \alpha_1)$, we have

$$\frac{d^2 z(\tau)}{d\tau^2} = \alpha_0[\wp_z(z, \tau) + \wp_z(z + \frac{\tau}{2}, \tau)] + \alpha_1[\wp_z(z + \frac{1}{2}, \tau) + \wp_z(z + \frac{1+\tau}{2}, \tau)]$$

$$= \frac{1}{4}\frac{d^2 z(\tau)}{d(\tau/2)^2} = \alpha_0 \wp_z(z, \frac{\tau}{2}) + \alpha_1 \wp_z(z + \frac{1}{2}, \frac{\tau}{2});$$

that is, $z(2\tau)$ is a solution to PVI with parameters $(4\alpha_0, 4\alpha_1, 0, 0)$. The converse statement is true as well. In this way we get the following bijections between the sets of solutions to (5.9):

(5.29) $\qquad (\alpha_0, \alpha_1, \alpha_0, \alpha_1) \leftrightarrow (4\alpha_0, 4\alpha_1, 0, 0),$

and in particular

(5.30) $\qquad (\alpha_0, 0, \alpha_0, 0) \leftrightarrow (4\alpha_0, 0, 0, 0).$

5.5.7. The symmetry group W. Now put $a_i^2 = 2\alpha_i, i = 0, \ldots, 3$. In [O2], Okamoto found out that the following group W of the transformations of the parameter space (a_i) can be birationally lifted to the group acting on the space of all solutions of all Painlevé VI equations. By definition, W is generated by

a) $(a_i) \mapsto (\varepsilon_i a_i)$, where $\varepsilon_i = \pm 1$.

b) Permutations of (a_i).

c) $(a_i) \mapsto (a_i + n_i)$, where $n_i \in \mathbf{Z}$ and $\sum_{i=0}^{3} n_i \equiv 0\,(2)$.

This result goes back to Schlesinger who discovered the general discrete symmetries of his equations. It is remarkable however that they act so neatly on a specific reduction represented by PVI. Explicit formulas are quite complicated even for the simplest shift $(a_i) \mapsto (a_i + 2\delta_{i0})$, and composition quickly makes them unmanageable. A neat invariant description of these symmetries in the language of isomonodromic deformations was recently provided in [AL2] based upon [AL1].

5.6. From Frobenius to Painlevé. Following [D2], Appendix E, we will now describe the map which produces a solution to (5.8) for any analytic potential of the form (5.6).

Let $\Phi(x_0, x_1, x_2)$ be the germ of analytic functions of the form (5.6), satisfying the Associativity Equations, for which (000) is a tame semisimple point with non–zero canonical coordinates, or equivalently, $M(2) \neq 0$. For $a, b = 0, 1, 2$ calculate consecutively the following functions of (x_0, x_1, x_2):

(5.31) $$G_{ab} := (a+b-1)\,\Phi_{ab} + \frac{1}{2}(r+1)\delta_{a+b,1},$$

(5.32) $$q = \frac{G_{11}G_{22} - G_{12}^2}{G_{22}},$$

(5.33) $$p = -\frac{G_{11}G_{22}}{G_{12}^3 + G_{02}G_{12}G_{22} - G_{11}G_{12}G_{22} - G_{22}^2 G_{01}}.$$

Denote by (u_1, u_2, u_3) the eigenvalues of the operator $E\circ$. Since they are local canonical coordinates, q and p are functions of u_i. Finally, put

(5.34) $$t = \frac{u_3 - u_1}{u_2 - u_1},\quad X(t) = \frac{q - u_1}{u_2 - u_1}.$$

The fact that locally X depends only on t and not on separate u_i follows from the equation $E\Phi = \Phi + (r+1)x_0 x_1$.

5.6.1. Claim. *The function $X(t)$ satisfies the PVI equation (5.8) with parameters*

(5.35) $$(\alpha, \beta, \gamma, \delta) = (\frac{9}{2}, 0, 0, \frac{1}{2}).$$

Moreover, we have

(5.36) $$\frac{dX}{dt} = \left(2p + \frac{1}{q - u_3}\right) \frac{\prod_{i=1}^{3}(q - u_i)}{(u_3 - u_2)(u_3 - u_1)}.$$

In fact, Dubrovin in [D2], Appendix E, deduces a more general statement applicable to the case $D \neq 0$ at semisimple points as well. We will restrict ourselves to comparing notation. Dubrovin's $y, x, t^1, t^2, t^3, g^{\alpha, \beta}, \mu$ are my

$$X, t, x_0, x_1, x_2, G_{3-\alpha, 3-\beta}, -1,$$

respectively. Our formulas (5.32) and (5.33) are Dubrovin's (E.8); (5.35) is obtained from the fact that Dubrovin's μ is -1 for $D = 0$. Finally, (5.34) and (5.36) are Dubrovin's (E.16).

Dubrovin also shows how to reconstruct the potential knowing $X(t)$ (this involves integration, which explains the discrepancy in the numbers of constants). Namely:

(i) Find q, p as functions of u_i from (E.16) (P is given by (E.10)).

(ii) Find $\log k$ from (E.13).

(iii) Find functions ψ_{kl} from (E.9) and (E.17).

(iv) Find flat coordinates as functions of u_i from (3.84b).

(v) Find the third derivatives of Φ with respect to the flat coordinates from (3.84c): $c_{\alpha\beta\gamma}$ are $\Phi_{\alpha\beta\gamma}$. They are expressed as functions of u_i, whereas we want the potential as a function of t^α, or x_i. So the last step:

(vi) Invert (iv) and input into (v).

5.6.2. Initial conditions for $\mathbf{P}^2$. We can now calculate the Painlevé initial conditions for $\mathbf{P}^2$ at the point $x_a = 0$. According to Theorem 4.9.1, we have (up to renumbering) $(u_1, u_2, u_3) = (3, 3\zeta, 3\zeta^2), \zeta = e^{2\pi i/3}$ at this point. After calculating (5.31), we obtain $q = p = 0$, again at the origin. Then (5.34) and (5.36) give

(5.37) $$t = \zeta + 1, \ X(\zeta + 1) = \frac{1}{1 - \zeta}, \ X'(\zeta + 1) = \frac{1}{3}.$$

Obviously, the elliptic curve $Y^2 = X(X-1)(X-\zeta-1)$ admits complex multiplication by ζ: the q-coordinate can be simply multiplied by ζ. The point $q = 0$ remains invariant, hence it must be of order three on this curve. (I do not see the meaning of the last condition $X'(t) = \frac{1}{3}$.)

It is interesting to remark that the point (5.35) in the parameter space of PVI in a sense also corresponds to the "half period". More precisely, the (a_i)-coordinates of this point are $(a_0, \ldots, a_3) = (3, 0, 0, 0)$. By the Schlesinger–Okamoto shift we can reduce this point to $(1, 0, 0, 0)$.

The point $(0, 0, 0, 0)$ corresponds to the equation $d^2z/d\tau^2 = 0$ trivially solvable with two arbitrary constants; all $X(t)$ can be expressed via the Weierstrass function. The same is true for the shifted point $(2, 0, 0, 0)$ by Okamoto. The $\mathbf{P}^2$-point lies exactly halfway in between.

CHAPTER III

Frobenius Manifolds and Moduli Spaces of Curves

§1. Formal Frobenius manifolds and $Comm_\infty$-algebras

In this chapter we return to the supergeometric setting of Chapter I, §1. Many constructions refer to the formal Frobenius manifolds, and we first reproduce their definition.

1.1. Formal Frobenius manifolds. Let k be a supercommutative **Q**-algebra, $H = \bigoplus_a k\Delta_a$ a free (**Z**$_2$-graded) k-module of finite rank, and $g : H \otimes H \to k$ an even symmetric pairing which is non-degenerate in the sense that it induces an isomorphism $g' : H \to H^t$ where H^t is the dual module.

Denote by $K = k[[H^t]]$ the completed symmetric algebra of H^t. In other words, if $\sum_a x^a \Delta_a$ is a generic even element of H, then K is the algebra of formal series $k[[x^a]]$.

1.1.1. Definition. *The structure of the formal Frobenius manifold on (H, g) is given by an even potential $\Phi \in K$, defined up to quadratic terms, and satisfying the Associativity Equations I.(1.6).*

In other words, the multiplication law $\Delta_a \circ \Delta_b = \sum_c \Phi_{ab}{}^c \Delta_c$ turns $H_K = K \otimes_k H$ into a supercommutative K-algebra.

1.1.2. Examples. a) If (M, g, Φ_M) is a Frobenius manifold over $k = \mathbf{R}$ or $\mathbf{C}$, and x is a point of M, put $H = T_{M,x}$ (the tangent superspace at x identified with the space of local flat tangent fields), Φ = the image of Φ_M in the completion of the local ring $\mathcal{O}_{M,x}$, and (x^a) a system of local flat coordinates vanishing at x.

More generally, we can start with a relative Frobenius manifold M/S where S is affine or Stein, and a section $x : S \to M$ with normal sheaf trivialized by the vertical flat vector fields. The completion along this section will be a formal Frobenius manifold over $k = \Gamma(S, \mathcal{O}_S)$.

b) Quantum cohomology, briefly described in Chapter I, 4.4, furnishes many examples of formal Frobenius structures on the cohomology modules ($H = H^*(V, k)$, g = Poincaré pairing); see e.g. potentials II.(4.13) of projective spaces.

In §5 below we reproduce the axiomatic theory of quantum cohomology from [KM1]. The constructive definition of quantum cohomology involves moduli spaces of algebraic curves with labeled points and maps of such curves. It turns out, however, that at least genus zero moduli spaces are intimately connected with formal Frobenius manifolds *of arbitrary origin*. We explain this connection in §§1–4 and apply it to the construction and study of the tensor product of Frobenius manifolds in §§6–8. The remaining part of this chapter discusses two more constructions of Frobenius manifolds related to the algebraic and symplectic geometry: via K. Saito's theory of singularities and via differential Gerstenhaber–Batalin–Vilkovyski algebras.

In this section we start by showing that the Taylor coefficients of any formal potential Φ can be interpreted as a family of multilinear composition laws on H furnishing a beautiful generalization of the usual commutative algebra. This opens the way to see this theory from the perspective of operads. Let (H,g) be as in 1.1.

1.2. Definition. *The structure of the cyclic $Comm_\infty$-algebra on (H,g) is a sequence of even polylinear maps $\circ_n : H^{\otimes n} \to H$, $n = 2, 3, \ldots$, satisfying the following conditions:*

a) Higher commutativity: $\circ_n$ are S_n-symmetric (in the sense of superalgebra). We will denote $\circ_n(\gamma_1 \otimes \cdots \otimes \gamma_n)$ by $(\gamma_1, \ldots, \gamma_n)$.

b) Cyclicity: the tensors

$$(1.1) \quad Y_{n+1} : H^{\otimes(n+1)} \to k, \quad Y_{n+1}(\gamma_1 \otimes \cdots \otimes \gamma_n \otimes \gamma_{n+1}) := g((\gamma_1, \ldots, \gamma_n), \gamma_{n+1})$$

are S_{n+1}-symmetric.

c) Higher associativity: for all $m \geq 0$ and $\alpha, \beta, \gamma, \delta_1, \ldots, \delta_m \in H$, we have

$$(1.2) \quad \sum_{\sigma : S_1 \coprod S_2 = \{1,\ldots,m\}} \varepsilon'(\sigma)((\alpha, \beta, \delta_i \,|\, i \in S_1), \gamma, \delta_j \,|\, j \in S_2)$$
$$= \sum_{\sigma : S_1 \coprod S_2 = \{1,\ldots,m\}} \varepsilon''(\sigma)(\alpha, (\beta, \gamma, \delta_i \,|\, i \in S_1), \delta_j \,|\, j \in S_2).$$

1.2.1. Comments. a) In (1.2) σ runs over all ordered partitions of $\{1, \ldots, m\}$ into two disjoint subsets. The signs $\varepsilon'(\sigma)$, $\varepsilon''(\sigma)$ are defined as follows: fix an initial ordering, say, $(\alpha, \beta, \gamma, \delta_1, \ldots, \delta_m)$, then calculate the sign of the permutation induced by σ on the odd arguments in (1.2).

b) For $m = 0$, (1.2) reads

$$(1.3) \quad ((\alpha, \beta), \gamma) = (\alpha, (\beta, \gamma)),$$

and for $m = 1$

$$((\alpha, \beta), \gamma, \delta) + (-1)^{\tilde{\gamma}\tilde{\delta}}((\alpha, \beta, \delta), \gamma) = (\alpha, (\beta, \gamma, \delta)) + (\alpha, (\beta, \gamma), \delta).$$

The general combinatorial structure of (1.2) can be memorized as follows: start with (1.3) and distribute $(\delta_1, \ldots, \delta_m)$ in all possible ways between the brackets at both sides, without introducing new brackets and retaining the initial ordering inside each bracketed group.

c) The term "cyclic" comes from cyclic cohomology. One could also say that g must be an invariant scalar product with respect to all multiplications: compare (1.1) to I.(1.2). Choosing $\circ_n = 0$ for all $n \geq 3$, we will get a conventional commutative algebra with invariant scalar product.

§1. FORMAL FROBENIUS MANIFOLDS AND $Comm_\infty$-ALGEBRAS

1.3. Abstract Correlation Functions. Clearly, if g is fixed, $\circ_n$ and Y_{n+1} uniquely determine each other. It will be useful to axiomatize the functional equations between Y_{n+1} which turn out to be equivalent to the higher associativity laws.

1.3.1. Definition. *A system of Abstract Correlation Functions (ACF) on (H,g) is a family of S_n-symmetric even polynomials $Y_n : H^{\otimes n} \to k$, $n = 3, 4, 5, \ldots$, satisfying the following coherence relations:*

for all $n \geq 4$, all pairwise distinct $i, j, k, l \in \{1, \ldots, n\}$, and all $\gamma_1, \ldots, \gamma_n \in H$, we have

$$(1.4) \quad \sum_{\sigma: ij\sigma kl} \varepsilon(\sigma)(Y_{|S_1|+1} \otimes Y_{|S_2|+1}) \left(\bigotimes_{p \in S_1} \gamma_p \otimes \Delta \otimes \left(\bigotimes_{q \in S_2} \gamma_q \right) \right) = (j \leftrightarrow k),$$

where $\Delta = \sum \Delta_a g^{ab} \otimes \Delta_b$.

Here σ runs over *stable* partitions of $\{1, \ldots, n\}$ (this means that $|S_i| \geq 2$), and the notation $ij\sigma kl$ means that either $i, j \in S_1$, $k, l \in S_2$, or $i, j \in S_2$, $k, l \in S_1$.

1.4. Correspondence between formal series, families of multiplications, and families of polynomials. Let $\Phi \in k[[H^t]]$ be a formal series. Disregarding terms of degree ≤ 2, write

$$(1.5) \quad \Phi = \sum_{n=3}^{\infty} \frac{1}{n!} Y_n,$$

where $Y_n \in (H^t)^{\otimes n}$ can also be considered as an even symmetric map $H^{\otimes n} \to k$. Having thus produced Y_n, we can define the symmetric polylinear multiplications $\circ_n$ satisfying (1.1). Clearly, both correspondences are bijective.

We can now formally state the main result of this section.

1.5. Theorem. *The correspondence of 1.4 establishes a bijection between the sets of the following structures on (H, g):*

a) Formal Frobenius manifolds.

b) Cyclic $Comm_\infty$-algebras.

c) Abstract Correlation Functions.

Proof. We start with the correspondence a) $\leftrightarrow$ c). The Associativity Equations for Φ can be written as

$$(1.6) \quad \forall a, b, c, d, \quad \sum_{ef} \Phi_{abe} g^{ef} \Phi_{fcd} = (a \leftrightarrow b \leftrightarrow c \leftrightarrow a),$$

where the subscripts label a basis of H. Representing Φ as in (1.5) and writing $\gamma = \sum_a x^a \Delta_a$ we see that (1.6) is equivalent to

$$\sum_{n_i \geq 3; e,f} \frac{1}{(n_1-3)!} Y_{n_1}(\gamma^{\otimes(n_1-3)} \otimes \Delta_a \otimes \Delta_b \otimes \Delta_e) g^{ef}$$

$$\times \frac{1}{(n_2-3)!} Y_{n_2}(\Delta_f \otimes \Delta_c \otimes \Delta_d \otimes \gamma^{\otimes(n_2-3)})$$

$$= \sum_{n_i \geq 3} \frac{1}{(n_1-3)!(n_2-3)!} (Y_{n_1} \otimes Y_{n_2})(\gamma^{\otimes(n_1-3)} \otimes \Delta_a \otimes \Delta_b \otimes \Delta \otimes \Delta_c \otimes \Delta_d \otimes \gamma^{\otimes(n_2-3)})$$

(1.7) $\qquad = (a \mapsto b \mapsto c \mapsto a).$

In order to deduce (1.7) from the coherence relations (1.4), we proceed as follows. Fix $n \geq 4$, consider in (1.7) only the terms with $n_1 + n_2 - 2 = n$, and multiply them by $(n_1 + n_2 - 6)!$ The resulting identity is a particular case of (1.4), corresponding to the following choices:

$$(\gamma_1, \ldots, \gamma_n) = (\gamma, \ldots, \gamma, \Delta_a, \Delta_b, \Delta_c, \Delta_d),$$

$$(i, j, k, l) = (n_1 + n_2 - 5, n_1 + n_2 - 4, n_1 + n_2 - 3, n_1 + n_2 - 2).$$

Since all the arguments except for the four deltas coincide, summation over the partitions in (1.4) will produce the binomial coefficient which we need. (Actually, we have $\gamma \in H_K$, but this does not violate (1.4).)

Arguing in reverse order, we can deduce (1.4) from (1.7). Then one first obtains (1.4) with a part of the γ's coinciding, belonging to H_K, and being generic even elements. An easy version of the polarization argument then gives the desired conclusion.

We now turn to the correspondence b) $\leftrightarrow$ c). The relation (1.1) can be rewritten as

(1.8) $\qquad (\gamma_1, \ldots, \gamma_n) = \sum_{ab} Y_{n+1}(\gamma_1 \otimes \cdots \otimes \gamma_n \otimes \Delta_a) g^{ab} \Delta_b.$

From here we deduce

$$((\gamma_1, \ldots, \gamma_{n_1}), \gamma_{n_1+1}, \ldots, \gamma_{n_1+n_2-1})$$

$$= \sum_{ab} Y_{n_2+1}(\sum_{cd} Y_{n_1+1}(\gamma_1 \otimes \cdots \otimes \gamma_{n_1} \otimes \Delta_c) g^{cd} \Delta_d \otimes \gamma_{n_1+1} \otimes \cdots \otimes \gamma_{n_1+n_2-1})$$

(1.9) $= \sum_{ab} (Y_{n_1+1} \otimes Y_{n_2+1})(\gamma_1 \otimes \cdots \otimes \gamma_{n_1} \otimes \Delta \otimes \gamma_{n_1+1} \otimes \cdots \otimes \gamma_{n_1+n_2-1} \otimes \Delta_a) g^{ab} \Delta_b.$

The associativity relations (1.2) will exactly match the coherence relations (1.7) rewritten via (1.9) if we put $m + 3 = n_1 + n_2$, $\alpha = \gamma_1, \beta = \gamma_2, \gamma = \gamma_{n_1+1}$; $i = 2, j = 1, k = n_1 + 1, l = m + 3$.

1.6. Identity. If a formal Frobenius manifold (H, g, Φ) admits a flat identity e, it can be identified with a basic element Δ_0. In the respective structure of the cyclic $Comm_\infty$-algebra, formula (2.3) of Chapter I transforms into the following definition of identity, perhaps slightly counter-intuitive:

(1.10) $\qquad (\Delta_0, \gamma_1, \ldots, \gamma_n) = \begin{cases} \gamma_1 & \text{for } n = 1, \\ 0 & \text{otherwise.} \end{cases}$

In fact, this formula for $n = 1$ is equivalent to the statement $(\Delta_0, \Delta_a) = \Delta_a$ for all a, or else $g((\Delta_0, \Delta_a), \Delta_b) = g_{ab}$ for all a, b. But in view of (1.1), the left hand side is the same as

$$Y_3(\Delta_0 \otimes \Delta_a \otimes \Delta_b) = \partial_0 \partial_a \partial_b \Phi(\gamma), \ \gamma = \sum_a x^a \Delta_a,$$

which is g_{ab} in view of I.(2.2).

1.7. The Euler operator. We will return to the discussion of E in the context of correlation functions below; cf. 4.10. Here it is probably worth noting that in the formal situation the grading induced by E interacts with the natural grading on K in which H is of degree 1. If in the semisimple decomposition I.(2.14) the term $\sum \partial_b$ is present, then the grading relation I.(2.7) connects Y_{n+1} to Y_n, otherwise they become decoupled. This last possibility occurs in quantum cohomology for manifolds with vanishing canonical class so that the general constraints of Frobenius manifolds become less stringent for such manifolds.

1.8. Semisimplicity. (H, g, Φ) is called (formally) semisimple if the k-algebra H with the structure constants $\Phi_{ab}{}^c(0)$ is isomorphic to k^n. One can prove then that H_K is isomorphic to K^n. The basic idempotents $e_i \in H_K$ have the same properties as in the geometric theory.

1.9. Why alternative descriptions? The formal version of Frobenius geometry is natural from the viewpoint of quantum cohomology: the relevant structure initially is formal, and only after some work can Φ be analytically continued and geometrized.

The reformulation in terms of $\circ_n$ suggests a non–trivial extension of the notion of commutative algebra and combined with operadic formalism (about which later) leads to an unexpected generalization of other classical structures. For example, one can introduce and study the notion of Lie_∞-algebras, given by a family of S_n-skew symmetric polylinear brackets $[\ ]_n : H^{\otimes n} \to H$, $n = 2, 3, \ldots$, satisfying the higher Jacobi identities: for all $k \geq 2$, $l \geq 0$, $a_1, \ldots, a_k, b_1, \ldots, b_l \in H$ the expression

$$\sum_{i<j} \varepsilon(i,j)[[a_i, a_j], a_1, \ldots, \widehat{a_i}, \ldots, \widehat{a_j}, \ldots, a_k, b_1, \ldots, b_l]$$

must vanish for $l = 0$ and be equal to $[[a_1, \ldots, a_k], b_1, \ldots, b_l]$ otherwise. This structure was called *gravity algebra* by E. Getzler (see [Ge1]). It is dual to $Comm_\infty$ in the same sense as Lie algebras are dual to the commutative ones (Quillen, Kontsevich, Kapranov, and Ginzburg).

It would be interesting to find and study a geometric counterpart of this structure (for which Lie_∞ would be a formal version), with an appropriate notion of potential.

Finally, the structure of Abstract Correlation Functions turns out to be a truncated version of an apparently much richer object, consisting of maps $I_n : H^{\otimes n} \to H^*(\overline{M}_{0n}, k)$, $n \geq 3$, where $\overline{M}_{0n}$ are the moduli spaces of stable curves of genus zero with n labeled points. These maps are constrained by the relations coming from the geometry of $\overline{M}_{0n}$ which extend and "explain" the formal identities (1.4).

The most remarkable fact is that this rich structure is in fact equivalent to its truncated version, thus to $Comm_\infty$ and formal Frobenius manifolds. On the other hand, it admits an intrinsic operation of tensor product, quite unexpected in either of the previous descriptions, geometric and formal alike.

§2. Pointed curves and their graphs

Moduli spaces (orbifolds, stacks) of curves with labeled points are stratified according to their degeneration type. In this section we review the combinatorial structure of this stratification. This study is continued in Chapter V.

2.1. Definition. *A prestable curve over a scheme T is a flat proper morphism $\pi : C \to T$ whose geometric fibers are reduced one-dimensional schemes with at most ordinary double points as singularities. Its genus is a locally constant function on T: $g(t) := \dim H^1(C_t, \mathcal{O}_{C_t})$.*

2.2. Definition. *Let S be a finite set. An S-pointed (equivalently, S-labeled) prestable curve over T is a family $(C, \pi, x_i \,|\, i \in S)$, where $\pi : C \to T$ is a prestable curve, and x_i are sections such that for any geometric point t of T we have $x_i(t) \neq x_j(t)$ for $i \neq j$ and $x_i(t)$ are smooth on C_t.*

Such a curve is called stable if it is connected, and the normalization of each irreducible component which has genus zero (resp. 1) carries at least three (resp. one) special points. A point is called special iff it is the inverse image of either the singular or labeled point of C.

2.2.1. Remark. Let $(C, \pi, x_i \,|\, i \in S)$ be an S-pointed prestable curve. It is stable iff automorphism groups of its geometric fibers fixing the labeled points are finite and there are no infinitesimal automorphisms.

2.3. Definition. *A (finite) graph τ is the data $(F_\tau, V_\tau, \partial_\tau, j_\tau)$ where F_τ is a (finite) set (of flags), V_τ is a finite set (of vertices), $\partial_\tau : F_\tau \to V_\tau$ is the boundary map, and $j_\tau : F_\tau \to F_\tau$ is an involution, $j_\tau^2 = id$.*

An isomorphism $\tau \to \sigma$ consists of two bijections $F_\tau \to F_\sigma$, $V_\tau \to V_\sigma$, compatible with ∂ and j.

Two-element orbits of j_τ form the set E_τ of edges, and one-element orbits form the set S_τ of tails.

It is convenient to think of graphs in terms of their geometric realizations. For each vertex $v \in V_\tau$ put $F_\tau(v) = \partial_\tau^{-1}(v)$ and consider the topological space "star of v" consisting of $|v| := |F_\tau(v)|$ semi-intervals having one common boundary point. These semi-intervals must be labeled by their respective flags. Then take the union of all stars and replace every two-element orbit of j_τ by a segment joining the respective vertices so that these two flags become halves of the edge, and tails become non-paired flags.

A graph τ is called connected (resp. simply connected) if its geometric realization $\|\tau\|$ is also.

2.4. Definition. *A modular graph is a graph τ together with a map $g : V_\tau \to \mathbf{Z}_{\geq 0}$, $v \mapsto g_v$. An isomorphism of two modular graphs is an isomorphism of the underlying graphs preserving the g-labels of vertices.*

A modular graph (τ, g) is called stable if $|v| \geq 3$ for all v with $g_v = 0$, and $|v| \geq 1$ for all v with $g_v = 1$.

2.5. Definition. *The (dual) modular graph (τ, g) of a prestable S-pointed curve $(C, \pi, x_i \,|\, i \in S)$ over an algebraically closed field consists of the following data:*

a) F_τ = the set of branches of C passing through special points.

b) V_τ = the set of irreducible components of C, g_v = the genus of the normalization of the component corresponding to v (sometimes denoted C_v).

c) $\partial_\tau(f) = v$, iff the branch f belongs to the component C_v.

d) $j_\tau(f) = \bar{f}$, $f \neq \bar{f}$, iff the two branches $f, \bar{f}$ intersect at a common double point. Therefore, edges of τ bijectively correspond to the singular points of C.

e) $j_\tau(f) = f$, iff f is a branch passing through a labeled point of C. Thus the tails of τ bijectively correspond to the labeled points of C and to the set S of their labels.

We will sometimes call the isomorphism class of (τ, g) *the combinatorial type* of C.

If C is stable, the combinatorial type of C is stable, and vice versa. Any modular graph represents the combinatorial type of some prestable labeled curve.

2.6. Proposition. *Let (τ, g) be the combinatorial type of a prestable S-pointed connected curve $(C, \pi, x_i \,|\, i \in S)$, g = genus of C, $n = |S|$. Then we have*

$$(2.1) \qquad g = \sum_{v \in V_\tau} g_v + \dim H_1(\|\tau\|),$$

$$(2.2) \qquad g - 1 = \sum_{v \in V_\tau} (g_v - 1) + |E_\tau|,$$

$$(2.3) \qquad 2|E_\tau| + n = \sum_{v \in V_\tau} |v|.$$

We postpone the proof to Chapter V, 1.2.2.

2.6.1. Corollary. *For any (g, n) with $2g - 2 + n > 0$ there exist only finitely many isomorphism classes of connected stable modular graphs of genus g with tails $\{1, \ldots, n\}$ (or simply stable (g, n)-graphs).*

More precisely, if (τ, g) is connected and stable, then:

a) $|V_\tau| \leq 2g - 2 + n$, with the equality sign exactly on graphs for which $(g_v, n) = (0, 3)$ or $(1, 1)$ for all vertices v.

b) For $\gamma \geq 2$,

$$\operatorname{card}\{v \,|\, g_v = \gamma\} \leq \frac{2g - 2 + n}{2\gamma - 2}.$$

c) $|E_\tau| \leq 3g - 3 + n$, with the equality sign exactly on graphs with $g = 0$, $(g_v, |v|) = (0, 3)$ for all vertices.

Proof. Adding (2.2) to one half of (2.3), we get:

$$\sum_{v \in V_\tau} \left(g_v - 1 + \frac{1}{2}|v|\right) = \frac{1}{2}n + g - 1 > 0.$$

Stability implies that $g_v - 1 + \frac{1}{2}|v| \geq \frac{1}{2}$ for $g_v = 0, 1$ and $\geq g_v - 1$ for $g_v \geq 2$. The first two assertions of the corollary now follow directly.

From (2.2) one sees that

$$|E_\tau| \leq g - 1 + \mathrm{card}\{v \mid g_v = 0\} \leq g - 1 + |V_\tau| \leq n + 3g - 3,$$

with the equality sign corresponding to all $g_v = 0$ and hence $|v| = 3$ in view of a).

2.6.2. Remarks. a) There exist infinitely many unstable (0,2), (0,1) and (0,0) graphs.

b) The stable modular graphs with $g_v = 0$ and $|v| = 3$ for all v describe maximally degenerate pointed curves. Such curves have no moduli because each component is a $\mathbf{P}^1$ with three special points on it.

2.6.3. Corollary. *a) Stable connected modular $(0,n)$-graphs are trees with vertices of valency ≥ 3.*

b) Any isomorphism of such graphs is uniquely defined by its restriction on tails.

c) $|V_\tau| - |E_\tau| = 1$ for such graphs.

2.7. Combinatorics of degeneration. Let $(\tau, g), (\sigma, h)$ be connected stable modular graphs with the same (or explicitly identified) set of tails $S = S_\tau = S_\sigma$. We will write $(\tau, g) \geq (\sigma, h)$ if there exists a family of stable curves with an irreducible base such that the generic geometric fiber has the combinatorial type (τ, g), some other geometric fiber has the type (σ, h), and the specialization of the structure sections induces the given identification of tails.

When the set S is fixed, this relation becomes a partial order, called specialization. If $(\tau, g) > (\sigma, h)$ and any intermediate (ρ, k) coincides with either (τ, g) or (σ, h), we will say that (σ, h) is a *codimension one* specialization of (τ, g).

Any codimension one specialization $(\tau, g) > (\sigma, h)$ can be uniquely specified by the data of one of the following two types:

a) *Splitting.* Choose a vertex $v \in V_\tau$ of genus $g_v \geq 0$, a decomposition $g_v = g'_v + g''_v$, and a partition of the set of the flags incident to v: $F_\tau(v) = F'_\tau(v) \cup F''_\tau(v)$, such that both subsets are j_τ-invariant. To obtain (σ, h), replace the vertex v in τ by two vertices v', v'' connected by an edge e, and put $g_{v'} = g'_v$, $g_{v''} = g''_v$, $F_\sigma(v') = F'_\tau(v) \cup \{e'\}$, $F_\sigma(v'') = F''_\tau(v) \cup \{e''\}$ where e', e'' are the two halves of e. The remaining vertices, flags, and incidence relations are the same for τ and σ.

Geometrically, this describes the following degeneration: the irreducible component C_v splits into two irreducible curves, among which the special points of C_v are distributed as specified by the partition of flags. The new edge e "is" the new singular point $C_{v'} \cap C_{v''}$.

b) *Acquisition of a loop/cusp.* Choose a vertex $v \in V_\tau$ of genus $g_v \geq 1$. Put $V_\sigma = V_\tau$, and keep all the g-labels of vertices the same except for g_v which is replaced by $g_v - 1$ in σ. Finally, add two new flags forming one j_σ-orbit (a loop) to $F_\tau(v)$.

Geometrically, this corresponds to a degeneration of C_v acquiring a new cusp. The genus of the normalization is thereby reduced by one.

Arbitrary combinatorial specialization of the stable modular graphs can be realized geometrically.

2.8. Stratified moduli spaces. For any (g,n) with $2g - 2 + n > 0$ there exist two basic types of moduli spaces: $M_{g,n}$ and $\overline{M}_{g,n}$. The first one classifies only smooth stable n–labeled curves, the second one arbitrary ones. The precise definition/construction of these spaces varies depending on the context. There are versions of the types "coarse moduli spaces", "orbifolds", "moduli stacks".

In all versions, however, the following intuitive picture can be made precise.

a) $M_{g,0}(g \geq 2), M_{1,1}, M_{0,3}$ are the basic smooth orbifolds of dimension $3g - 3, 1, 0$ respectively. Each of them carries the universal curve $C \to M$.

b) $M_{g,n}$ is the n-th (resp. $(n-1)$-th, $(n-3)$-th) relative power of the respective $C \to M$, with partial diagonals (and eventually incidence loci with 1 or 3 basic structure sections) deleted.

We can similarly define $M_{g,S}$ parametrizing S–marked curves.

c) For any stable connected n–labeled graph (τ, g) put

$$M_{(\tau,g)} := \left(\prod_{v \in V_\tau} M_{g_v, F_\tau(v)} \right) / G,$$

where G is the automorphism group of (τ, g) identical on tails. This is the moduli space of stable n-labeled curves of the combinatorial type (τ, g). In fact, deforming such a curve is equivalent to independently deforming its irreducible components, keeping track of special points and their incidence relations.

d) Finally, we have a decomposition of $\overline{M}_{g,n}$ into pairwise disjoint locally closed strata indexed by the isomorphism classes of n–graphs:

$$\overline{M}_{g,n} = \coprod_{(\tau,g)} M_{(\tau,g)} = \coprod_{(\tau,g)} \left(\prod_{v \in V_\tau} M_{(g_v, F_\tau(v))} \right)/G.$$

The stratum $M_{(\sigma,h)}$ belongs to the closure of $M_{(\tau,g)}$ exactly when $(\tau,g) > (\sigma,h)$.

In the next section we treat the genus zero case in more detail. An essential simplification is due to the fact that stable n–trees have no non–trivial n–automorphisms (that is, automorphisms identical on leaves.) Therefore moduli spaces of genus zero are actually smooth manifolds.

§3. Moduli spaces of genus 0

This section is a report on the structure of the moduli spaces of curves of genus zero elaborating the general discussion of the previous section. We give precise statements but often omit or only sketch the proofs which can be found in [Kn1] and [Ke]. We work in the category of schemes over an arbitrary field (in most cases, Spec **Z** would do as well).

3.1. Theorem. *a) For any $n \geq 3$, there exists a universal n-pointed stable curve $(\overline{\pi}_n : \overline{C}_{0n} \to \overline{M}_{0n}; x_i, i = 1, \ldots, n)$ of genus zero. This means that any such curve over a scheme T is induced by a unique morphism $T \to \overline{M}_{0n}$.*

b) $\overline{M}_{0n}$ is a smooth irreducible projective algebraic variety of dimension $n - 3$.

c) *For any stable n-tree τ, there exists a locally closed reduced irreducible subscheme $D(\tau) \subset \overline{M}_{0n}$ parametrizing exactly curves of the combinatorial type τ. Its codimension equals the cardinality of the set of edges $|E_\tau|$, that is, the number of singular points of any curve of the type τ. This subscheme depends only on the n-isomorphism class of τ.*

d) *$\overline{M}_{0n}$ is the disjoint union of all $D(\tau)$. The closure of any of the strata $D(\tau)$ is the union of all strata $D(\sigma)$ such that $\tau > \sigma$ in the sense of 2.7.*

Let σ_n be the one-vertex n-tree. We will denote $D(\sigma_n)$ by M_{0n} and the induced stable curve by $\pi_n : C_{0n} \to M_{0n}$. It classifies the irreducible pointed curves. Its geometric points are systems of n pairwise distinct points on $\mathbf{P}^1$ *considered up to a common fractional linear transformation.*

The codimension one strata are labeled by the isomorphism classes of stable one-edge n-graphs σ. Each such class can be identified with an *unordered* partition $\{1,\ldots,n\} = S_1 \coprod S_2$; stability means that $|S_i| \geq 2$ for $i = 1, 2$. The curve $\overline{C}_{0n}$ over $D(\sigma)$ has two components, and the partition $S_1 \coprod S_2$ corresponds to the distribution of the structure sections x_i between these components. Of course, with obvious modifications we can replace $\{1,\ldots,n\}$ here by any finite set S.

3.2. Examples. Figures 1 and 2 show the structure of $\overline{M}_{0n}$ with its canonical stratification, and the structure of $\overline{C}_{0n}$, for $n = 3, 4, 5$.

$M_{03} = \overline{M}_{03}$ is simply a point, and $\overline{C}_{0n}$ is $\mathbf{P}^1$ endowed with three points labeled by 1,2,3 because the fractional linear group acts simply transitively on the ordered triples.

$\overline{M}_{04}$ is also $\mathbf{P}^1$ with three labeled points, but this time the labels are one-edge stable trees with tails $\{1,2,3,4\}$ corresponding to the divisorial strata, and M_{04} is the complement to these three points. Furthermore, $\overline{C}_{04}$ is a surface fibered over $\mathbf{P}^1$ and endowed with four labeled sections. In addition to them, there are six components of degenerate fibers. One can check that all the ten curves are exceptional of the first kind, forming a configuration well known in the theory of the Del Pezzo surfaces. In fact, $\overline{C}_{04}$ is isomorphic to the (rigid) Del Pezzo surface of degree 5, which can be obtained by blowing up four points of $\mathbf{P}^2$ in general position. It is known that S_5, and not just S_4 (renumbering sections), acts on such a surface. In our context this can be explained by the fact that $\overline{C}_{04}$ can be identified with $\overline{M}_{05}$ (non-canonically) or rather with some $\mathbf{M}_{0S}$, $|S| = 5$, canonically; the reader is invited to describe S.

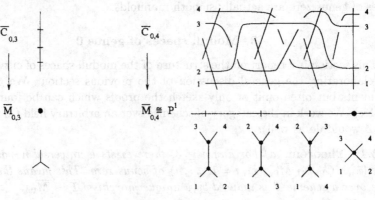

FIGURE 1

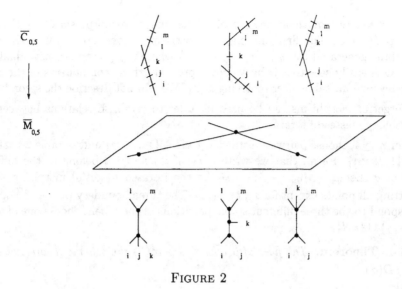

FIGURE 2

Of course, M_{05} is the complement to the ten boundary divisors marked by the stable 5-trees with one edge. Each of these divisors contains three 0-dimensional strata marked by the stable 5-trees with two edges.

We see an emerging pattern: $\overline{C}_{0,n}$ is isomorphic to $\overline{M}_{0,n+1}$. It can be explained by the following considerations.

Consider a stable pointed curve $(C, x_1, \ldots, x_{n+1})$ of genus 0 over a field. We will say that $(C, x_1, \ldots, x_n)$ is obtained from it by forgetting the point x_{n+1}. However, $(C, x_1, \ldots, x_n)$ may well be unstable. This will happen precisely when the component of C supporting x_{n+1} has only one additional labeled point, say x_j. In this case we can contract this component to its intersection point x'_j with some other component of C, thus getting the n-pointed curve $(C', x_1, \ldots, x_{j-1}, x'_j, \ldots, x_n)$. We will call the last step *stabilization*, and the resulting construction (forgetting plus stabilization whenever necessary) *stable forgetting*.

3.3. Theorem. *a) There is a canonical morphism $\rho_{n+1} : \overline{M}_{0,n+1} \to \overline{M}_{0n}$ which acts on the isomorphism classes of $(n+1)$-pointed curves by stably forgetting the last point.*

b) There exists a canonical isomorphism $\mu_n : \overline{M}_{0,n+1} \to \overline{C}_{0n}$ commuting with projections to $\overline{M}_{0n}$.

The first statement is not obvious because it is not clear that collapsing of unstable components can be performed uniformly over a base.

F. Knudsen ([Kn1], §2) proves both statements in the following way. He remarks that not only $\overline{M}_{0,n+1}$ but $\overline{C}_{0n}$ as well represents a natural functor, namely

$$\overline{C}_{0n}(T) = \{T\text{-families of stable } (0,n)\text{-curves with an extra section } \Delta\}/(\text{iso}).$$

No restriction is imposed on this extra section. The universal family is $\overline{C}_{0n} \times_{\overline{M}_{0n}} \overline{C}_{0n}$ fibered over $\overline{C}_{0n}$ via the second projection, with relative diagonal as Δ.

Therefore it suffices to produce a functorial bijection between the T-families of the types $(C, x_1, \ldots, x_n, x_{n+1})$ and $(D, y_1, \ldots, y_n, \Delta)$, respectively. This bijection is defined via two mutually inverse birational maps: a morphism $C \to D$ and a blow up $D \to C$. The first one maps C to the projective spectrum of the sheaf of algebras generated by $\omega_{C/T}(x_1 + \cdots + x_n)$ where $\omega_{C/T}$ is the relative dualizing sheaf. One easily sees that it blows down precisely those components of the fibers which become unstable after removing x_{n+1}. We will not describe the second map.

Forgetful morphisms can be used in order to establish relations between the cohomology classes of strata.

For $n \geq 4$, choose pairwise distinct $i, j, k, l \in \{1, \ldots, n\}$ and a stable 2-partition σ of $\{1, \ldots, n\}$. Recall that we write $ij\sigma kl$ if i, j and k, l belong to the different parts of σ. Let $\mu : \overline{M}_{0n} \to \overline{M}_{0,\{ijkl\}}$ be the iterated forgetful morphism stably forgetting all points except for x_i, x_j, x_k, x_l. The three boundary points of $\overline{M}_{0,\{ijkl\}}$ correspond to the three different stable partitions of the labels; choose one of them, say $\{i, j\} \cup \{k, l\}$.

3.4. Theorem. *The fiber of μ over this point is the scheme theoretical union*
$$\bigcup_{\sigma: ij\sigma kl} D(\sigma).$$

For a proof, see [Kn1], Theorem 2.7, and [Ke], p. 552, Fact 3.

3.4.1. Corollary. *Let $[D(\sigma)]$ be the cohomology (or Chow) class of $D(\sigma)$. Then for any quadruple $i, j, k, l \in \{1, \ldots, n\}$ we have*

$$(3.1) \qquad \sum_{ij\sigma kl} [D(\sigma)] - \sum_{kj\tau il} [D(\tau)] = 0.$$

In fact, (3.1) is the difference of two fibers of the forgetful morphism.

In order to state the second corollary, we introduce some notation. For two unordered stable partitions $\sigma = \{S_1, S_2\}$ and $\tau = \{T_1, T_2\}$ of S put

$$a(\sigma, \tau) := \text{ the number of non--empty pairwise}$$
$$\text{distinct sets among } S_a \cap T_b, \ a, b = 1, 2.$$

Clearly, $a(\sigma, \tau) = 2, 3$, or 4. Moreover, $a(\sigma, \tau) = 2$ iff $\sigma = \tau$, and $a(\sigma, \tau) = 4$ iff there exist pairwise distinct $i, j, k, l \in S$ such that simultaneously $ij\sigma kl$ and $ik\tau jl$. If $a(\sigma, \tau) = 3$, we sometimes call σ and τ *compatible*. A family of 2-partitions $\{\sigma_1, \ldots, \sigma_m\}$ is called *good*, if for all $i \neq j$, σ_i and σ_j are compatible.

3.4.2. Corollary. *If $a(\sigma, \tau) = 4$, then*

$$(3.2) \qquad \overline{D}(\sigma) \cap \overline{D}(\tau) = \emptyset.$$

In fact, $\overline{D}(\sigma)$ and $\overline{D}(\tau)$ belong to two different fibers of an appropriate forgetful morphism to $\mathbf{P}^1$.

3.5. The ring structure of $H^*(\overline{M}_{0S})$. Keel [Ke] has shown that the dual classes of $[\overline{D}(\tau)]$ generate the ring $H^*(\overline{M}_{0S})$, whereas (3.1) and (3.2) generate the ideal of relations.

More precisely, for a given finite set S of cardinality ≥ 3, consider a family of independent commuting variables D_σ indexed by stable unordered 2-partitions of S. Put $F_S = k[D_\sigma]$ ($F_S = k$ for $|S| = 3$). This is a graded polynomial ring, $\deg D_\sigma = 1$. Define the ideal $I_S \subset F_S$ generated by the following elements:

a) For each ordered quadruple $i, j, k, l \in S$

(3.3) $$R_{ijkl} := \sum_{ij\sigma kl} D_\sigma - \sum_{kj\tau il} D_\tau \in I_S.$$

b) For each pair σ, τ with $a(\sigma, \tau) = 4$:

(3.4) $$D_\sigma D_\tau \in I_S.$$

Finally, put $H_S^* = K[D_\sigma]/I_S$.

3.5.1. Theorem (Keel [Ke]). *The map*

$$D_\sigma \longmapsto \text{dual class of } \overline{D}(\sigma)$$

induces the isomorphism of rings (doubling the degrees)

(3.5) $$H_S^* \xrightarrow{\sim} H^*(\overline{M}_{0S}, k) = A^*(\overline{M}_{0S})_k.$$

Here A^* is the Chow ring.

Keel's presentation (3.5) in principle solves the problem of algorithmic calculations in the cohomology ring. In practice, however, even the most basic properties of this ring are not obvious for H_S^*, e.g., the facts that $H_S^i = 0$ for $i > |S| - 3$, $\dim H_S^{|S|-3} = 1$, and the Poincaré pairing is perfect duality.

In the next section we will need more precise information about the homogeneous components not only of H_S^*, but of I_S as well. The remaining part of this subsection is devoted to the preparatory work. We keep the notation of Theorem 3.5.1.

The monomial $D_{\sigma_1} \ldots D_{\sigma_a} \in F_S$ is called good if the family of 2-partitions $\{\sigma_1, \ldots, \sigma_a\}$ is good, i.e. $a(\sigma_i, \sigma_j) = 3$ for $i \neq j$. Notice that the relevant divisors are then pairwise distinct. In particular, D_σ and 1 are good.

Consider a stable S-tree τ. Any edge $e \in E_\tau$ defines a stable partition $\sigma(e)$: if one cuts e, the tails of the resulting two trees (except for halves of e) form $\sigma(e)$.

3.5.2. Proposition. *a) The monomial*

$$m(\tau) := \prod_{e \in E_\tau} D_{\sigma(e)}$$

is good.

b) For any $0 \leq r \leq |S| - 3$, the map $\tau \longmapsto m(\tau)$ establishes a bijection between the set of good monomials of degree r in F_S and stable S-trees τ with $|E_\tau| = r$ modulo S-isomorphisms. There are no good monomials of degree $> |S| - 3$.

Proof. a) Let f be a flag of a tree τ whose boundary is the vertex v. It defines a subtree of τ which we will call *the branch of f*. If f is itself a tail, its branch consists of f and v. In general, it comprises all vertices, flags, and edges that can be reached (in geometric realization) by a no–return path starting with (v, f). Denote by $S(f)$ the set of leaves on this branch (or the set of their labels).

Now let $e \neq e' \in E_\tau$. There exists a sequence of pairwise distinct edges $e = e'_0, e'_1, \ldots, e'_r, e'_{r+1} = e'$, $r \geq 0$, such that e'_j and e'_{j+1} have a common vertex v_j. Let u be the remaining vertex of e, and w that of e'. Let S' be the set of all tails of τ belonging to the branches starting at u but not with a flag belonging to e; similarly, let S'' be the set of all tails of τ belonging to the branches that start at w but not with a flag belonging to e'. Finally, let T be the set of all tails on the branches at $v_0, \ldots, v_r$ not starting with the flags in $e'_0, \ldots, e'_{r+1}$ (we identify tails with their labels). Since τ is stable, all three sets S', S'', and T are non-empty, and

$$\sigma(e) = \{S', S'' \coprod T\}, \ \sigma(e') = \{S' \coprod T, S''\}.$$

It follows that $a(\sigma(e), \sigma(e')) = 3$ so that $m(\tau)$ is a good monomial.

b) For $r = 0, 1$ the assertion is clear. Assume that for some $r \geq 1$ the map $\tau \longmapsto m(\tau)$ is surjective on good monomials of degree r. We will prove then that it is surjective in degree $r + 1$.

Let m' be a good monomial of degree $r + 1$. Choose a divisor D_σ of m' which is *extremal* in the following sense: one element, say S_1, of the partition $\sigma = \{S_1, S_2\}$ is minimal in the set of all elements of all 2-partitions σ' such that $D_{\sigma'}$ divides m'. Put $m' = D_\sigma m$. Since m is good of degree r, we have $m = m(\tau)$ for some stable S-tree τ. We will show that $m' = m(\tau')$, where τ' is obtained from τ by inserting a new edge with tails marked by S_1 at an appropriate vertex $v \in V_\tau$. In other words, τ' is a codimension one specialization of τ in the sense of 2.7.

First we must find v in τ. To this end, consider any edge $e \in E_\tau$ and the respective partition $\sigma(e) = \{S'_e, S''_e\}$. Since m' is good, we have $a(\{S_1, S_2\}, \{S'_e, S''_e\}) = 3$. As S_1 is minimal, one sees that exactly one of the sets $\{S'_e, S''_e\}$ strictly contains S_1. Let it be S''_e. Orient e by declaring that the direction from the vertex (corresponding to) S'_e to S''_e is positive. We claim that with this orientation of all edges, for any $w \in V_\tau$ there can be at most one edge outgoing from w. In fact, if τ contains a vertex w with two positively oriented flags f_1 and f_2, then S_1 must be contained in the two subsets of S, branches $S(f_1)$ and $S(f_2)$. But their intersection is empty.

It follows that there exists exactly one vertex $v \in V_\tau$ having no outgoing edges. Moreover, S_1 is contained in the set of labels of the tails at v by construction. If we now define τ' by inserting a new edge e' at v so that $\sigma(e') = \sigma$, we will clearly have $m' = m(\tau')$. If $r \leq |S| - 4$, the tree τ' cannot be unstable because, first, $|S_1| \geq 2$, and second, at least two more flags converge at v: otherwise the unique incoming edge would produce the partition $\{S_1, S_2\} = \sigma$ which would mean that D_σ already divides $m(\tau)$.

For $r = |S| - 3$, this argument shows that m' cannot exist because all the vertices of τ have valency three.

It remains to check that if $m(\tau_1) = m(\tau_2)$, then τ_1 and τ_2 are S–isomorphic.

Assume that this has been checked in degree $\leq r$ and that $\deg \tau_1 = \deg \tau_2 = r + 1$. Choose an extremal divisor D_σ of $m(\tau_1) = m(\tau_2)$ as above and contract the respective edges of τ_1, τ_2, getting the trees τ'_1, τ'_2. Since $m(\tau'_1) = m(\tau'_2) = m(\tau_i)/D_\sigma$, τ'_1 and τ'_2 are S–isomorphic by the inductive assumption. This isomorphism respects the marked vertices v'_1, v'_2 corresponding to the contracted edges because as we have seen they are uniquely defined. Hence it extends to an S-isomorphism $\tau_1 \to \tau_2$.

3.5.3. Remark. Since the boundary divisors intersect transversally, the image of $m(\tau)$ in $H^*(\overline{M}_{0S})$ is the dual class of $D(\tau)$.

3.6. Multiplication formulas. In this subsection we will show that good monomials modulo I_S span H_S and therefore, dual classes of strata span $H^*(\overline{M}_{0S})$.

This will follow from the more precise formulas (3.6)–(3.9), allowing one to express recursively a product of good monomials modulo I_S as a linear combination of good monomials.

Let σ, τ be two stable S-trees, and $|E_\sigma| = 1$. We have to consider the following alternatives.

a) $D_\sigma m(\tau)$ *is a good monomial.* Then

$$(3.6) \qquad D_\sigma m(\tau) = m(\tau'),$$

where τ is obtained from τ' by contracting the edge in $E_{\tau'}$ whose 2-partition coincides with that of σ.

More generally, if $m(\sigma)m(\tau)$ is a good monomial, then

$$(3.7) \qquad m(\sigma)m(\tau) = m(\sigma \times \tau),$$

where the direct product is the categorical one in the category of S-trees and S-morphisms, to be described later. We can identify $E_{\sigma \times \tau}$ with $E_\sigma \coprod E_\tau$, and $p_1 : \sigma \times \tau \to \sigma$ (resp. $p_2 : \rho \times \tau \to \tau$) contracts edges of the second factor (resp. of the first one).

b) *There exists a divisor $D_{\sigma'}$ of $m(\tau)$, $|E_{\sigma'}| = 1$, such that $a(\sigma, \sigma') = 4$.* Then

$$(3.8) \qquad D_\sigma m(\tau) \equiv 0 \bmod I_S$$

in view of (3.4).

c) D_σ *divides* $m(\tau)$. Then let $e \in E_\tau$ be the edge corresponding to σ; v_1, v_2 its vertices, and (v_i, e) the corresponding flags.

We will write several different expressions for $D_\sigma m(\tau)$ mod I_S, corresponding to various choices of unordered pairs of distinct flags $\{\bar{\imath}, \bar{\jmath}\} \subset F_\tau(v_1) \setminus \{(v_1, e)\}$, $\{\bar{k}, \bar{l}\} \subset F_\tau(v_2) \setminus \{(v_2, e)\}$. For each choice, put

$$T_1 = F_\tau(v_1) \setminus \{\bar{\imath}, \bar{\jmath}, (v_1, e)\},$$
$$T_2 = F_\tau(v_2) \setminus \{\bar{k}, \bar{l}, (v_2, e)\}.$$

Notice that because of stability the set of such choices is non-empty.

3.6.1. Proposition. *For every such choice we have*

$$(3.9) \qquad D_\sigma m(\tau) \equiv - \sum_{\substack{T \subset T_1 \\ |T| \geq 1}} m(tr_{T,e}(\tau)) - \sum_{\substack{T \subset T_2 \\ |T| \geq 1}} m(tr_{T,e}(\tau)) \bmod I_S,$$

where $tr_{T,e}(\tau)$ is the tree obtained from τ by "transplanting all branches starting in T to the middle point of the edge e" (see Figure 3). (An empty sum is zero.)

98   III. FROBENIUS MANIFOLDS AND MODULI SPACES OF CURVES

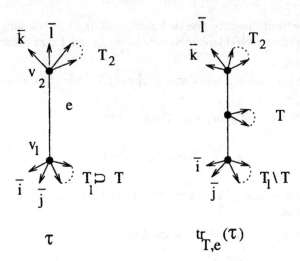

FIGURE 3. Transplants: arrows correspond to branches

Remark. We can also describe $tr_{T,e}(\tau)$ as a result of inserting an extra edge instead of the vertex v_1 (resp. v_2) and putting the branches T to the common vertex of the new edge and e. There exists a well-defined edge in $tr_{T,e}(\tau)$ whose contraction produces τ.

Proof. We choose pairwise distinct labels on the chosen branches $i \in S(\bar{i})$, $j \in S(\bar{j})$, $k \in S(\bar{k})$, $l \in S(\bar{l})$ and then calculate the element (see (3.3))

$$(3.10) \qquad R_{ijkl} \cdot m(\tau) = \left(\sum_{ij\rho kl} D_\rho - \sum_{kj\rho il} D_\rho \right) m(\tau) \equiv 0 \bmod I_S.$$

Clearly, $ij\sigma kl$, so that for all terms D_ρ of the second sum in (1.5) we have $a(\sigma, \rho) = 4$ so that $D_\rho m(\tau) \in I_S$. Among the terms of the first sum, there is one D_σ. If $ij\rho kl$ and $\rho \neq \sigma$, then D_ρ cannot divide $m(\tau)$. Otherwise ρ would correspond to an edge $e' \neq e$, but the 2-partition of such an edge cannot break $\{i,j,k,l\}$ into $\{i,j\}$ and $\{k,l\}$ as a glance to a picture of τ shows. It follows that $D_\rho m(\tau) = m(\rho \times \tau)$ as in (3.7). The projection $\rho \times \tau \to \tau$ contracts the extra edge onto a vertex that can be only one of the ends of e, otherwise, as above, the condition $ij\rho kl$ cannot hold. It should be clear by now that $\rho \times \tau$ must be one of the trees $tr_{T,e}(\tau)$, and that each tree of this kind can be uniquely represented as $\rho \times \tau$ for some ρ with $ij\rho kl$. But from (3.10) it follows that

$$D_\sigma m(\tau) \equiv - \sum_{\substack{ij\rho kl \\ \rho \neq \sigma}} D_\rho m(\tau) \bmod I_S,$$

which is (3.9).

3.7. Integral over the fundamental class. The functional

$$\int_{\overline{M}_{0,S}} : H^*(\overline{M}_{0,S}) \to k$$

is given by

$$m(\tau) \longmapsto \begin{cases} 1 & \text{if } \deg m(\tau) = |S| - 3, \\ 0 & \text{otherwise.} \end{cases}$$

Notice that $\deg m(\tau) = |S| - 3$ iff $|v| = 3$ for all $v \in V_\tau$, and $\overline{D}(\tau)$ is a point in this case.

§4. Formal Frobenius manifolds and Cohomological Field Theories

4.1. Definition. *In the notation of 1.1, the structure of the (tree level) Cohomological Field Theory (CohFT) on (H, g) is given by a family of even linear maps (correlators)*

$$(4.1) \qquad I_n : H^{\otimes n} \to H^*(\overline{M}_{0n}, k), \quad n = 3, 4, \ldots,$$

satisfying the following conditions:

a) $\mathbf{S}_n$-covariance (with respect to the natural action of $\mathbf{S}_n$ on both sides of (4.1)).

b) Splitting, or compatibility with restriction to the boundary divisors: for any stable ordered partition $\sigma : \{1, \ldots, n\} = S_1 \coprod S_2$, $n_i = |S_i|$, and the respective map

$$\varphi_\sigma : \overline{M}_{0,n_1+1} \times \overline{M}_{0,n_2+1} \to \overline{D}(\sigma) \subset \overline{M}_{0n}$$

we have

$$(4.2) \quad \varphi_\sigma^*(I_n(\gamma_1 \otimes \ldots \otimes \gamma_n)) = \varepsilon(\sigma)(I_{n_1+1} \otimes I_{n_2+1})\left(\bigotimes_{p \in S_1} \gamma_p \otimes \Delta \otimes \left(\bigotimes_{q \in S_2} \gamma_q\right)\right),$$

where $\Delta = \sum \Delta_a g^{ab} \otimes \Delta_b$ is the Casimir element, and $\varepsilon(\sigma)$ is the sign of the permutation induced on the odd arguments $\gamma_1, \ldots, \gamma_n$.

Let (H, g, I_*) be a CohFT. Its correlation functions are polylinear functionals

$$(4.3) \qquad Y_n : H^{\otimes n} \to k, \quad Y_n(\gamma_1 \otimes \cdots \otimes \gamma_n) := \int_{\overline{M}_{0n}} I_n(\gamma_1 \otimes \cdots \otimes \gamma_n),$$

where the integral denotes the value of the top-dimensional component of I_n on the fundamental cycle of $\overline{M}_{0n}$; cf. 3.7 above.

4.2. Proposition. *Correlation functions of a CohFT satisfy the axioms of Abstract Correlation Functions (Definition 1.3.1).*

Proof. Clearly, functionals (4.3) are symmetric, because I_n are $\mathbf{S}_n$-covariant.

In order to check (1.4), look at (3.1), this time interpreted as the linear relation between the homology classes of the boundary divisors. This implies

$$(4.4) \qquad \sum_{\sigma : ij\sigma kl} \int_{\overline{M}_{0,n_1+1} \times \overline{M}_{0,n_2+1}} \varphi_\sigma^*(I_n(\gamma_1 \otimes \ldots \otimes \gamma_n)) = (j \leftrightarrow k).$$

Substituting (4.2) into (4.4), we obtain (1.4).

We can now state the central result of this section:

4.3. Theorem. *Each ACF is the system of correlation functions of the unique CohFT. Thus, the following notions are equivalent:*

a) *Formal Frobenius manifolds.*

b) *Cohomological Field Theories.*

Before proving this theorem, we will discuss two related themes.

4.4. Tensor product. Let $\{H^{(1)}, g^{(1)}, I_n^{(1)}\}$ and $\{H^{(2)}, g^{(2)}, I_n^{(2)}\}$ be two CohFT's. Define $H = H^{(1)} \otimes H^{(2)}$ and $g = g^{(1)} \otimes g^{(2)}$. Put

$$I_n(\gamma_1^{(1)} \otimes \gamma_1^{(2)} \otimes \ldots \otimes \gamma_n^{(1)} \otimes \gamma_n^{(2)})$$
$$(4.5) \qquad := \varepsilon(\gamma^{(1)}, \gamma^{(2)}) I_n^{(1)}(\gamma_1^{(1)} \otimes \ldots \otimes \gamma_n^{(1)}) \cup I_n^{(2)}(\gamma_1^{(2)} \otimes \ldots \otimes \gamma_n^{(2)}),$$

where $\varepsilon(\gamma^{(1)}, \gamma^{(2)})$ is our standard sign in superalgebra, and $\cup$ is the cup product in $H^*(\overline{M}_{0n}, k)$.

CLAIM. (H, g, I_*) *is a CohFT*.

This can be shown by a straightforward check.

Thanks to Theorem 4.3, this tensor product operation can be defined on $Comm_\infty$-algebras and formal Frobenius manifolds. But (4.5) cannot be trivially restricted to these structures. In fact, they are directly formulated in terms of the top components of I_n, whereas the tensor product involves components of all degrees.

We will continue the study of this tensor product after the end of the proof of Theorem 4.3.

4.5. Complete Cohomological Field Theories. In this subsection only we will denote the metric by h, reserving g for genus. The complete, as opposed to tree level, CohFT structure on (H, h) is given by a family of maps

$$(4.6) \qquad I_{g,n} : H^{\otimes n} \to H^*(\overline{M}_{g,n}, k)$$

indexed by all stable pairs (g, n). They must satisfy the following extension of the genus zero axioms:

a) S_n-covariance for all g.

b) Splitting: for any g_1, g_2, $g_1 + g_2 = g$, and σ as above, such that $(g_i, n_i + 1)$ are stable, we must have
(4.7)
$$\varphi^*(I_{g,n}(\gamma_1 \otimes \ldots \otimes \gamma_n)) = \varepsilon(\sigma)(I_{g_1,n_1+1} \otimes I_{g_2,n_2+1}) \left(\bigotimes_{p \in S_1} \gamma_p \otimes \Delta \otimes \left(\bigotimes_{q \in S_2} \gamma_q \right) \right),$$

where $\varphi : \overline{M}_{g_1,n_1+1} \times \overline{M}_{g_2,n_2+1} \to \overline{M}_{g,n}$ is the respective boundary morphism corresponding to the degeneration described in 2.7a).

c) Acquiring a cusp: for $g \geq 1$

$$(4.8) \qquad \psi^*(I_{g,n}(\gamma_1 \otimes \ldots \otimes \gamma_n)) = I_{g-1,n+2}(\gamma_1 \otimes \ldots \otimes \gamma_n \otimes \Delta),$$

where $\psi : \overline{M}_{g-1,n+2} \to \overline{M}_{g,n}$ is the boundary morphism described in 2.7b).

§4. COHOMOLOGICAL FIELD THEORIES

The theory of Gromov–Witten invariants actually furnishes such a structure on the cohomology spaces of projective algebraic and symplectic manifolds. Hence it is very important to study the complete CohFT's. The tensor product formula extends to the complete case and plays the role of the Künneth formula for the Gromov–Witten invariants.

Technically, the difference between the tree level and the complete case reflects our very incomplete understanding of the topology of $\overline{M}_{g,n}$ for $g \geq 1$.

A further extension of this theory involves the so-called *gravitational descendants*. See Chapter VI for more details.

The proof of Theorem 4.3 is contained in 4.6–4.9 below. We shall first show that if a CohFT with a given system of correlation functions exists at all, then it is unique.

4.6. Proposition. *Let (H, g, I_*) be a CohFT with correlation functions (Y_n). Then for any stable n-tree τ we have*

$$(4.9) \qquad \int_{\overline{D}(\tau)} I_n(\gamma_1 \otimes \ldots \otimes \gamma_n) = \left(\bigotimes_{v \in V_\tau} Y_{F_\tau(v)} \right) \left(\bigotimes_{i \in S_\tau} \gamma_i \otimes \Delta^{\otimes E_\tau} \right).$$

Since the homology classes of $\overline{D}(\tau)$ span $H_*(\overline{M}_{0n}, k)$ (cf. 3.6), this establishes the uniqueness.

Proof. Let us explain the meaning of (4.9). We use the extension of the formalism of direct products to the arbitrary finite sets S. Then, say, $Y_n(\gamma_1 \otimes \cdots \otimes \gamma_n)$ can be replaced by $Y_S(\bigotimes_{i \in S} \gamma_i)$, and the argument of $\bigotimes_{v \in V_\tau} Y_{F_\tau(v)}$ must be some linear combination of the elements of the form $\bigotimes_{v \in V_\tau} (\bigotimes_{f \in F_\tau(v)} \alpha_f)$, $\alpha_f \in H$. If f is a tail marked by i, we choose $\alpha_f = \gamma_i$ in (4.9). Otherwise f is a half of an edge $\{f, \overline{f}\}$, and each such edge contributes Δ.

The formula (4.2) furnishes a particular case of (4.9) for the one-edge tree τ. But we can iterate (4.2), refining the inclusion $\overline{D}(\tau) \subset \overline{M}_{0n}$ to a sequence of codimension one boundary embeddings and using (4.2) at each step. A contemplation will convince the reader that (4.9) will be the final answer, independent on the chosen refinement. This proves Proposition 4.6.

It remains to establish that if (Y_n) is an arbitrary ACF, then the formulas (4.9) actually define a CohFT. The only problem is to check that for any $n \geq 3$ and $\gamma_1, \ldots, \gamma_n \in H$, there exists a cohomology class $I_n(\gamma_1 \otimes \cdots \otimes \gamma_n) \in H^*(\overline{M}_{0n}, k)$ which as a linear functional on $[\overline{D}(\tau)]$ is defined by (4.9). Then it will be automatically S_n-invariant, and will satisfy (4.2).

In other words, it remains to show that all linear relations between $[\overline{D}(\tau)]$ are also satisfied by the right hand sides of (4.9). Again, for the codimension one case this is a built-in property: Keel's relations (3.3) between $[\overline{D}(\tau)]$ are precisely reflected in the quadratic relations (1.4) postulated for any ACF. To deal with arbitrary codimension, we will start with a generalization of Keel's relations.

4.7. Basic linear relations. As in 3.6, we will work with classes of boundary strata in $H^*(\overline{M}_{0S})$, represented by the classes of good monomials in F_S mod I_S.

Let $|S| \geq 4$. Consider a system $(\tau, v, \overline{i}, \overline{j}, \overline{k}, \overline{l})$ where τ is an S-tree, $v \in V_\tau$ is a vertex with $|v| \geq 4$, and $\overline{i}, \overline{j}, \overline{k}, \overline{l} \in F_\tau(v)$ are pairwise distinct flags (taken in this order). Put $T = F_\tau(v) \setminus \{\bar{i},\bar{j},\bar{k},\bar{l}\}$. For any ordered 2-partition of T, $\alpha = \{T_1, T_2\}$ (one or both T_i can be empty), we can define two trees $\tau'(\alpha)$ and $\tau''(\alpha)$. The first one is obtained by inserting a new edge e at $v \in V$ with branches $\{\bar{i},\bar{j},T_1\}$ and $\{\bar{k},\bar{l},T_2\}$ at its edges. The second one corresponds similarly to $\{\bar{k},\bar{j},T_1\}$ and $\{\bar{i},\bar{l},T_2\}$. We recall that $S(\bar{i})$ is the set of labels of tails belonging to the branch of $\bar{i}$: see Figure 4.

4.7.1. Proposition. *We have*

$$(4.10) \qquad R(\tau, v, \bar{i}, \bar{j}, \bar{k}, \bar{l}) := \sum_\alpha [m(\tau'(\alpha)) - m(\tau''(\alpha))] \in I_S.$$

Proof. Choose $i \in S(\bar{i})$, $j \in S(\bar{j})$, $k \in S(\bar{k})$, $l \in S(\bar{l})$, and calculate $R_{ijkl} m(\tau) \in I_S$, where R_{ijkl} is defined by (3.3). Consider for instance the summands $D_\sigma m(\tau)$ for $ij\sigma kl$.

From the picture of τ it is clear that D_σ does not divide $m(\tau)$. If $D_\sigma m(\tau)$ does not vanish modulo I_S, we must have $D_\sigma m(\tau) = m(\sigma \times \tau)$, and $\sigma \times \tau$ is of the type $\tau'(\alpha)$. Similarly, the summands of $D_\sigma m(\tau)$ with $kj\sigma il$ are of the type $m(\tau''(\alpha))$.

4.8. Theorem. *All linear relations modulo I_S between good monomials of degree $r+1$ are spanned by the relations (4.10) for $|E_\tau| = r$.*

Proof. For $r = 0$ this holds by the definition of I_S. Generally, denote by H_{*S} the linear space, generated by the symbols $\mu(\tau)$ for all S–isomorphism classes τ of stable S–trees satisfying the analog of the relations (4.10),

$$(4.11) \qquad r(\tau, v, \bar{i}, \bar{j}, \bar{k}, \bar{l}) := \sum_\alpha [\mu(\tau'(\alpha)) - \mu(\tau''(\alpha))] = 0.$$

Denote by 1 the symbol $\mu(\rho)$ where ρ is the one-vertex tree.

4.8.1. Main Lemma. *There exists on H_{*S} a structure of the H_S^*–module given by the following multiplication formulas reproducing (3.7), (3.8), and (3.9):*

$$(4.12) \qquad D_\sigma \mu(\tau) = \mu(\sigma \times \tau),$$

if $D_\sigma m(\tau)$ is a good monomial;

$$(4.13) \qquad D_\sigma \mu(\tau) = 0,$$

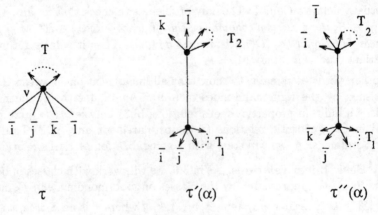

FIGURE 4

§4. COHOMOLOGICAL FIELD THEORIES

if there exists a divisor $D_{\sigma'}$ of $m(\tau)$ such that $a(\sigma, \sigma') = 4$;

$$(4.14) \qquad D_\sigma \mu(\tau) = - \sum_{\substack{T \subset T_1 \\ |T| \geq 1}} \mu(tr_{T,e}(\tau)) - \sum_{\substack{T \subset T_2 \\ |T| \geq 1}} \mu(tr_{T,e}(T)),$$

if D_σ divides $M(\tau)$, and e corresponds to σ. The notation in (4.14) is the same as in (3.9).

Deduction of Theorem 4.8 from the Main Lemma. Since the monomials $m(\tau)$ satisfy (4.10), there exists a surjective linear map $a : H_{*S} \to H_S^*$, $\mu(\tau) \mapsto m(\tau)$. On the other hand, from (4.12) it follows that $m(\sigma)\mu(\tau) = \mu(\sigma \times \tau)$ if $m(\sigma)m(\tau)$ is a good monomial. Hence we have a linear map $b : H_S^* \to H_{*S}$: $m(\tau) \mapsto \mu(\tau) = m(\tau)1$ inverse to a. Therefore $\dim H_{*S} = \dim H_S^*$ so that Theorem 4.8 follows.

We now start proving the Main Lemma.

4.8.2. (4.14) is well defined. The right hand side of (4.14) formally depends on the choice of $\bar{i}, \bar{j}, \bar{k}, \bar{l}$. We first check that different choices give the same answer modulo (4.11). It is possible to pass from one choice to another by replacing one flag at a time. So let us consider $\bar{i}' \neq \bar{i}, \bar{j}, \bar{k}, \bar{l}$ and write the difference of the right hand sides of the relations (4.14) written for $(\tau, v, \bar{i}, \bar{j}, \bar{k}, \bar{l})$ and $(\tau, v, \bar{i}', \bar{j}, \bar{k}, \bar{l})$. The terms corresponding to those T that do not contain $\{\bar{i}, \bar{i}'\}$ cancel. This includes all terms with $T \subset T_2$. The remaining sum can be rewritten as

$$(4.15) \qquad - \sum_{T \subset T_1 \setminus \{\bar{i}, \bar{i}', \bar{j}\}} \left[\mu(tr_{T \cup \{\bar{i}'\}}(\tau)) - \mu(tr_{T \cup \{\bar{i}\}}(\tau)) \right],$$

where T can now be empty.

We contend that (4.15) is of the type (4.11). More precisely, consider any of the trees $tr_{T \cup \{\bar{i}'\}}(\tau)$, $tr_{T \cup \{\bar{i}\}}(\tau)$ and contract the edge whose vertices are incident to the flags $\bar{i}, \bar{j}, \bar{i}'$. We will get a tree σ and its vertex $v \in V_\tau$. The pair (σ, v) up to a canonical isomorphism does not depend on the transplants we started with. In $F_\sigma(v)$ there are flags $\bar{i}, \bar{j}, \bar{i}'$ and one more flag whose branch contains both k and l and which we denote $\bar{h}$. Then (4.15) is $-r(\sigma, v, \bar{i}, \bar{j}, \bar{i}', \bar{h})$. This is illustrated by Figure 5.

4.8.3. Operators D_σ on H_{*S} pairwise commute. We have to prove the identities

$$(4.16) \qquad D_{\sigma_1}(D_{\sigma_2}\mu(\tau)) = D_{\sigma_2}(D_{\sigma_1}\mu(\tau)).$$

Consider several possibilities separately.

i) *There exists a divisor D_σ of $m(\tau)$ such that $a(\sigma_1, \sigma) = 4$, so that $D_{\sigma_1}\mu(\tau) = 0$.*

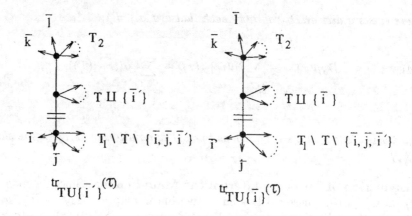

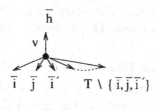

FIGURE 5. The edge $\frac{\ }{\ }$ is contracted to v.

If $D_{\sigma_2}\mu(\tau) = 0$ as well, (4.16) is true. If $D_{\sigma_2}\mu(\tau) = \mu(\sigma_2 \times \tau)$, then D_σ divides $m(\sigma_2 \times \tau)$, and (4.16) is again true. Finally, if D_{σ_2} divides $m(\tau)$, then $\sigma_2 \neq \sigma$ (otherwise $m(\tau)$ would not be a good monomial). Hence the transplants $tr_{T,e}(\tau)$ involved in the formula of the type (4.14) which we can use to calculate $D_{\sigma_2}\mu(\tau)$ will all contain an edge corresponding to σ so that $D_{\sigma_1}(tr_{T,e}(\tau)) = 0$, and (4.16) again holds.

The same argument applies to the case when $D_{\sigma_2}\mu(\tau) = 0$.

From now on we may and will assume that for any divisor D_σ of $m(\tau)$ we have $a(\sigma, \sigma_1) \leq 3$, $a(\sigma, \sigma_2) \leq 3$, and that $\sigma_1 \neq \sigma_2$.

ii) $a(\sigma_1, \sigma_2) = 4$ and D_{σ_2} divides $m(\tau)$.

Then D_{σ_1} does not divide $m(\tau)$, so that $D_{\sigma_1}\mu(\tau) = \mu(\sigma_1 \times \tau)$, and $D_{\sigma_2}(D_{\sigma_1}\mu(\tau)) = 0$. On the other hand, $D_{\sigma_2}\mu(\tau)$ is a sum of transplants to the midpoint of the edge, corresponding to σ_2. Each such transplant has an edge giving the 2-partition σ_2, so that $D_{\sigma_1}(D_{\sigma_2}\mu(\tau)) = 0$.

The case $a(\sigma_1, \sigma_2) = 4$ and $D_{\sigma_1}/m(\tau)$ is treated in the same way.

Hence from this point on we can and will in addition assume that $a(\sigma_1, \sigma_2) = 3$.

iii) D_{σ_1} does not divide $m(\tau)$.

If D_{σ_2} does not divide $m(\tau)$ as well, then $D_{\sigma_1}(D_{\sigma_2}\mu(\tau)) = D_{\sigma_1}\mu(\sigma_2 \times \tau) = \mu(\sigma_1 \times \sigma_2 \times \tau) = D_{\sigma_2}(D_{\sigma_1}\mu(\tau))$. If D_{σ_2} divides $m(\tau)$, we will use a carefully chosen formula of the type (4.14) for the calculation of $D_{\sigma_2}\mu(\tau)$. Namely, let v_1 be the (unique) vertex of τ which gets replaced by an edge in $\sigma_1 \times \tau$, and let e_2 be the edge of τ corresponding to D_{σ_2}. Let u_2, u_1 be the vertices of e_2 such that u_1 can be joined to v_1 by a path not passing by e_2.

Consider first the subcase $u_1 \neq v_1$. Choose some $\bar{i}, \bar{j} \in F_\tau(u_2)$ and $\bar{k}, \bar{l} \in F_\tau(u_1)$ in such a way that $\bar{l}$ starts a path leading from u_1 to v_1. Use these $\bar{i}, \bar{j}, \bar{k}, \bar{l}$ in a formula of the type (4.14) to calculate $D_{\sigma_2}\mu(\tau)$ and then $D_{\sigma_1}(D_{\sigma_2}\mu(\tau))$, that will insert an edge instead of the vertex v_1 which survives in all the transplants entering $D_{\sigma_2}\mu(\tau)$. Then calculate $D_{\sigma_2}(D_{\sigma_1}\mu(\tau))$ by first inserting the edge at v_1, and then constructing the transplants not moving $\bar{i}, \bar{j}, \bar{k}, \bar{l}$. Since by our choice of $\bar{l}$ we never transplant the branch containing v_1, the two calculations will give the same result.

Now let $v_1 = u_1$. Let $\{S_1, S_2\}$ be the 2-partition of S corresponding to σ_1. Since $\sigma_1 \times \tau$ exists, $\{S_1, S_2\}$ is induced by a partition of $F_\tau(v_1) = \bar{S}_1 \coprod \bar{S}_2$. We denote by $\bar{S}_2$ the part to which the flag $(v_1 = u_1, e_2)$ belongs. Let

$$\bar{T} = \bar{S}_2 \setminus (\{(v_1 = u_1, e_2)\} \coprod F_\tau(u_2)).$$

This set is non–empty because otherwise e_2 would correspond to $\{S_1, S_2\}$, and we would have $\sigma_1 = \sigma_2$. Take $\bar{i}, \bar{j} \in F_\tau(u_2)$, $\bar{k} \in \bar{S}_1$, and $\bar{l} \in \bar{T}$: see Figure 6.

Now consider $D_{\sigma_2}(D_{\sigma_1}\mu(\tau))$ and $D_{\sigma_1}(D_{\sigma_2}\mu(\tau))$. To calculate the first expression we form a sum of transplants of $\sigma_1 \times \tau$. To calculate the second one, we form transplants of τ, and then insert an edge at $v_1 = u_1$.

The transplants corresponding to the branches at u_2 will be the same in both expressions. The transplants corresponding to the subsets $T \subset \bar{T} \setminus \{\bar{l}\}$ will also be the same. In addition, the second expression will contain the terms $-D_{\sigma_1}(\mu(tr_{T,e_2}(\tau)))$ where $T \cap \bar{S}_1 \neq \emptyset$. But each such term vanishes. In fact, consider the 2–partition $\rho = \{R_1, R_2\}$ of S corresponding to the edge of $tr_{T,e_2}(\tau)$ containing the flag $(v_1 = u_1, e_2)$, and let $k, l \in R_1$. A glance at the third tree of Figure 6 shows that $a(\rho, \sigma_1) = 4$, because if $\bar{t} \in T \cap \bar{S}_1$, $t \in S(\bar{t})$, then $kt\sigma_1 il$ and $kl\rho it$. Hence the extra terms are irrelevant.

The case when D_{σ_2} does not divide $m(\tau)$ is treated in the same way. It remains to consider the last possibility.

iv) D_{σ_1} and D_{σ_2} divide $m(\tau)$, $a(\sigma_1, \sigma_2) = 3$.

Denote by e_1 (resp. e_2) the edge corresponding to σ_1 (resp. σ_2). Let u_1, u_2 (resp. v_1, v_2) be the vertices of e_1 (resp. e_2) numbered in such a way that there is a path from u_2 to v_1 not passing through e_1, e_2 (the case $u_2 = v_1$ is allowed). To calculate the multiplication by D_{σ_1} choose $\bar{i}, \bar{j} \in F_{u_1}(\tau) \setminus \{(u_1, e_1)\}$, $\bar{l}$ on the path from u_2 to v_1 if $u_2 \neq v_1$, and $\bar{l} = (v_1, e_2)$ if $u_2 = v_1$; $\bar{k} \in F_\tau(v_2) \setminus \{\bar{l}\}$. To calculate

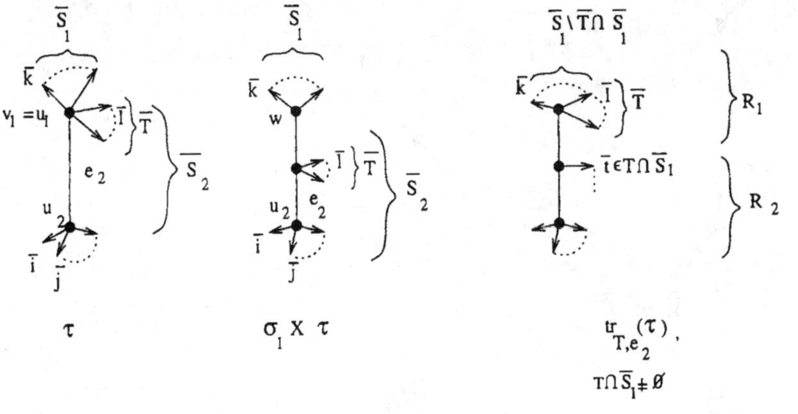

FIGURE 6

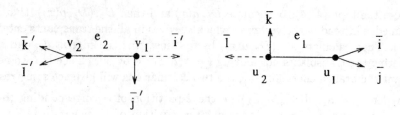

FIGURE 7

the product by D_{σ_2}, choose similarly $\bar{k}', \bar{l}' \in F_\tau(v_2) \setminus \{(v_2, e_2)\}$, $\bar{i}' \in F_\tau(v_1)$ on the path from v_1 to u_2, if $v_1 \neq u_2$, and $\bar{i}' = (u_2, e_1)$ if $v_1 = u_2$, $\bar{j}' \in F_\tau(v_1)$ (see Figure 7).

The critical choice here is that of $\bar{l}$ and $\bar{i}'$. It ensures that by calculating $D_{\sigma_1}(D_{\sigma_2}\mu(\tau))$ and $D_{\sigma_2}(D_{\sigma_1}\mu(\tau))$ we will get the same sum of transplanted trees. This ends the proof of (4.16).

4.8.4. Compatibility with I_S–generating relations. If $D_{\sigma_1} D_{\sigma_2} = 0$ because $a(\sigma_1, \sigma_2) = 4$, one sees that $D_{\sigma_1}(D_{\sigma_2}\mu(\tau)) = 0$ looking through various subcases in 4.8.3. It remains to show that $R_{ijkl}\mu(\tau) = 0$ where R_{ijkl} is defined by (3.3).

Consider the smallest connected subgraph in τ containing the flags i, j, k, l. Figure 8 gives the following exhaustive list of alternatives. Paths from i to j and from k to l: i) have at least one common edge; ii) have exactly one common vertex; iii) do not intersect.

Consider them in turn.

i) Let e be an edge common to the paths ij and kl. Denote by ρ the respective 2-partition. Then $ik\rho jl$ or $il\rho kj$. Therefore any summand of R_{ijkl} annihilates D_ρ so that $R_{ijkl}\mu(\tau) = 0$ in view of (4.13).

ii) Let v be the vertex common to the paths ij and kl. Then exactly the same calculation as in the proof of Proposition 4.7.1 shows that

$$R_{ijkl}\mu(\tau) = \sum_\alpha [\mu(\tau'(\alpha)) - \mu(\tau''(\alpha))] = 0$$

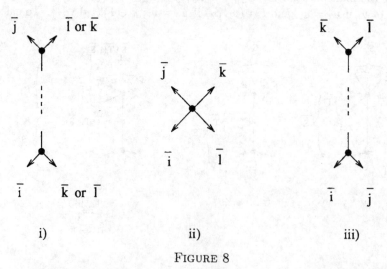

FIGURE 8

Figure 9

(notation as in (4.10) and (4.11)).

iii) This is the most complex case. Let us draw a more detailed picture of τ in the neighborhood of the subgraph we are considering (Figure 9).

Let v_1 be the vertex on the path ij which is connected by a sequence of edges $e_1, \ldots, e_m$ ($m \geq 1$) with the vertex v_m on the path kl so that e_a has vertices (v_a, v_{a+1}) in this order. Let T_a be the set of flags at v_a which do not coincide with $\bar{i}, \bar{j}, \bar{k}, \bar{l}$, and do not belong to e_{a-1}, e_a.

Consider any summand D_σ of R_{ijkl}. If $jk\sigma il$, then $D_\sigma \mu(\tau) = 0$ because each edge e_a determines a partition ρ of S such that $ij\rho kl$. From now on we assume that $ij\sigma kl$. Then $D_\sigma \mu(\tau)$ can be non–zero if one of the two following alternatives holds:

a) For some v_a, there exists a partition $T_a = T_a' \coprod T_a''$ (with $|T_a'| \geq 1, |T_a''| \geq 1$, except for the case $a = 1$ where T_1' can be empty, and $a = m$ where T_m'' can be empty) such that the two sets

$$S_1 = S(\bar{i}) \coprod S(\bar{j}) \coprod S(T_1') \coprod \cdots \coprod S(T_a'),$$

$$S_2 = S(T_a'') \coprod S(T_{a+1}) \coprod \cdots \coprod S(T_m) \coprod S(\bar{k}) \coprod S(\bar{l})$$

form the 2-partition corresponding to σ. In this case

$$D_\sigma \mu(\tau) = \mu(\sigma \times \tau),$$

and $\sigma \times \tau$ is obtained by inserting a new edge at v_a and by distributing T_a' and T_a'' at different vertices of this edge.

b) For some e_a, the two sets

$$S_1 = S(\bar{i}) \coprod S(\bar{j}) \coprod (\coprod_{i \leq a} S(T_i)),$$

$$S_2 = (\coprod_{i \geq a+1} S(T_i)) \coprod S(\bar{k}) \coprod S(\bar{l})$$

form the 2-partition corresponding to σ.

In this case D_σ divides $m(\tau)$, and in order to calculate $D_\sigma \mu(\tau)$ using a formula of the type (4.14) we must first choose two pairs of flags at the two vertices of v_a.

Contributions from a) and b) come with opposite signs, and we contend that they completely cancel each other.

To see the pattern of the cancellation look first at case a) at v_1. It brings (with positive sign) the contributions corresponding to the following trees. Form all the partitions $T_1 = T_1' \coprod T_1''$ such that $T_1'' \neq \emptyset$, where $T_1 = F_\tau(v_1) \setminus \{\bar{i}, \bar{j}, (v_1, e_1)\}$.

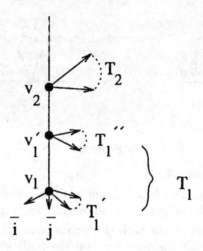

FIGURE 10

Transplant all T_1''-branches to the midpoint of e_1. Denote the new vertex v_1'. The result is drawn as Figure 10.

Now consider the terms of type b) for the edge e_1. If $m = 2$, we choose for the calculation of $D_{\sigma_1}\mu(\tau)$ (where σ_1 corresponds to e_1) the flags $\bar{i}, \bar{j}, \bar{k}, \bar{l}$. If $m > 2$, we choose the flags $\bar{i}, \bar{j}, (v_2, e_1), t \in T_2$. Then we get the sum of two contributions. One will consist of the trees obtained by transplanting branches at v_1. They come with negative signs and exactly cancel the previously considered terms of type a). If $m = 2$, the second group will cancel the terms of type a) coming from v_2.

Consider a somewhat more difficult case $m > 2$. Then this second group of terms comes from the trees indexed by the partitions $T_2 = T_2' \coprod T_2'', t \in T_2'', T_2' \neq \emptyset$. Branches corresponding to T_2' are transplanted to the midpoint v_1' of the edge e_1. These terms come with negative signs: see Figure 11.

These trees in turn cancel with those coming from the terms of type a) at the vertex v_2 with positive sign. However, there will be additional terms of type a) for which $t \in T_2'$. They will cancel with one group of transplants contributing to $D_{\sigma_2}\mu(\tau)$ where σ_2 corresponds to the edge e_2 of Figure 9, if for the calculation of

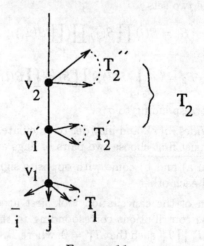

FIGURE 11

$D_{\sigma_2}\mu(\tau)$ one uses (4.14) with the following choice of flags: $(v_2, e_1), t$ at one end, (v_3, e_1), some $t' \in T_3$ at the other end (this last choice must be replaced by $\bar{k}, \bar{l}$, if $m = 3$).

The same pattern continues until all the terms cancel.

4.8.5. Compatibility with relations (4.11). By this time we have checked that the action of any element of F_S/I_S on the individual generators $\mu(\tau)$ of H_{*S} is well defined modulo the span I_{*S} of relations (4.11). It remains to show that the subspace in $\bigoplus_\tau k\mu(\tau)$ spanned by these relations is stable with respect to this action. But the calculation in the proof of Proposition 4.7.1 shows that

$$r(\tau, v, \bar{i}, \bar{j}, \bar{k}, \bar{l}) \equiv m(\tau) r_{ijkl} \bmod I_{*S},$$

where r_{ijkl} is obtained from R_{ijkl} by replacing $m(\sigma)$ with $\mu(\sigma)$. To multiply this by any element of H_S^* we can first multiply it by $m(\tau)$, then represent the result as a linear combination of good monomials, and finally multiply each good monomial by r_{ijkl}. The result will lie in I_{*S}.

This finishes the proof of the Main Lemma and Theorem 4.8.

4.9. End of the proof of Theorem 4.3. According to the remark in the last paragraph of 4.6, it remains to show the following. Start with an ACF $Y_n : H^{\otimes n} \to k$, $n \geq 3$ (Definition 1.3.1), and extend these polynomial maps to all stable trees τ by putting

(4.17) $\quad Y(\tau): H^{\otimes n} \to k, \ Y(\tau)(\gamma_1 \otimes \cdots \otimes \gamma_n) := \int_{\overline{D}(\tau)} I_n(\gamma_1 \otimes \cdots \otimes \gamma_n).$

For the one–vertex tree we get Y_n; cf. (4.3). Recall that by iterating the splitting identity (4.2) we can express $Y(\tau)$ through Y_n:

(4.18) $\quad Y(\tau)\left(\bigotimes_{i \in S_\tau} \gamma_i\right) = \left(\bigotimes_{v \in V_\tau} Y_{F_\tau(v)}\right)\left(\bigotimes_{i \in S_\tau} \gamma_i \otimes \Delta^{\otimes E_\tau}\right),$

and then $Y(\tau)$ will satisfy the following version of the relations (4.10):

(4.19) $\quad \sum_\alpha Y(\tau'(\alpha)) = \sum_\beta Y(\tau''(\beta)).$

The notation is explained in the first paragraph of 4.7. We start with a tree τ, in which a vertex v and four tails $\bar{i}, \bar{j}, \bar{k}, \bar{l}$ are marked. The trees τ', τ'' are obtained from τ by inserting an edge at v. This can be done in many ways, parametrized by 2–partitions of $F_\tau(v)$. They induce 2–partitions of $\{\bar{i}, \bar{j}, \bar{k}, \bar{l}\}$. We put to the left those which break this quadruple into $\{\bar{i}, \bar{j}\} \cup \{\bar{k}, \bar{l}\}$, and to the right those which break it into $\{\bar{i}, \bar{l}\} \cup \{\bar{k}, \bar{j}\}$. The remaining partitions do not contribute.

To prove (4.19), rewrite every summand using (4.18). Look at the factor Δ corresponding to the inserted edge, and represent it as $\sum \Delta_a g^{ab} \otimes \Delta_b$. After some fumbling with indices, one can recognize in the obtained expression a linear combination of identities (4.10) written for various arguments and the one–vertex tree with flags S_τ.

4.10. Tensor product revisited. We now return to the situation of 4.4. The correlation functions of $I_n^{(j)}$ will be denoted $Y_n^{(j)}, Y^{(j)}(\tau)$ and the same superscript will be added to other objects related to one of the two factors. The respective data for the tensor product are denoted simply H, g, I_n, Y_n, E, etc.

Since the cup product in $H^*(\overline{M}_{0n}, k)$ is induced by the diagonal map, we can rewrite (4.5) in terms of correlators Y_n as follows:

$$Y_n(\gamma_1^{(1)} \otimes \gamma_1^{(2)} \otimes \ldots \otimes \gamma_n^{(1)} \otimes \gamma_n^{(2)})$$
$$(4.20) := \varepsilon(\gamma^{(1)}, \gamma^{(2)})(Y^{(1)} \otimes Y^{(2)})(\Delta_{\overline{M}_{0n}})(\gamma_1^{(1)} \otimes \ldots \otimes \gamma_n^{(1)} \otimes \gamma_1^{(2)} \otimes \ldots \otimes \gamma_n^{(2)}).$$

The right hand side here must be interpreted as follows: represent the class of the diagonal $\Delta_{\overline{M}_{0n}}$ as a linear combination of classes $\overline{D}(\tau) \otimes \overline{D}(\sigma)$, and replace the respective summand by $Y^{(1)}(\tau) \otimes Y^{(2)}(\sigma)$. With this notation, the formula for the potential of the tensor product takes the form

$$(4.21) \qquad \Phi(\gamma) = \sum_{n \geq 3} \frac{1}{n!} (Y^{(1)} \otimes Y^{(2)})(\Delta_{\overline{M}_{0n}})(\gamma^{\otimes n}).$$

We will use this formula in order to extend the tensor product to formal Frobenius manifolds with flat identities and Euler fields.

Let E be an Euler field for (H, g, I_n). Denote by E_0 the flat vector field which at zero coincides with E and put $E = E_1 + E_0$. In flat coordinates (x^a) on H vanishing at the origin we have

$$(4.22) \qquad E = \sum_{a,b} d_a^b x^a \partial_b + \sum_a r^a \partial_a = E_1 + E_0$$

for some $d_a^b, r^a \in k$ (cf. I.2.2 and I.2.4). To avoid some extra signs, we will assume that d_a^b are even. The formula I.(2.5) becomes

$$(4.23) \qquad \forall a, b : \sum_c d_a^c g_{cb} + \sum_c d_b^c g_{ac} = D g_{ab}.$$

Here we will call D the weight of E. The spectrum point d_0 is defined by I.(2.6) and I.(2.16) and called *the weight* of e. (This terminology differs from that of I.5.4.2 where we worked with weak Frobenius manifolds.)

For the two factors as above we use the notation $E^{(j)}, E_0^{(j)}, D^{(j)}$, etc. Moreover, if their flat coordinates are labeled by some sets $A^{(1)}$, resp. $A^{(2)}$, the elements of these sets are denoted a', b', resp. a'', b'', etc. The flat coordinates in $H^{(1)} \otimes H^{(2)}$ dual to $\partial_{a'} \otimes \partial_{b''}$ are denoted $x^{a'b''}$ and the coefficients in (4.22) become $d_{a'}^{(1)b'}$, etc. We have $x^{a'b''} = (-1)^{a'b''} x^{a'} \otimes x^{b''}$ where the notation for signs is explained in I.1.1.2.

Put $F^{(1)} := \sum x^{a'} \partial_{a'}$ and similarly for $F^{(2)}$.

4.10.1. Theorem. *The tensor product of two formal Frobenius manifolds with flat identities $e^{(i)}$ of the same weight d_0 and with Euler fields $E^{(i)}$ of weights $D^{(i)}$ admits $e^{(1)} \otimes e^{(2)}$ as the flat identity of weight d_0 and*

$$(4.24) \quad E := E_1^{(1)} \otimes F^{(2)} + F^{(1)} \otimes E_1^{(2)} + E_0^{(1)} \otimes e^{(2)} + e^{(1)} \otimes E_0^{(2)} - d_0 F^{(1)} \otimes F^{(2)}$$

as the Euler field of weight $D^{(1)} + D^{(2)} - 2d_0$.

§4. COHOMOLOGICAL FIELD THEORIES

Proof. In flat coordinates (4.24) becomes

$$E = \sum_{b',b''} \left[\sum_{a'} d_{a'}^{(1)b'} x^{a'b''} + \sum_{a''} d_{a'}^{(2)b''} x^{b'a''} - d_0 x^{b'b''} \right] \partial_{b'b''}$$

(4.25)
$$+ \sum_{a'} r^{(1)a'} \partial_{a'0} + \sum_{a''} r^{(2)a''} \partial_{0a''}.$$

We must first check that E is conformal; cf. I.2.2. Writing I.(2.5) in the basis of flat decomposable vector fields on the tensor product, we find using (4.25):

$$g([\partial_{a'a''}, E], \partial_{b'b''}) + g(\partial_{a'a''}, [\partial_{b'b''}, E])$$

$$= (-1)^{b'a''} \left(\sum_{c'} d_{a'}^{(1)c'} g_{c'b'}^{(1)} g_{a''b''}^{(2)} + \sum_{c''} d_{a''}^{(2)c''} g_{c'b'}^{(1)} g_{c''b''}^{(2)} \right.$$

$$\left. + \sum_{c'} d_{b'}^{(1)c'} g_{a'c'}^{(1)} g_{a''b''}^{(2)} + \sum_{c''} d_{b''}^{(2)c''} g_{a'b'}^{(1)} g_{a''c''}^{(2)} \right)$$

$$- 2(-1)^{b'a''} d_0 g_{a'b'}^{(1)} g_{a''b''}^{(2)} = (D^{(1)} + D^{(2)} - 2d_0) g_{a'a'',b'b''}$$

in view of (4.23) and

$$g_{a'a'',b'b''} := g(\partial_{a'} \otimes \partial_{a''}, \partial_{b'} \otimes \partial_{b''}) = (-1)^{b'a''} g_{a'b'}^{(1)} g_{a''b''}^{(2)}.$$

Before checking the quasihomogeneity property of the tensor product potential expressed by I.2.2.2, we must first express it in the language of the tree correlation functions (4.17). From now on, we will slightly simplify the calculations, avoiding cumbersome signs by working in the pure even case. The result is valid without this assumption.

We will extend $Y(\tau)$ by linearity to formal linear combinations of trees. For an S-labeled stable tree of genus zero τ and some $s \notin S$ denote by $\pi^*(\tau)$ the sum of all trees obtained from τ by putting an additional tail at some vertex of τ and labeling it with s. This is the combinatorial counterpart of the pullback morphism of the Chow rings corresponding to the morphism "forget x_s and stabilize" $\overline{M}_{0,S \cup \{s\}} \to \overline{M}_{0,S}$.

4.10.2. Lemma. *Let $Y(\tau)$ be the tree correlation functions corresponding to the potential*

$$\Phi(\gamma) = \sum_n \frac{1}{n!} Y_n(\gamma^{\otimes n}).$$

Then the quasi-homogeneity condition

(4.26) $\quad E\Phi = (D + d_0)\Phi + $ *a quadratic polynomial in flat coordinates*

with E given by (4.22) is equivalent to the family of conditions

$$\sum_{i=1}^{n} \sum_a d_{a_i}^a Y(\tau)(\partial_{a_1} \otimes \cdots \otimes \widehat{\partial_{a_i}} \otimes \cdots \otimes \partial_{a_n} \otimes \partial_a) + \sum_a r^a Y(\pi^*(\tau))(\partial_{a_1} \otimes \cdots \otimes \partial_{a_n} \otimes \partial_a)$$

(4.27)
$$= (D + d_0 + |E_\tau| d_0) Y(\tau)(\partial_{a_1} \otimes \cdots \otimes \partial_{a_n}).$$

Here τ runs over n–trees whose tails are additionally labeled by ∂_{a_i}, E_τ is the set of edges, and ∂_a labels the extra tail in each summand of $\pi^*(\tau)$.

Proof. First of all, directly from the definitions we see that (4.26) is equivalent to the following subfamily of (4.27) corresponding to the one–vertex trees:

$$\sum_a \left(\sum_{i=1}^n d_{a_i}^a Y_n(\partial_{a_1} \otimes \cdots \otimes \widehat{\partial_{a_i}} \otimes \cdots \otimes \partial_{a_n} \otimes \partial_a) + r^a Y_{n+1}(\partial_{a_1} \otimes \cdots \otimes \partial_{a_n} \otimes \partial_a) \right)$$

$$(4.28) \qquad = (D + d_0) Y_n(\partial_{a_1} \otimes \cdots \otimes \partial_{a_n}).$$

Now put

$$(4.29) \qquad \widetilde{Y}(\tau) := \bigotimes_{v \in V_\tau} Y_{F_\tau(v)} : H^{\otimes F_\tau} \to k.$$

Then we have for any tree τ whose flags f are labeled by some $\partial_{a(f)}$:

$$\sum_a \left(\sum_{f \in F_\tau} d_{a(f)}^a \widetilde{Y}(\tau)((\bigotimes_{f' \in F_\tau \setminus \{f\}} \partial_{a(f')}) \otimes \partial_a) + r^a \widetilde{Y}(\pi^*(\tau))((\bigotimes_{f \in F_\tau} \partial_{a(f)}) \otimes \partial_a) \right)$$

$$(4.30) \qquad = |V_\tau|(D + d_0) \widetilde{Y}(\tau)(\bigotimes_{f \in F_\tau} \partial_{a(f)}).$$

To check this, rewrite the rhs of (4.30) in the following form:

$$\sum_{v \in V_\tau} (D + d_0) \prod_{v' \in V_\tau} Y_{F_\tau(v')}\left(\bigotimes_{f \in F_\tau(v')} \partial_{a(f)} \right)$$

$$= \sum_{v \in V_\tau} (D + d_0) Y_{F_\tau(v)}\left(\bigotimes_{f \in F_\tau(v)} \partial_{a(f)} \right) \prod_{v' \in V_\tau \setminus \{v\}} Y_{F_\tau(v')}\left(\bigotimes_{f \in F_\tau(v')} \partial_{a(f)} \right).$$

Now replace each term $(D + d_0) Y_{F_\tau(v)}(\bigotimes_{f \in F_\tau(v)} \partial_{a(f)})$ in the last sum by the (appropriately modified) left hand side of (4.28). Some reshuffling produces the right hand side of (4.30).

Finally, (4.27) can now be deduced by expressing each value of $Y(\tau)$ in (4.27) as a special value of $\widetilde{Y}(\tau)$ with the help of (4.18). Namely, let $\Delta = \sum \partial_p g^{pq} \partial_q$. The calculations become slightly better organized if we start with the following expression:

$$\sum_{i=1}^n \sum_a d_{a_i}^a Y(\tau)(\partial_{a_1} \otimes \cdots \otimes \widehat{\partial_{a_i}} \otimes \cdots \otimes \partial_{a_n} \otimes \partial_a) + |E_\tau| D Y(\tau)(\partial_{a_1} \otimes \cdots \otimes \partial_{a_n})$$

$$= \sum_{i=1}^n \sum_a d_{a_i}^a Y(\tau)(\partial_{a_1} \otimes \cdots \otimes \widehat{\partial_{a_i}} \otimes \cdots \otimes \partial_{a_n} \otimes \partial_a)$$

$$(4.31) \quad +|E_\tau| D \sum_\phi \left(\bigotimes_{v \in V_\tau} Y_{F_\tau(v)} \right) (\partial_{a_1} \otimes \cdots \otimes \partial_{a_n} \otimes \bigotimes_{e \in E_\tau} (\partial_{\phi_1(e)} g^{\phi_1(e)\phi_2(e)} \otimes \partial_{\phi_2(e)})),$$

§4. COHOMOLOGICAL FIELD THEORIES

where ϕ runs over all labelings of internal flags of τ by ∂_a's, and $\phi_1(e), \phi_2(e)$ refer to the two halves of the edge e. We may assume that each edge is oriented in order to make this notation unambiguous.

Now with the help of (4.23) and a somewhat tedious rewriting we can represent this expression as

$$= (|E_\tau| + 1)(D + d_0)Y(\tau)\left(\bigotimes_{i=1}^{n}\partial_{a_i}\right) - \sum_a r^a Y(\pi^*(\tau))\left(\bigotimes_{i=1}^{n}\partial_{a_i} \otimes \partial_a\right).$$

This finishes the proof of the lemma.

We now need similar information for the identity. According to 1.6, ∂_0 is the flat identity iff

(4.32) $\quad Y_3(\partial_a, \partial_b, \partial_0) = g_{ab}$ and $Y_n(\partial_{a_1} \otimes \cdots \otimes \partial_{a_{n-1}} \otimes \partial_0) = 0$ for all $n > 3$.

In order to extend this to the tree correlation functions, we introduce the following notation. Let τ be a stable tree with a distinguished tail which in expressions involving correlation functions will always be labeled by ∂_0. Denote by $\pi_*(\tau)$ the tree obtained by forgetting this tail and subsequent stabilization, if the resulting tree is not stable; otherwise put $\pi_*(\tau) = 0$. Similar to $\pi^*(\tau)$ above, this is the combinatorial counterpart of the flat pushforward of the Chow groups of moduli spaces.

4.10.3. Lemma. *If ∂_0 is the flat identity, then for any stable n-tree with n-th tail marked by ∂_0 we have*

(4.33) $\quad Y(\tau)(\partial_{a_1} \otimes \cdots \otimes \partial_{a_{n-1}} \otimes \partial_0) = Y(\pi_*(\tau))(\partial_{a_1} \otimes \cdots \otimes \partial_{a_{n-1}}).$

Proof. From (4.32) it follows that if the vertex v carrying the n-th tail remains stable after forgetting this tail, then both sides of (4.33) vanish. Otherwise this vertex has valency three, and it contributes to the values of $Y(\tau)$ metric coefficients $Y_3(\partial_a, \partial_b, \partial_0) = g_{ab}$. The statement now can be checked by straightforward calculation. Two cases should be distinguished depending on whether v carries one or two tails. We leave the details to the reader.

Finally, we need the following simple identities. Let π be the forgetful map (of the n-th point) $\overline{M}_{0n} \to \overline{M}_{0,n-1}$ as above. Then

(4.34) $\quad (\pi, \pi)_*(\Delta_{\overline{M}_{0n}}) = 0, \ (id, \pi)_*(\Delta_{\overline{M}_{0n}}) = (\pi, id)^*(\Delta_{\overline{M}_{0,n-1}}).$

The first identity follows from dimensional considerations, and the second one from the projection formula.

We can now finish the proof of Theorem 4.10.1.

Clearly, $e^{(1)} \otimes e^{(2)} = \partial_0^{(1)} \otimes \partial_0^{(2)}$ is a flat field. Furthermore

(4.35) $\quad \begin{aligned} & Y_3(\partial_{a'}^{(1)} \otimes \partial_{a''}^{(2)} \otimes \partial_{b'}^{(1)} \otimes \partial_{b''}^{(2)} \otimes \partial_0^{(1)} \otimes \partial_0^{(2)}) \\ &= Y_3^{(1)}(\partial_{a'}^{(1)} \otimes \partial_{b'}^{(1)} \otimes \partial_0^{(1)}) Y_3^{(2)}(\partial_{a''}^{(2)} \otimes \partial_{b''}^{(2)} \otimes \partial_0^{(2)}) = g_{a'a'',b'b''}.\end{aligned}$

Moreover, in view of (4.20), (4.33) and (4.34) we have for $n > 3$:

$$Y_n \left(\bigotimes_{i=1}^{n-1} (\partial_{a_i'}^{(1)} \otimes \partial_{a_i''}^{(2)}) \otimes (\partial_0^{(1)} \otimes \partial_0^{(2)}) \right)$$

$$= (Y^{(1)} \otimes Y^{(2)})((\pi,\pi)_* \Delta_{\overline{M}_{0n}}) \left(\bigotimes_{i=1}^{n-1} (\partial_{a_i'}^{(1)} \otimes \partial_{a_i''}^{(2)}) \right) = 0.$$

This shows that $e^{(1)} \otimes e^{(2)}$ is the identity.

It remains to check (4.26) for the tensor product potential. Let $E = E_1 + E_0$ be the tensor product Euler field defined by (4.24) and (4.25), $D = D^{(1)} + D^{(2)} - 2d_0$. We have

$$E_1(Y_n(\gamma^{\otimes n})) = E_1((Y^{(1)} \otimes Y^{(2)})(\Delta_{\overline{M}_{0n}})(\gamma^{\otimes n}))$$
$$= (D + d_0)(Y^{(1)} \otimes Y^{(2)})(\Delta_{\overline{M}_{0n}})(\gamma^{\otimes n})$$
$$- \sum_{a'} r^{(1)a'} (Y^{(1)} \otimes Y^{(2)})((\pi, id)^* \Delta_{\overline{M}_{0n}})(\gamma^{\otimes n} \otimes \partial_{a'}^{(1)})$$

(4.36)
$$- \sum_{a''} r^{(2)a''} (Y^{(1)} \otimes Y^{(2)})((id, \pi)^* \Delta_{\overline{M}_{0n}})(\gamma^{\otimes n} \otimes \partial_{a''}^{(2)}).$$

To check this, one represents the class of the diagonal as a linear combination of $\overline{D}(\sigma) \otimes \overline{D}(\tau)$ with $|E_\sigma| + |E_\tau| = n - 3$ and applies (4.27) to both tensor factors of each summand.

Similarly, in view of (4.33),

$$E_0(Y_{n+1}(\gamma^{\otimes n+1})) = (n+1) \sum_{a'} r^{(1)a'} (Y^{(1)} \otimes Y^{(2)})(\Delta_{\overline{M}_{0,n+1}})(\gamma^{\otimes n} \otimes \partial_{a'}^{(1)} \otimes \partial_0^{(2)})$$

$$+ (n+1) \sum_{a''} r^{(2)a''} (Y^{(1)} \otimes Y^{(2)})(\Delta_{\overline{M}_{0,n+1}})(\gamma^{\otimes n} \otimes \partial_0^{(1)} \otimes \partial_{a''}^{(2)})$$

$$= \sum_{a'} r^{(1)a'} (Y^{(1)} \otimes Y^{(2)})((id, \pi)_* \Delta_{\overline{M}_{0,n+1}})(\gamma^{\otimes n} \otimes \partial_{a'}^{(1)})$$

(4.37)
$$+ \sum_{a''} r^{(2)a''} (Y^{(1)} \otimes Y^{(2)})((\pi, id)_* \Delta_{\overline{M}_{0,n+1}})(\gamma^{\otimes n} \otimes \partial_{a''}^{(2)}).$$

Combining (4.35) and (4.36) and taking into account (4.34), one gets the desired equation (4.26).

4.10.4. The spectrum of the tensor product. Let $D, \{d_a\}$ be the spectrum of the Frobenius manifold M. We consider the spectrum as a family of numbers with assigned multiplicities. We put $d = 2 - D$ and $q_a = 1 - d_a$ and call the family $d, \{q_a\}$ the *d–spectrum* of M. If M is the quantum cohomology of V, we have $d = \dim V$ and the multiplicity of q_a is $h^{2q_a}(V)$. If M is the tensor product as above, its d–spectrum in self–explanatory notation is

(4.38) $$d = d^{(1)} + d^{(2)}, \ \{q_{a'} + q_{b''}\}.$$

Of course, this agrees with the Künneth formula.

§5. Gromov–Witten invariants and quantum cohomology: Axiomatic theory

5.1. Introduction. This section introduces Gromov–Witten (GW) invariants and quantum cohomology in the axiomatic form, following [KM1]. The main goal is to show how to pass from GW–invariants to formal Frobenius manifolds. We also introduce the notion of Frobenius manifolds of qc–type which axiomatizes some formal properties of quantum cohomology.

Let V be a projective algebraic manifold with canonical class $K_V = c_1(\omega_V)$, and H^* a cohomology theory with coefficients in a $\mathbf{Q}$–algebra. An appropriate counting of parametrized curves of genus g with n labeled points lying in a given homology (or numerical) equivalence class β in V leads to the construction of maps

$$(5.1) \qquad I^V_{g,n,\beta} : H^*(V)^{\otimes n} \to H^*(\overline{M}_{g,n})$$

defined for all stable pairs (g,n). These maps, called *Gromov–Witten* or *GW–invariants*, were constructed in this generality (in motivic cohomology, and for ground fields of characteristic zero) in [Beh], following the initial insight of physicists (E. Witten, C. Vafa, R. Dijkgraaf, E. and H. Verlinde), the preliminary works [KM1], [BehM], [BehF] in the framework of the algebraic geometry, Gromov's work, and the construction [RT1] of GW–invariants in the symplectic geometry.

If we consider only the case $g = 0, n \geq 3$, we get the tree level GW–invariants.

The construction of GW–invariants is not very effective, especially because it involves cohomological or analytic machinery needed to implement the notion of "general position" in a too rigid geometric category. Besides direct calculations in a rather restricted range of special cases, general methods for understanding GW–invariants are based on studying degenerations.

5.2. Axioms for GW–invariants: The first group. We keep the conventions of 5.1. Denote by $B = B(V)$ the semigroup of the numerical equivalence classes β in V such that $(\beta, L) \geq 0$ for any ample divisor class L. We will call a family of maps (5.1) *a system of GW–invariants for V*, if it has the properties i)–vii) listed below.

i) Effectivity. $I^V_{g,n,\beta} = 0$ for $\beta \notin B$.

ii) $\mathbf{S}_n$–covariance. The symmetric group $\mathbf{S}_n$ acts upon $H^*(V, \mathbf{Q})^{\otimes n}$ (considered as a superspace via $\mathbf{Z}$ mod 2–grading) and upon $\overline{M}_{g,n}$ via renumbering of marked points. The maps $I^V_{g,n,\beta}$ must be compatible with this action.

iii) Degeneration. In §2 we briefly described the combinatorics of the degeneration of stable pointed curves; cf. in particular 2.7, where the two types of the boundary divisors in $\overline{M}_{g,n}$ are characterized, corresponding to the splitting of curve and to the acquisition of cusp. We will treat both types in turn.

Fix g_1, g_2 and n_1, n_2 such that $g = g_1 + g_2$, $n = n_1 + n_2$, $n_i + 2g_i - 2 \geq 0$. Also fix two complementary subsets $S = S_1, S_2$ of $\{1,\ldots,n\}$, $|S_i| = n_i$. Denote by $\varphi_S : \overline{M}_{g_1, S_1 \cup \{*\}} \times \overline{M}_{g_2, S_2 \cup \{*\}} \to \overline{M}_{g,n}$ the canonical map which assigns to the stable pointed curves $(C_i; x^{(i)}_*, x^{(i)}_j \mid j \in S_i), i = 1, 2$, their union $C_1 \cup C_2$, with $x^{(1)}_*$ identified to $x^{(2)}_*$.

Denote by $\Delta \in H^*(V)^{\otimes 2}$ the dual class of the diagonal.

The first part of the Degeneration Axiom now reads:

$$\varphi_S^*(I_{g,n,\beta}^V(\gamma_1 \otimes \cdots \otimes \gamma_n))$$

$$(5.2) \quad = \varepsilon(S) \sum_{\beta=\beta_1+\beta_2} (I_{g_1,n_1+1,\beta_1}^V \otimes I_{g_2,n_2+1,\beta_2}^V)\left(\left(\bigotimes_{j \in S_1} \gamma_j\right) \otimes \Delta \otimes \left(\bigotimes_{j \in S_2} \gamma_j\right)\right),$$

where $\varepsilon(S)$ is the sign of the permutation (S_1, S_2) on $\{\gamma_j\}$ of odd dimension.

The sum in (5.2) is finite thanks to the Effectivity Axiom.

Notice that because of the $\mathbf{S}_n$-covariance it suffices to check (5.2) for $S_1 = \{1, \ldots, n_1\}$.

For the second part of the Degeneration Axiom, denote by $\psi : \overline{M}_{g-1,n+2} \to \overline{M}_{g,n}$ the map corresponding to gluing together the last two marked points. Then

$$(5.3) \quad \psi^*(I_{g,n,\beta}^V(\gamma_1 \otimes \cdots \otimes \gamma_n)) = I_{g-1,n+2,\beta}^V(\gamma_1 \otimes \cdots \otimes \gamma_n \otimes \Delta).$$

Again, it is easy to combine this with $\mathbf{S}_n$-covariance in order to get the restriction formula valid for the boundary map gluing two arbitrary marked points.

5.2.1. From GW–invariants to Cohomological Field Theories and formal Frobenius manifolds.
Comparing (5.1)–(5.3) with the notions introduced in §4, especially in 4.5, one sees that this structure is close to that of the Complete Cohomological Field Theory (CCohFT). To make this correspondence precise, consider a topological algebra Λ over the coefficient ring of the cohomology theory H^* endowed with a multiplicative character

$$(5.4) \quad B \to \Lambda: \quad \beta \mapsto q^\beta.$$

Assume in addition that all the series

$$(5.5) \quad I_{g,n}^V := \sum_\beta q^\beta I_{g,n,\beta}^V : \ H^*(V)^{\otimes n} \to H^*(\overline{M}_{g,n}, \Lambda)$$

converge. Then we have:

CLAIM. *The Λ-linear extension of (5.5) defines a structure of CCohFT on the metric Λ-module $(H^*(V, \Lambda), g)$ where g is the Poincaré form.*

We leave the easy check to the reader.

Restricting ourselves to $g = 0, n \geq 3$, that is, starting with a tree level system of GW–invariants, we get in this way formal Frobenius manifolds.

Two choices of Λ are the most common ones. First, one can take a localization of the completed semigroup algebra of B, that is, a version of the Novikov ring. It is topologically spanned by the monomials $q^\beta = q_1^{b_1} \ldots q_r^{b_r}$ where $\beta = (b_1, \ldots, b_r)$ is a basis of the numerical class group, and $(q_1, \ldots, q_r)$ are independent formal variables. Here the convergence of (5.5) becomes a formal property. Second, one can take $\Lambda = \mathbf{C}$ and $q^\beta = \exp(-(L, \beta))$ where $(L, *)$ is the scalar product with an ample sheaf or, more generally, a linear function on B. The convergence for exponents with sufficiently positive real part on B may present a non–trivial problem, but whenever it is established, it allows us to pass from the formal Frobenius manifold to a Frobenius manifold in the category of complex analytic spaces.

5.3. Axioms for GW–invariants: The second group. Denote by $1 \in H^*(V)$ the identity in the cohomology ring, that is, the dual class to $[V]$. For $\gamma \in H^i$, put $|\gamma| = i$.

iv) Identity. Let $\pi_n : \overline{M}_{g,n} \to \overline{M}_{g,n-1}$ be the map that forgets the n–th marked point and then stabilizes the resulting curve. Then we must have

$$(5.6) \qquad I^V_{g,n,\beta}(\gamma_1 \otimes \cdots \otimes \gamma_{n-1} \otimes 1) = \pi_n^*(I^V_{g,n-1,\beta}(\gamma_1 \otimes \cdots \otimes \gamma_{n-1})).$$

v) Dimension. The map of $\mathbf{Z}$–graded modules $I^V_{g,n,\beta}$ must be homogeneous of degree $2(K_V, \beta) + (2g-2)\dim V$, that is,

$$(5.7) \qquad |I^V_{g,n,\beta}(\gamma_1 \otimes \cdots \otimes \gamma_n)| = \sum_{i=1}^n |\gamma_i| + 2(K_V, \beta) + (2g-2)\dim V.$$

vi) Divisor. Let $\delta \in H^2(V)$. Then

$$(5.8) \qquad \pi_{n*}(I^V_{g,n,\beta}(\gamma_1 \otimes \cdots \otimes \gamma_{n-1} \otimes \delta)) = (\beta, \delta)\, I^V_{g,n-1,\beta}(\gamma_1 \otimes \cdots \otimes \gamma_{n-1}).$$

vii) Mapping to a Point. By this catchphrase we describe the situation when $\beta = 0$. Let $\pi : \mathcal{C} \to \overline{M}_{g,n}$ be the universal curve. Denote by $\mathcal{E} = \mathcal{E}_{g,n}$ the locally free sheaf $R^1\pi_*\mathcal{O}_\mathcal{C}$ of rank g on $\overline{M}_{g,n}$. Let p_1, p_2 be the two projections of $V \times \overline{M}_{g,n}$. Then we must have

$$(5.9) \qquad I^V_{g,n,0}(\gamma_1 \otimes \cdots \otimes \gamma_n) = p_{2*}(p_1^*(\gamma_1 \cup \cdots \cup \gamma_n) \cup c_G(\mathcal{T}_V \boxtimes \mathcal{E})),$$

where $G := g \dim V$ is the rank of $\mathcal{E} \boxtimes \mathcal{T}_V$.

It is a nice exercise to check that the Dimension Axiom is compatible with the remaining ones.

Below we will prove that on the tree level, the Identity Axiom together with a corollary of the Mapping to a Point Axiom implies that the vector field 1 is the flat identity on the corresponding formal Frobenius manifold in the sense of I.2.1, whereas a combination of the Divisor and Dimension Axioms implies that this Frobenius manifold is equipped with the Euler field of the form described in I.4.4.

5.3.1. Potential. Put

$$(5.10) \qquad \langle \gamma_1 \ldots \gamma_n \rangle_{g,\beta} = \langle I^V_{g,n,\beta} \rangle (\gamma_1 \otimes \cdots \otimes \gamma_n) := \int_{\overline{M}_{g,n}} I^V_{g,n,\beta}(\gamma_1 \otimes \cdots \otimes \gamma_n).$$

The potential of the Frobenius manifold defined by (5.5) for $g = 0$ is

$$(5.11) \qquad \Phi(\Gamma) = \langle e^\Gamma \rangle_0 := \sum_{n \geq 3} \sum_\beta \frac{q^\beta}{n!} \langle I^V_{0,n,\beta} \rangle (\Gamma^{\otimes n}).$$

Here Γ is the even generic point of the linear superspace $H^*(V, \Lambda)$ over $\operatorname{Spec}\Lambda$.

More precisely, choose a homogeneous basis $\{\Delta_a \,|\, a = 0, \ldots, D\}$ of $H^*(V)$ and put $g_{ab} = (\Delta_a, \Delta_b) = \int_V \Delta_a \wedge \Delta_b$, $(g^{ab}) = (g_{ab})^{-1}$. Putting $\Gamma = \sum_a x^a \Delta_a$, we get

$$\Phi(\Gamma) = \sum_{n=n_0+\cdots+n_D \geq 3} \sum_\beta \frac{\varepsilon(n_0,\ldots,n_D) q^\beta}{n_0!\ldots n_D!}$$

(5.12)
$$\times \langle I^V_{0,n,\beta} \rangle (\Delta_0^{\otimes n_0} \otimes \cdots \otimes \Delta_D^{\otimes n_D})(x^0)^{n_0}\ldots(x^D)^{n_D},$$

where ε is the standard sign acquired in the supercommutative algebra

$$S(H^*(V))[x^0,\ldots,x^D]$$

after reshuffling

$$\prod_{a=0}^D (x^a \Delta_a)^{n_a} = \varepsilon \prod_{a=0}^D \Delta_a^{n_a} \prod_{a=0}^D (x^a)^{n_a}.$$

5.3.2. Lemma. *We have*

(5.13)
$$\langle I^V_{g,n,\beta} \rangle(\gamma_1 \otimes \cdots \otimes \gamma_{n-1} \otimes 1)$$
$$= \begin{cases} 0 \text{ if } \beta \neq 0, \text{ or } \beta = 0, (g,n) \notin \{(0,3),(1,1),(\geq 2,0)\}, \\ \int_V \gamma_1 \cup \gamma_2 \text{ if } \beta = 0, g = 0, n = 3. \end{cases}$$

Proof. Assume first that $\beta \neq 0$, and

(5.14)
$$\langle I^V_{g,n,\beta} \rangle(\gamma_1 \otimes \cdots \otimes \gamma_{n-1} \otimes 1) \neq 0$$

for some homogeneous γ_i. Choose $\delta \in H^2(V)$ with $(\beta,\delta) \neq 0$. Then from (5.8) and (5.14) it follows that

(5.15)
$$\langle I^V_{g,n+1,\beta} \rangle(\gamma_1 \otimes \cdots \otimes \gamma_{n-1} \otimes \delta \otimes 1) \neq 0.$$

In view of (5.6), $I^V_{g,n+1,\beta}(\gamma_1 \otimes \cdots \otimes \gamma_{n-1} \otimes \delta \otimes 1)$ is the lift of

(5.16)
$$I^V_{g,n,\beta}(\gamma_1 \otimes \cdots \otimes \gamma_{n-1} \otimes \delta)$$

which therefore also must be non-vanishing. But according to the Dimension Axiom, the degree of $I^V_{g,n,\beta}(\gamma_1 \otimes \cdots \otimes \gamma_{n-1} \otimes \delta)$ is greater by two than that of $I^V_{g,n,\beta}(\gamma_1 \otimes \cdots \otimes \gamma_{n-1} \otimes 1)$, which is already maximal in view of (5.14). Hence (5.14) and (5.16) cannot be simultaneously different from zero.

Now let $\beta = 0$. Then (5.6) shows that (5.14) can hold only if $(g,n) = (0,3)$, $(1,1)$ or $(\geq 2, 0)$, because a lifted class cannot have maximal dimension. To calculate these numbers, one can apply (5.9). For $g = 0$, the term $c_G(\mathcal{E} \boxtimes \mathcal{T}_V)$ is the identity class, so that (5.9) becomes the second line of (5.13).

5.3.3. Corollary. *1 is the flat identity.*

Proof. Choose $\Delta_0 = 1$. From (5.12) and (5.13) it is clear that only terms involving $I_{0,3,0}$ may contain x^0. Moreover,

$$\partial_0 \partial_b \partial_c \langle I^V_{0,3,\beta} \rangle(\gamma \otimes \gamma \otimes \gamma) = 6\langle I^V_{0,3,\beta} \rangle(1 \otimes \Delta_b \otimes \Delta_c).$$

This vanishes if $\beta \neq 0$ and is $6g_{bc}$ otherwise. It remains to compare this with Chapter I, (2.1) and (2.2).

5.3.4. Proposition. *(i) The vector field*

$$E = \sum_a (1 - \frac{|\Delta_a|}{2})x^a \partial_a + \sum_{b: |\Delta_b|=2} r^b \partial_b, \quad (5.17)$$

where r^b are defined by

$$c_1(\mathcal{T}_V) = -K_V = \sum_{b: |\Delta_b|=2} r^b \Delta_b, \quad (5.18)$$

is an Euler field with $d_0 = 1$, $D = 2 - \dim V$.

(ii) There exists a unique formal function Φ differing from (5.11) only by terms of degree ≤ 2 which is representable as a formal Fourier series in $q^\beta e^{(\beta,\delta)}$, $\beta \in B$, with coefficients which are formal series of the remaining coordinates, having the following properties. Put $\Phi = \Psi + c$ where c is the constant ($\beta = 0$) term of the Fourier series. Then, assuming $d_0 = 1$ and denoting by $E(0)$ the anticanonical class summand of E, we have:

a) $E\Psi = (D+1)\Psi$.

b) c is a cubic form with $(E - E(0))c = (D+1)c$, the classical cubic self-intersection index divided by 6.

Proof. In terms of Chapter I, 2.2, E is obviously conformal. We have $d_0 = 1$ because $\Delta_0 = 1$ is the identity. We will check the formula I.(2.7) which reads here

$$E\Phi = (3 - \dim V)\Phi + \text{a quadratic polynomial in } x^a \quad (5.19)$$

simultaneously with construction of the normalized potential. According to Lemma 5.3.2, the part of Φ corresponding to $\beta = 0$ is $\frac{1}{6}\langle\gamma^3\rangle$. Hence it is a linear combination of $x^a x^b x^c$ with $\Delta_a + \Delta_b + \Delta_c = 2 \dim V$ which obviously satisfies (5.19). It remains to consider the $\beta \neq 0$ part. We may add to it a quadratic polynomial without influencing the validity or otherwise of (5.19). The following correction terms simplify the calculation.

The maps $\langle I^V_{0,n,\beta}\rangle : H^*(V)^{\otimes n} \to \Lambda$ are defined for $n \geq 3$ and satisfy

$$\langle I^V_{0,n,\beta}\rangle(\alpha \otimes \delta) = (\beta, \delta)\langle I^V_{0,n-1,\beta}\rangle(\alpha) \quad (5.20)$$

for $\delta \in H^2(V)$: this follows from the Divisor Axiom. It is easy to check that there exists a unique polylinear extension of $\langle I^V_{0,n,\beta}\rangle$ to all $n \geq 0$ satisfying (5.20). In fact, it suffices to put

$$\langle I^V_{0,n,\beta}\rangle(\alpha) = (\beta, \delta)^{-m} \langle I^V_{0,n+m,\beta}\rangle(\alpha \otimes \delta^{\otimes m})$$

for any (m, δ) with $m + n \geq 3$ and invertible (β, δ). Now put

$$\Psi(\Gamma) := \sum_{n \geq 0} \sum_{\beta \neq 0} \frac{q^\beta}{n!} \langle I^V_{0,n,\beta}\rangle(\Gamma^{\otimes n}). \quad (5.21)$$

Clearly, $\Psi + c$ differs from the initial Φ by terms of degree ≤ 2. Moreover, c is a linear combination of $x^a x^b x^c$ with $|\Delta_a| + |\Delta_b| + |\Delta_c| = 2 \dim V$ so that $(E - E(0))c =$

$(D+1)c$. As for $E\Psi$, we have for $\Gamma = \gamma_0 + \delta$, where δ is the divisorial part,

$$\Psi(\gamma_0 + \delta) = \sum_{i,k \geq 0} \sum_{\beta \neq 0} \frac{q^\beta}{i!\, k!} \langle I_{0,n}(V,\beta) \rangle (\gamma_0^{\otimes i} \otimes \delta^{\otimes k})$$

(5.22)
$$= \sum_{i \geq 0} \sum_{\beta \neq 0} \frac{q^\beta e^{(\beta,\delta)}}{i!} \langle I_{0,n}(V,\beta) \rangle (\gamma_0^{\otimes i}).$$

Let us now apply E to any summand in (5.22). The $E(0)$ part acts only upon $e^{(\beta,\delta)}$ and multiplies it by $(c_1(V), \beta)$. The $E - E(0)$ part multiplies any monomial $x_{a_1} \ldots x_{a_n}$ in non–divisorial coordinates by $\sum_i (1 - |\Delta_{a_i}|/2)$. From the definition we see that β can furnish a non–zero contribution to such a term only if

$$\dim I_{g,n}(V,\beta) = \dim V - 3 + (c_1(V), \beta) + n = \sum_{i=1}^n \frac{|\Delta_{a_i}|}{2}.$$

Hence every non–vanishing term of (5.22) is an eigenvector of E with eigenvalue $D + 1 = 3 - \dim V$.

This proves the proposition.

Notice in conclusion that $q^\beta e^{(\beta,\delta)}$ is the universal character of B together with q^β. We have introduced q^β only to achieve the formal convergence. If it holds without q^β, we can forget about it. Moreover, if the formal Fourier series actually converges for δ lying somewhere in the complexified ample cone, $\Psi(x)$ has a free abelian symmetry group: translations by an appropriate discrete subgroup in the space $H^2(V, i\mathbf{R})$. Conversely, in the analytic category this condition is necessary for the existence of the appropriate Fourier series.

5.4. Potentials of qc–type. Based upon the analysis above, we will introduce the following definition. Its goal is to provide an intermediate step in the problem of checking whether a given formal Frobenius manifold is quantum cohomology. We must be able at least to detect the following structures.

5.4.1. Definition. *Let $(M = \mathrm{Spf}\, k[[H^t]], g, \Phi_0)$ be a formal Frobenius manifold over a $\mathbf{Q}$–algebra k with flat identity, Euler field E, and spectrum $D, \{d_a\}$ in k. Assume that $-\mathrm{ad}\, E$ is semisimple on flat vector fields and denote by $H(d_a)$ the eigenspace corresponding to d_a. We consider H to be a free $\mathbf{Z}_2$–graded k–module of flat vector fields, and H^t as the dual module of flat coordinates vanishing at the origin. Assume that $H^2 = H(0), H_2 = H(D)$ are pure–even.*

Moreover, assume that there exists a semigroup $B \subset H_2$ with finite decomposition and indecomposable zero, generating a lattice, and the cubic form c on H, such that by eventually changing terms of degree ≤ 2 in Φ_0 we can obtain the potential of the form

$$\Phi = \Psi + c,\ E\Psi = (D+1)\Psi,\ (E - E(0))c = (D+1)c,$$

(5.23)
$$\Psi(\gamma_0 + \delta) = \sum_{i \geq 0} \sum_{\beta \in B \setminus \{0\}} \frac{e^{(\beta,\delta)}}{i!} I_\beta(\gamma_0^{\otimes i})$$

in which all summands in the last sum are eigenvectors of E with eigenvalue $D+1$. Here Γ is a generic even element of H, δ its H^2-component, $\gamma_0 = \Gamma - \delta$. The coefficient $I_\beta(\gamma_0^{\otimes i})$ is a form in non-divisorial coordinates.

A formal Frobenius manifold satisfying these conditions will be called of qc-type.

A flat identity e in this language is an element $e \in H$ which considered as a derivation satisfies
$$e\,\Psi = 0,\ e\,c = g.$$

5.4.2. Correlators of qc-manifolds. Let M be a formal Frobenius manifold of qc-type. Recall that $\Phi_{ab}{}^c$ are the structure constants of the quantum multiplication. On qc-manifolds there are two useful specializations of this structure.

a) The "small quantum multiplication" obtained by restricting $\Phi_{ab}{}^c$ to $\gamma_0 = 0$.

b) The cup multiplication $\cup$ obtained by formally putting $e^{(\beta,\gamma)} = 0$ for all $\beta \neq 0$ ("large volume limit"). In other words, this is the multiplication for which c can be written as
$$c(\gamma) = \frac{1}{6} g(\gamma, \gamma \cup \gamma).$$

We now define correlators $\langle \ldots \rangle : H^{\otimes n} \to k$ as $\mathbf{S}_n$-invariant polylinear functions whose values are derivatives of Φ at zero. In other words, for a basis $\{\Delta_a\}$ of H and dual coordinates $\{x_a\}$ as above, we have
$$\Phi(x) = \sum_{n, a_1, \ldots, a_n} \epsilon(a) \frac{x_{a_1} \ldots x_{a_n}}{n!} \langle \Delta_{a_1} \ldots \Delta_{a_n} \rangle.$$

In the qc-case we can write
$$\langle \Delta_{a_1} \ldots \Delta_{a_n} \rangle = \sum_{\beta \in B \setminus \{0\}} \langle \Delta_{a_1} \ldots \Delta_{a_n} \rangle_\beta + \langle \Delta_{a_1} \ldots \Delta_{a_n} \rangle_0,$$

where the first sum comes from Ψ and the second, non-vanishing only for triple arguments, from c.

Looking at (5.23) one sees that small quantum multiplication depends only on the triple correlators of non-divisorial elements of the basis.

5.4.3. Claim. *The correlators of the Frobenius manifolds of qc-type satisfy the following Divisor Identity: if $\delta \in H^2$, $\beta \neq 0$,*
$$\langle \delta \gamma_1 \ldots \gamma_n \rangle_\beta = (\delta, \beta) \langle \gamma_1 \ldots \gamma_n \rangle_\beta.$$

Reading the proof of (5.22) backwards, one sees that this property follows from (5.23).

This formula allows us to extend the definition of the correlators to $n \leq 2$ arguments.

§6. Formal Frobenius manifolds of rank one and Weil–Petersson volumes of moduli spaces

6.1. Notation. The rank of a CohFT on (H, g) over a field k is the (super)dimension of H. In this section we consider the case $\dim H = 1$. To slightly simplify notation let us assume that all square roots exist in k. Then $H = k\Delta_0$, $g(\Delta_0, \Delta_0) = 1$, and $\Delta = \Delta_0^{\otimes 2} \in H^{\otimes 2}$. The basic vector Δ_0 is defined up to a sign.

We will consider its choice as a rigidification and without further ado call such rigidified theories simply CohFT's of rank one.

A structure of CohFT on (H, Δ_0) boils down to the sequence of cohomology classes (generally non–homogeneous)

$$(6.1) \qquad c_n := I_n(\Delta_0^{\otimes n}) \in H^*(\overline{M}_{0n}, k)^{S_n}, \ n \geq 3,$$

satisfying the identities

$$(6.2) \qquad \varphi_\sigma^*(c_n) = c_{n_1+1} \otimes c_{n_2+1},$$

where $\varphi_\sigma : \overline{M}_{0,n_1+1} \times \overline{M}_{0,n_2+1} \to \overline{M}_{0n}$ is the embedding of the boundary divisor corresponding to a partition σ (see (4.2)). Put $c_n = \sum_{i=0}^{n-3} c_n^{(i)}$, $c_n^{(i)} \in H^{2i}(\overline{M}_{0n})$. Changing the sign of Δ_0 leads to $c_n \mapsto (-1)^n c_n$.

The tensor product formula (4.5) becomes

$$(6.3) \qquad \{c_n'\} \otimes \{c_n''\} = \{c_n' \cup c_n''\},$$

if we agree that $(H', \Delta_0') \otimes (H'', \Delta_0'') = (H' \otimes H'', \Delta_0' \otimes \Delta_0'')$.

Here are some simple consequences of (6.2) and (6.3).

6.1.1. *The theory $c_n = c_n^{(0)} = \mathbf{1}_{0n}$ for $n \geq 3$ is the identity with respect to the tensor product.*

6.1.2. *The theories $c_n(t) = t^{n-2}\mathbf{1}_{0n}$, $t \in k^*$, form a group isomorphic to k^*.*

6.1.3. *The theory $\{c_n\}$ is invertible iff $c_3^{(0)} \neq 0$. Any invertible theory is a tensor product of one of the type $\{c_n(t)\}$ and one with $c_n^{(0)} = 1$ for all n, and this decomposition is unique.*

6.1.4. *Assume that $c_n^{(0)} = 1$ and put $\lambda_n = \log c_n \in H^*(\overline{M}_{0n}, k)^{S_n}$. Then (6.2) becomes*

$$(6.4) \qquad \varphi_\sigma^*(\lambda_n) = \lambda_{n_1+1} \otimes 1 + 1 \otimes \lambda_{n_2+1},$$

and (6.3) becomes

$$(6.5) \qquad \{\lambda_n'\} \otimes \{\lambda_n''\} = \{\lambda_n' + \lambda_n''\}.$$

Vice versa, any sequence of classes $\lambda_n \in H^*(\overline{M}_{0n}, k)^{S_n}$ satisfying (6.4) gives rise to a CohFT of rank 1, $c_n = \exp \lambda_n$. We can say that $\{\lambda_n\}$ forms a *logarithmic CohFT of rank 1*.

6.1.5. *There is a canonical bijection between the set of the isomorphism classes of CohFT's of rank 1 and the set of infinite sequences $(C_3, C_4, \ldots) \in k^\infty$ given by*

$$(6.6) \qquad C_n = \int_{\overline{M}_{0n}} c_n = \int_{\overline{M}_{0n}} c_n^{(n-3)}.$$

In fact, this is a particular case of Theorem 4.3, because any formal series in one variable $\Phi(x) = \sum \dfrac{C_n}{n!} x^n$ satisfies the associativity equations.

We will call the theory *normalized*, if $c_3 = c_3^0 = 1$. Equivalently, $c_n = 1_n +$ terms of dimension $< n - 3$, or else $C_3 = 1$. Thus the space $\mathbf{CohFT}_1(\mathbf{k})$ of all normalized (hence invertible) rank one CohFTs is canonically isomorphic to $\frac{1}{6}x^3 + x^4 k[[x]]$ and has canonical coordinates $C_n, n \geq 4$. Our goal is to describe the tensor product in this infinite–dimensional space. To this end we will find another coordinate system in which the tensor product becomes addition, and calculate the transition functions between the two systems.

6.2. Mumford classes. Consider the universal curve $p_n : C_n \to \overline{M}_{0n}$ and its structure sections $s_i : \overline{M}_{0n} \to C_n$, $i = 1, \ldots, n$. Let $x_i \subset C_n$ be the image of s_i, and ω the relative dualizing sheaf on C_n. For $a = 1, 2, \ldots$ put

$$(6.7) \qquad \kappa_n(a) := p_{n*}\left(c_1(\omega(\sum_{i=1}^{n} x_i))^{a+1}\right) \in H^{2a}(\overline{M}_{0n}, \mathbf{Q})^{S_n}.$$

It is proved in [AC1] that for any $a \geq 1$ (in fact, $a = 0$ as well) $\{\kappa_n(a) \mid n \geq 3\}$ satisfy (3.4), i.e. form a logarithmic field theory. (This follows from V.1.2 below.) Hence we can construct an infinite–dimensional family of invertible theories of rank one putting

$$(6.8) \qquad I_n(\Delta_0^{\otimes n}) := \kappa_n[s_1, s_2, \ldots] := \exp\left(\sum_{a=1}^{\infty} s_a \kappa_n(a)\right), \; n \geq 3.$$

6.2.1. Theorem. (s_a) *form a coordinate system on the space* $\mathbf{CohFT}_1(\mathbf{k})$ *defining its group isomorphism with* K_+^{∞}.

Proof. The sum in the rhs effectively stops at $a = n - 3$ (cf. (6.7)). The C_n-coordinate of the theory (3.8) is therefore

$$\int_{\overline{M}_{0n}} \kappa_n[s_1, s_2, \ldots] = s_{n-3} \int_{\overline{M}_{0n}} \kappa_n(n-3) + P_n(s_1, \ldots, s_{n-4}), \; n \geq 4,$$

where P_n is a universal polynomial. Hence it remains to check that the coefficient at s_{n-3} does not vanish. But we will actually show below that it equals 1.

6.3. Potential. Potential of the theory (6.8) is

$$(6.9) \qquad \Phi(x; s_1, s_2, \ldots) = \sum_{n=3}^{\infty} \frac{x^n}{n!} \int_{\overline{M}_{0n}} \sum_{\sum a m(a) = n-3} \prod_{a \geq 1} \kappa_n(a)^{m(a)} \frac{s_a^{m(a)}}{m(a)!}.$$

Here $\mathbf{m} = (m(1), m(2), \ldots)$ runs over sequences of non–negative integers vanishing for sufficiently large a.

In shorter versions of expressions like (6.9) we will use notation of the type

$$|\mathbf{m}| := \sum_{a \geq 1} a m(a), \quad \|\mathbf{m}\| := \sum_{a \geq 1} m(a), \quad \mathbf{m}! := \prod_{a \geq 1} m(a)!,$$

$$(6.10) \qquad \kappa_n^{\mathbf{m}} = \prod_{a \geq 1} \kappa_n(a)^{m(a)}, \quad \mathbf{s}^{\mathbf{m}} = \prod_{a \geq 1} s_a^{m(a)},$$

where $\mathbf{s} = (s_1, s_2, \ldots)$ is a family of independent formal variables or complex numbers. In particular we put

$$(6.11) \qquad V(\mathbf{m}) := \frac{1}{(\sum_{a \geq 1} a m(a))!} \int_{\overline{M}_{0n}} \prod_{a \geq 1} \frac{\kappa_n(a)^{m(a)}}{m(a)!} = \int_{\overline{M}_{0n}} \frac{\kappa^{\mathbf{m}}}{\mathbf{m}! |\mathbf{m}|!} \in \mathbf{Q}.$$

Then the second x-derivative of (6.9) can be written as

$$(6.12) \qquad \Phi''(x;\mathbf{s}) = \sum_{\mathbf{m}} V(\mathbf{m}) \frac{x^{|\mathbf{m}|+1}}{|\mathbf{m}|+1} \mathbf{s}^{\mathbf{m}}.$$

6.4. Calculation of $V(\mathbf{m})$. Integrals $V(\mathbf{m})$ are called in [KaMZ] the higher Weil–Petersson volumes of the moduli spaces. (Here we restrict ourselves to the genus zero case; for arbitrary genera, see 6.9–6.12 below.)

In order to calculate them, we have to introduce the classes

$$\psi_{n,i} := \xi_i^*(c_1(\omega_{C/\overline{M}_{0n}})) \in H^2(\overline{M}_{0n}),$$

where $\xi_i : \overline{M}_{0n} \to C$, $i = 1, \ldots, n$, are the universal sections. The integrals of their monomials according to Witten are denoted

$$(6.13) \qquad \langle \tau_{a_1} \ldots \tau_{a_n} \rangle = \int_{\overline{M}_{0n}} \psi_{n,1}^{a_1} \ldots \psi_{n,n}^{a_n}.$$

It is well known (see recursive formulas below and in VI.3) that they do not vanish exactly for $a_1 + \cdots + a_n = n - 3$, and then are equal to

$$(6.14) \qquad \langle \tau_{a_1} \ldots \tau_{a_n} \rangle = \frac{(n-3)!}{a_1! \ldots a_n!}.$$

More generally, we will consider the relative integrals of type (6.13). For $k \geq l$, denote by $\pi_{k,l} : \overline{M}_{0k} \to \overline{M}_{0l}$ the morphism forgetting the last $k - l$ points. For any $a_1, \ldots, a_p \geq 0$ define

$$(6.15) \quad \kappa_n(a_1, \ldots, a_p) := \pi_{n+p,n*}(\psi_{n+p,n+1}^{a_1+1} \cdots \psi_{n+p,n+p}^{a_p+1}) \in H^{2(a_1+\cdots+a_p)}(\overline{M}_{0n}, \mathbf{Q}).$$

Notice that whenever $a_1 + \cdots + a_p = \dim \overline{M}_{0n} = n - 3$, we also have $(a_1 + 1) + \cdots + (a_p + 1) = \dim \overline{M}_{0,n+p}$, and then
(6.16)
$$\int_{\overline{M}_{0n}} \kappa_n(a_1, \ldots, a_p) = \int_{\overline{M}_{0,n+p}} \psi_{n+p,n+1}^{a_1+1} \cdots \psi_{n+p,n+p}^{a_p+1} = \langle \tau_0^n \tau_{a_1+1} \ldots \tau_{a_p+1} \rangle.$$

For $p = 1$ we get an apparent ambiguity of notation: (6.15) specializes to

$$\kappa_n(a) = \pi_{n+1,n*}(\psi_{n+1,n+1}^{a+1}) \in H^{2a}(\overline{M}_{0n}, \mathbf{Q})$$

which formally differs from (6.7). However one can check that the two definitions actually coincide. For more general results, see Chapters V and VI.

6.4.1. Theorem. *For any $n, a_1, \ldots, a_p \geq 0$ we have*

$$(6.17) \quad \kappa_n(a_1) \ldots \kappa_n(a_p) = \sum_{k=1}^{p} \frac{(-1)^{p-k}}{k!} \sum_{\{1,\ldots,p\}=S_1 \sqcup \cdots \sqcup S_k} \kappa_n(\sum_{j \in S_1} a_j, \ldots, \sum_{j \in S_k} a_j).$$

The inner sum is taken over partitions with non-empty S_i. Equivalently, for any $\mathbf{m} \in \mathbf{N}^\infty \setminus \{\overline{0}\}, p = \|\mathbf{m}\|$,

$$(6.18) \qquad \frac{(-1)^p}{\mathbf{m}!} \kappa_n^{\mathbf{m}} = \sum_{k=1}^{p} \frac{(-1)^k}{k!} \sum_{\mathbf{m}=\mathbf{m}_1+\cdots+\mathbf{m}_k, \mathbf{m}_i \neq 0} \frac{\kappa_n(|\mathbf{m}_1|, \ldots, |\mathbf{m}_k|)}{\mathbf{m}_1! \ldots \mathbf{m}_k!}.$$

The proof consists of a geometric and a combinatorial part.

6.4.2. Lemma. *We have*

$$（6.19） \quad \kappa_n(a_1,\ldots,a_p) = \sum_{\sigma \in \mathbf{S}_p} \prod_{o \in o(\sigma)} \kappa_n\Big(\sum_{j \in o} a_j\Big),$$

where $o(\sigma)$ denotes the set of the cycles of σ acting on $\{1,\ldots,p\}$, i.e. the orbits of the cyclic group $\langle \sigma \rangle$.

This identity formally follows from another geometric identity going back to Witten,

$$（6.20） \quad \pi_{n+1,n*}(\psi_{n+1,1}^{a_1} \cdots \psi_{n+1,n}^{a_n} \psi_{n+1,n+1}^{a_{n+1}+1}) = \psi_{n,1}^{a_1} \cdots \psi_{n,n}^{a_n} \kappa_n(a_{n+1}),$$

combined with the formula

$$（6.21） \quad \kappa_n(a) = \pi_{n,n-1}^*(\kappa_{n-1}(a)) + \psi_{n,n}^a.$$

We will prove (6.20) and (6.21) (in fact, their generalizations to the spaces of stable maps) in Chapter VI, §3, and omit the deduction of (6.19), referring the reader to [AC1].

The passage from (6.19) to (6.17) and (6.18) is a formal inversion result which we will prove here in an axiomatized form.

Let R be a commutative $\mathbf{Q}$-algebra generated by some elements $\kappa(a)$ where a runs over all elements of an additive semigroup A.

6.4.3. Lemma. *Define elements $\kappa(a_1,\ldots,a_p) \in R$ for $p \geq 2$, $a_i \in A$ recursively by*

$$（6.22） \quad \kappa(a_1,\ldots,a_p) = \kappa(a_1,\ldots,a_{p-1})\kappa(a_p) + \sum_{i=1}^{p-1} \kappa(a_1,\ldots,a_i+a_p,\ldots,a_{p-1}).$$

Then $\{\kappa(a_1,\ldots,a_p)\,|\,p \geq 1\}$ span R as a linear space. They can be expressed via monomials in $\kappa(a)$ by the following universal identity (coinciding with (6.19)):

$$（6.23） \quad \kappa(a_1,\ldots,a_p) = \sum_{\sigma \in \mathbf{S}_p} \prod_{o \in o(\sigma)} \kappa\Big(\sum_{j \in o} a_j\Big).$$

In particular, $\kappa(a_1,\ldots,a_p)$ are symmetric in $a_1,\ldots,a_p$.

Conversely, monomials in $\kappa(a)$ can be expressed via these elements by the universal formula (coinciding with (6.17)):

$$（6.24） \quad \kappa(a_1)\ldots\kappa(a_p) = \sum_{k=1}^{p} \frac{(-1)^{p-k}}{k!} \sum_{\{1,\ldots,p\}=S_1 \amalg \cdots \amalg S_k} \kappa\Big(\sum_{j \in S_1} a_j,\ldots,\sum_{j \in S_k} a_j\Big).$$

If $A = \{1,2,3,\ldots\}$, we also have

$$（6.25） \quad \frac{(-1)^p}{\mathbf{m}!}\kappa^{\mathbf{m}} = \sum_{k=1}^{p} \frac{(-1)^k}{k!} \sum_{\mathbf{m}=\mathbf{m}_1+\cdots+\mathbf{m}_k,\,\mathbf{m}_i \neq 0} \frac{\kappa(|\mathbf{m}_1|,\ldots,|\mathbf{m}_k|)}{\mathbf{m}_1!\ldots\mathbf{m}_k!},$$

where $p = \|\mathbf{m}\|$ as in (6.18) and $\kappa^{\mathbf{m}} = \kappa(1)^{m(1)}\kappa(2)^{m(2)}\ldots$.

Furthermore, $\kappa(a)$ are algebraically independent iff $\kappa(a_1,\ldots,a_p)$ are linearly independent. In this case R is graded by A via $\deg \kappa(a_1,\ldots,a_p) = a_1 + \cdots + a_p$.

Example. The elements $\kappa(a_1,\ldots,a_p)$ are given for $p = 2$ by

$$\kappa(a,b) = \kappa(a)\kappa(b) + \kappa(a+b), \quad \kappa(a)\kappa(b) = \kappa(a,b) - \kappa(a+b),$$

and for $p = 3$ by

$$\kappa(a,b,c) = \kappa(a)\kappa(b)\kappa(c) + \kappa(a+b)\kappa(c) + \kappa(a+c)\kappa(b)$$
$$+ \kappa(b+c)\kappa(a) + 2\kappa(a+b+c),$$
$$\kappa(a)\kappa(b)\kappa(c) = \kappa(a,b,c) - \kappa(a+b,c) - \kappa(a+c,b)$$
$$- \kappa(b+c,a) + \kappa(a+b+c).$$

Proof of Lemma 6.4.3. The following identity shows by induction on p that (6.22) and (6.23) are equivalent:

$$\sum_{\sigma \in \mathbf{S}_{p+1}} \prod_{o \in o(\sigma)} \kappa(\sum_{j \in o} a_j) = \kappa(a_{p+1}) \left[\sum_{\sigma \in \mathbf{S}_p} \prod_{o \in o(\sigma)} \kappa(\sum_{j \in o} a_j) \right]$$
(6.26)
$$+ \sum_{i=1}^{p} \sum_{\tau \in \mathbf{S}_p} \prod_{o \in o(\tau)} \kappa(\sum_{j \in o} a_j + \delta_{ij} a_{p+1}).$$

To convince yourself of the validity of (6.26) look at the following bijective map from the lhs monomials in $\kappa(a)$ to the rhs monomials. If a lhs monomial is indexed by $\sigma \in \mathbf{S}_{p+1}$ for which $\sigma(p+1) = p+1$, we get it in the rhs for $\tau =$ restriction of σ to $\{1,\ldots,p\}$. Otherwise $p+1$ belongs to an orbit of σ of cardinality ≥ 2; deleting $p+1$ from this cycle, we get a permutation $\tau \in \mathbf{S}_p$ and a number $i = \sigma(p+1) \leq p$ producing exactly the needed monomial in the second sum.

One can similarly pass from (6.22) to (6.24) and backwards. For example, assume that (6.24) is already proved for some p. Then we have

$$(\kappa(a_1)\ldots\kappa(a_p))\kappa(a_{p+1})$$

$$= \left(\sum_{k=1}^{p} \frac{(-1)^{p-k}}{k!} \sum_{\substack{\{1,\ldots,p\} = S_1 \amalg \cdots \amalg S_k \\ S_i \neq \emptyset}} \kappa(\sum_{j \in S_1} a_j, \ldots, \sum_{j \in S_k} a_j) \right) \kappa(a_{p+1})$$

$$= \sum_{k=1}^{p} \frac{(-1)^{p-k}}{k!} \sum_{\substack{\{1,\ldots,p\} = S_1 \amalg \cdots \amalg S_k \\ S_i \neq \emptyset}} \left(\kappa(\sum_{j \in S_1} a_j, \cdots \sum_{j \in S_k} a_j, a_{p+1}) \right.$$

(6.27)
$$\left. - \sum_{i=1}^{p} \kappa(\sum_{j \in S_1}(a_j + \delta_{ij} a_{p+1}), \ldots, \sum_{j \in S_k}(a_j + \delta_{ij} a_{p+1})) \right).$$

Now essentially the same combinatorics as above govern a correspondence between the summands in (6.27) and those in the rhs of (6.24) for $p+1$ arguments which is

$$\sum_{k=1}^{p+1} \frac{(-1)^{p+1-k}}{k!} \sum_{\substack{\{1,\ldots,p+1\}=S_1\amalg\cdots\amalg S_k \\ S_i\neq\emptyset}} \kappa(\sum_{j\in S_1} a_j, \ldots, \sum_{j\in S_k} a_j).$$

Namely, any ordered k-partition $\{1,\ldots,p\} = S_1 \amalg \cdots \amalg S_k$ can be enhanced to $k+1$ ordered $(k+1)$-partitions of $\{1,\ldots,p+1\}$ containing $\{p+1\}$ as a separate part, and to k ordered k-partitions of $\{1,\ldots,p+1\}$ for which $p+1$ is put into one of the S_i's.

It remains to rewrite (6.24) in the form (6.25), when $A = \{1,2,3,\ldots\}$. To this end, notice that if $\delta_{a_1} + \cdots + \delta_{a_p} = \mathbf{m}$, we have

$$\kappa(a_1)\ldots\kappa(a_p) = \prod_{a\geq 1} \kappa(a)^{m(a)} = \kappa^{\mathbf{m}}, \qquad p = \|\mathbf{m}\|,$$

and $m(a) = \mathrm{card}\{j \mid a_j = a\}$. Any set partition $\{1,\ldots,p\} = S_1 \amalg \cdots \amalg S_k$, $S_i \neq \emptyset$, produces a vector partition

$$\mathbf{m} = \mathbf{m}_1 + \cdots + \mathbf{m}_k, \quad \mathbf{m}_i = (m_i(a)) \neq 0, \quad m_i(a) = \mathrm{card}\{j \mid a_j = a\}.$$

We have $\sum_{j\in S_i} a_j = |\mathbf{m}_i|$, and each $\mathbf{m} = \mathbf{m}_1 + \cdots + \mathbf{m}_k$ comes from

$$\prod_{a\geq 1} \frac{m(a)!}{m_1(a)!\ldots m_k(a)!} = \frac{\mathbf{m}!}{\mathbf{m}_1!\ldots\mathbf{m}_k!}$$

set partitions. This finishes the proof of Lemma 6.4.3 and Theorem 6.4.1.

As a corollary, we get the explicit formula for $V(\mathbf{m})$:

$$(6.28) \qquad V(\mathbf{m}) = \sum_{k=1}^{p} (-1)^{p-k} \binom{|\mathbf{m}|+k}{k} \sum_{\substack{\mathbf{m}=\mathbf{m}_1+\cdots+\mathbf{m}_k \\ \mathbf{m}_i\neq 0}} \frac{1}{\prod_{k=1}^{p}(|\mathbf{m}_i|+1)!\mathbf{m}_i!}.$$

To check it, combine (6.18), (6.16), and (6.14).

6.5. Theorem. *In the ring of formal series of one variable with coefficients in $\mathbf{Q}[\mathbf{s}] = \mathbf{Q}[s_1, s_2, \ldots]$ we have the following inversion formula:*

$$(6.29) \qquad y = \sum_{|\mathbf{m}|\geq 0} V(\mathbf{m}) \frac{x^{|\mathbf{m}|+1}}{|\mathbf{m}|+1} \mathbf{s}^{\mathbf{m}} \iff x = \sum_{|\mathbf{m}|\geq 0} \frac{y^{|\mathbf{m}|+1}}{(|\mathbf{m}|+1)!} \frac{(-\mathbf{s})^{\mathbf{m}}}{\mathbf{m}!}.$$

Proof. From (6.28) we have for any $\mu \geq 1$:

$$\sum_{\mathbf{m}:|\mathbf{m}|=\mu} V(\mathbf{m}) \mathbf{s}^{\mathbf{m}} = \sum_{k=1}^{\infty} (-1)^k \binom{\mu+k}{k} \left(\sum_{|\mathbf{m}|>0} \frac{(-\mathbf{s})^{\mathbf{m}}}{(|\mathbf{m}|+1)!\mathbf{m}!}\right)^k \Bigg|_{\mathrm{degree}\,\mu}$$

$$= \left(\sum_{\mathbf{m}} \frac{(-\mathbf{s})^{\mathbf{m}}}{(|\mathbf{m}|+1)!\mathbf{m}!}\right)^{-\mu-1} \Bigg|_{\mathrm{degree}\,\mu}$$

$$= \mathrm{coeff.\ of\ } y^{\mu} \mathrm{\ in\ } \left(\frac{x(y)}{y}\right)^{-\mu-1},$$

where
$$x = x(y; s) := \sum_{|\mathbf{m}| \geq 0} \frac{y^{|\mathbf{m}|+1}}{(|\mathbf{m}|+1)!} \frac{(-\mathbf{s})^{\mathbf{m}}}{\mathbf{m}!} \in \mathbf{Q}[\mathbf{s}][[y]]$$
is the power series occurring on the right hand side of (6.29), and we have used the binomial identity
$$\sum_{k=1}^{\infty} (-1)^k \binom{\mu+k}{k} z^k = (1+z)^{-(\mu+1)} - 1.$$
But for any power series $x(y) = \sum_{r \geq 1} b_r y^r$, $b_1 \neq 0$, we have
$$\text{coeff. of } y^{\mu} \text{ in } \left(\frac{x(y)}{y}\right)^{-\mu-1} = \text{res}_{y=0}\left(\frac{1}{y^{\mu+1}}\left(\frac{y}{x(y)}\right)^{\mu+1} dy\right)$$
$$= \text{res}_{x=0}\left(\frac{1}{x^{\mu+1}} \frac{dy(x)}{dx} dx\right) = \text{coeff. of } x^{\mu} \text{ in } \frac{dy(x)}{dx} = \text{coeff. of } \frac{x^{\mu+1}}{\mu+1} \text{ in } y(x),$$
where $y(x)$ is the power series obtained by the formal inversion of $x = x(y)$. Applying this to our situation, we find that the inverse series of $x(y)$ is given by
$$y(x) = \sum_{\mu \geq 0} \frac{x^{\mu+1}}{\mu+1} \sum_{|\mathbf{m}|=\mu} V(\mathbf{m}) \mathbf{s}^{\mathbf{m}}$$
which is the left hand side of (6.29).

Combining now Theorem 6.2.1, (6.12), and (6.29) and using the formal Laplace transform, we arrive at the following result.

6.6. Theorem. *There is a natural bijection between the potentials of the normalized rank one CohTs $\Phi(x) \in \frac{x^3}{6} + x^4 k[[x]]$ and formal series $U(\eta) \in 1 + \eta k[[\eta]]$ given by any of the two equivalent prescriptions:*

(i) $U(\eta) = \sum_{n=0}^{\infty} B_n \eta^n$ *where*
$$x = \sum B_n \frac{y^{n+1}}{(n+1)!}$$
is the inverse power series of $y = \Phi''(x)$.

(ii) $U(\eta) = \int_0^{\infty} e^{-\Phi''(\eta x)/\eta} dx$.

Under this correspondence, the tensor product of CohfTs corresponds to the product of the series U.

Thus the coefficients of $-\log U(\eta)$ are the canonical coordinates in the sense of Lie group theory.

6.7. Explicit formulas. Substituting the generic potential (with coefficients C_n, $C_3 = 1$) into 6.6 (ii), expanding, and integrating term by term gives the explicit formula

(6.30) $$B_n = \sum_{\substack{n_4, n_5, \ldots \geq 0 \\ n_4 + 2n_5 + \cdots = n}} \frac{(2n_4 + 3n_5 + \cdots)!}{2!^{n_4} 3!^{n_5} \cdots n_4! n_5! \cdots} (-C_4)^{n_4} (-C_5)^{n_5} \cdots$$

§6. FORMAL FROBENIUS MANIFOLDS OF RANK ONE

for the coefficients of $U(\eta)$ in terms of the coefficients of $\Phi(x)$, and the same argument applied to the inverse power series gives the reciprocal formula

$$(6.31) \qquad C_n = \sum_{\substack{n_1,n_2,\dots \geq 0 \\ n_4+2n_5+\cdots = n-3}} \frac{(2n_1+3n_2+\cdots)!}{2!^{n_1}\, 3!^{n_2}\cdots n_1!\, n_2!\cdots}\, (-B_1)^{n_1}\,(-B_2)^{n_2}\cdots .$$

From this we obtain the explicit law for the tensor product of two normalized invertible CohFT's in terms of the coefficients of their potential functions:

$$C_4 = C_4' + C_4'',$$
$$C_5 = C_5' + 5 C_4' C_4'' + C_5'',$$
$$C_6 = C_6' + (8\, {C_4'}^2 + C_5')\, C_4'' + C_4'\, (8\, {C_4''}^2 + C_5'') + C_6'',$$
$$C_7 = C_7' + (35\, C_4' C_5' + 14\, C_6')\, C_4'' + (61\, {C_4'}^2\, {C_4''}^2 + 33\, {C_4'}^2 C_5'' + 33\, C_5' {C_4''}^2$$
$$+ 19\, C_5' C_5'') + C_4'\, (35\, C_4'' C_5'' + 14\, C_6'') + C_7'',\dots .$$

Finally, we observe that the values of the genus 0 Weil-Petersson volumes $V(\mathbf{m})$ can be calculated numerically from the closed formula (6.28), or the generating function formula (6.29). Here are the values up to $|\mathbf{m}| = 5$, expressed in terms of the generating function:

$$F(x,\mathbf{s}) = 1 + s_1 x + \left(5\frac{s_1^2}{2} + s_1\right)\frac{x^2}{2} + \left(61\frac{s_1^3}{6} + 9 s_1 s_2 + s_3\right)\frac{x^3}{6}$$
$$+ \left(1379\frac{s_1^4}{24} + 161\frac{s_1^2 s_2}{2} + 14 s_1 s_3 + 19\frac{s_2^2}{2} + s_4\right)\frac{x^4}{24}$$
$$+ \left(49946\frac{s_1^5}{120} + 4822\frac{s_1^3 s_2}{6} + 344\frac{s_1^2 s_3}{2} + 470\frac{s_1 s_2^2}{2} + 20 s_1 s_4 + 34 s_2 s_3 + s_5\right)\frac{x^5}{120}$$
$$+ O(x^6).$$

6.8. Theorem. *The tensor product of two convergent (at some non-zero point) rank one potentials is convergent.*

Proof. We start with the case when both theories are invertible. In view of 6.1.3 and 6.1.2 we may assume that $C_3^{(1)} = C_3^{(2)} = 1$. Hence we can apply Theorem 6.6. It shows that the series $U^{(1)}(\eta)$ and $U^{(2)}(\eta)$ converge somewhere, hence their product converges. But this product is inverse to the second derivative of the tensor product potential which therefore converges as well.

To treat the general case, put

$$C_n = P_n(C_3^{(1)},\dots,C_n^{(1)},C_3^{(2)},\dots,C_n^{(2)}),$$

where P_n are the universal polynomials. In view of Theorem 6.6, we can calculate these polynomials at $C_3^{(1)} = C_3^{(2)} = 1$. Again using 6.1.2 and 6.1.3 we will show that these restrictions can be uniquely extended to P_n because of grading constraints.

More precisely, define the *degree* of $C_{i_1}^{(j)}\dots C_{i_n}^{(j)}$ as $i_1 + \dots + i_n$ and the *length* as n. Monomials in both series of coefficients have bidegree and bilength. If P_n contains a monomial of bidegree $(k^{(1)}, k^{(2)})$ and bilength $(l^{(1)}, l^{(2)})$, then

$$(6.32) \qquad k^{(1)} - 2 l^{(1)} = k^{(2)} - 2 l^{(2)} = n - 2,\quad l^{(1)} + l^{(2)} = n - 1.$$

This can be checked either by using Theorem 6.6, or by directly applying the formula
$$C_n = (Y^{(1)} \otimes Y^{(2)})(\Delta_{\overline{M}_{0n}})(\partial^{\otimes n}).$$
In fact, the Künneth components of the diagonal $\overline{D}(\tau^{(1)}) \otimes \overline{D}(\tau^{(2)})$ produce monomials of bidegree $(|F_{\tau^{(1)}}|, |F_{\tau^{(2)}}|)$ and bilength $(|V_{\tau^{(1)}}|, |V_{\tau^{(2)}}|)$.

From (6.32) one sees that to reconstruct P_n from its restriction to $C_3^{(1)} = C_3^{(2)} = 1$, it suffices to multiply each monomial of this restriction by the appropriate monomial $C_3^{(1)i} C_3^{(2)j}$ which is unique.

Convergence follows from this formally.

6.9. Complete invertible Cohomological Field Theories. Keeping the notation of 6.1, we will now consider the structures of complete CohFT's on (H, Δ_0). Recall that by definition one such structure is given by the family of maps $I_{g,n} : H^{\otimes n} \to H^*(\overline{M}_{g,n}, k^*)^{S_n}$ defined for all stable pairs (g, n) and satisfying the axioms (4.7) and (4.8), extending (6.2). They form a commutative semigroup with respect to the tensor product $\{c'_{g,n}\} \otimes \{c''_{g,n}\} = \{c'_{g,n} \cup c''_{g,n}\}$. We will be mostly interested in the group of invertible theories.

The statements 6.1.1–6.1.3 readily generalize to this case. The theory $c_{g,n} = \mathbf{1}_{g,n}$ is the identity, the theories $c_{g,n}(t) = t^{2g-2+n} \mathbf{1}_{g,n}$ form a subgroup isomorphic to k^*, and invertibility is equivalent to $c_{0,3} \neq 0$. Moreover, any invertible theory is a tensor product of one of the type $\{c_{g,n}(t)\}$ and one with $c_{g,n}^{(0)} = 1$ for all g, n, and this decomposition is unique.

The statement 6.1.4 generalizes as well, however, we have no analog of the statement 6.1.5. We will reformulate the emerging difficulty as the problem of explicit description of the space L of "logarithmic" complete CohFTs, defined below.

6.10. The space L. By definition, $L \subset \prod_{g,n} H^*(\overline{M}_{g,n}, k)$ is the linear space formed by all families $l = \{l_{g,n} \in H^*(\overline{M}_{g,n}, k)\}$ satisfying the following conditions:

(i) $l_{g,n} \in H^{ev}(\overline{M}_{g,n}, k)^{S_n}$.

(ii) For every boundary morphism

(6.33) $\qquad b : \overline{M}_{g_1, n_1+1} \times \overline{M}_{g_2, n_2+1} \to \overline{M}_{g,n}, \; g = g_1 + g_2, n_1 + n_2 = n,$

we have

(6.34) $\qquad b^*(l_{g,n}) = l_{g_1, n_1+1} \otimes 1 + 1 \otimes l_{g_2, n_2+1}.$

(iii) For any boundary morphism

(6.35) $\qquad b' : \overline{M}_{g-1, n+2} \to \overline{M}_{g,n}$

we have

(6.36) $\qquad b'^*(l_{g,n}) = l_{g-1, n+2}.$

Here the boundary morphisms b (resp. b') glue together a pair of marked points situated on different connected components (resp. on the same one) of the appropriate universal curves.

The discussion above can be completed and summarized in the following proposition.

6.10.1. Proposition. *(a) L is graded by codimension in the following sense. Put $L^{(a)} = L \cap \prod_{g,n} H^{2a}(\overline{M}_{g,n}, K)$. Then $L = \prod_{a \geq 1} L^{(a)}$.*

(b) Let (H, h) be a one-dimensional vector space over k endowed with the metric h and an even basic vector Δ_0 such that $h(\Delta_0, \Delta_0) = 1$. Let P be the group of the invertible CohFT structures on (H, Δ_0) with respect to the tensor multiplication. Then the following map $K^* \times L \to P$ is a group isomorphism:

$$(6.37) \qquad (t, \{l_{g,n}\}) \mapsto \{c_{g,n} = t^{2g-2+n} \exp(l_{g,n})\}.$$

Notice that the statement on the grading means that if one puts $l_{g,n} = \sum_a l_{g,n}^{(a)}$, $l_{g,n}^{(a)} \in H^{2a}(\overline{M}_{g,n}, k)$, then $l \in L$ is equivalent to $l^{(a)} \in L$ for all a, where $l^{(a)} = \{l_{g,n}^{(a)}\}$.

6.10.2. Classes $\kappa(a)$ and $\mu(a)$. Put

$$(6.38) \qquad \kappa_{g,n}(a) = \pi_*(c_1(\omega_{g,n}(\sum_{i=1}^n x_i))^{a+1}) \in H^{2a}(\overline{M}_{g,n}, K)^{S_n},$$

where $\omega_{g,n}$ is the relative dualizing sheaf of the universal curve $\pi = \pi_{g,n} : C_{g,n} \to \overline{M}_{g,n}$, and x_i are the structure sections of π.

Furthermore, with the same notation put

$$(6.39) \qquad \mu_{g,n}(a) = \mathrm{ch}_a(\pi_* \omega_{g,n}).$$

Classes $\mu_{g,n}(a)$ vanish for all even a: see [Mu3].

We have $\kappa(a), \mu(a) \in L^{(a)}$. As we have already mentioned, for κ this was noticed in [AC1], and for μ in [KabKi]. For odd a, these elements are linearly independent, for example, because $\mu_{g,n}(a)$ are lifted from $\overline{M}_{g,0}$ for $g \geq 2$, or from $\overline{M}_{1,1}$, whereas $\kappa_{g,n}(a)$ are not lifted.

Problem. *Describe the whole space L.*

Probably, it is generated by κ and μ in small codimensions: this must not be too difficult to check, using the results of [AC2].

6.11. Three generating functions. Witten's *total free energy* of two-dimensional gravity ([W1]) is the formal series

$$(6.40) \qquad F(t_0, t_1, \dots) = \sum_{g=0}^{\infty} F_g(t_0, t_1, \dots) = \sum_{g=0}^{\infty} \sum_{\substack{|\mathbf{m}|=3g-3+n \\ \|\mathbf{m}\|=n}} \langle \tau_0^{m_0} \tau_1^{m_1} \dots \rangle \prod_{i=0}^{\infty} \frac{t_i^{m_i}}{m_i!}.$$

Here $\langle \tau_0^{m_0} \tau_1^{m_1} \dots \rangle$ is the intersection index generalizing (6.13) and defined as follows. Consider a partition $d_1 + \dots + d_n = 3g - 3 + n$ such that m_0 of the summands d_i are equal to 0, m_1 are equal to 1, etc., and put

$$(6.41) \qquad \langle \tau_0^{m_0} \tau_1^{m_1} \dots \rangle = \langle \tau_{d_1} \dots \tau_{d_n} \rangle = \int_{\overline{M}_{g,n}} \psi_1^{d_1} \dots \psi_n^{d_n},$$

where now $\psi_i = \psi_{g,n;i} = c_1(x_i^* \omega_{g,n})$.

The generating function for *higher Weil–Petersson volumes* was introduced in [KaMZ]. It extends (6.9) to all genera:
(6.42)
$$K(x, s_1, \dots) = \sum_{g=0}^{\infty} K_g(x, s_1, \dots) = \sum_{g=0}^{\infty} \sum_{|\mathbf{m}|=3g-3+n} \langle \kappa(1)^{m_1} \kappa(2)^{m_2} \dots \rangle \frac{x^n}{n!} \prod_{a=1}^{\infty} \frac{s_a^{m_a}}{m_a!}.$$

Here
(6.43)
$$\langle \kappa(1)^{m_1} \dots \kappa(3g-3+n)^{m_{3g-3+n}} \rangle = \int_{\overline{M}_{g,n}} \kappa_{g,n}(1)^{m_1} \dots \kappa_{g,n}(3g-3+n)^{m_{3g-3+n}}.$$

Actually, (6.42) is a specialization of the general notion of the *potential* of a complete CohFT $(H, h; \{I_{g,n}\})$ which is a formal series in coordinates of H and is defined as $\langle \exp(\gamma) \rangle$ where γ is the generic even element of H and the functional $\langle * \rangle$ as above is the integration over fundamental classes. Considering variable invertible CohFT's, we will get the potential as a formal function on $H \times K^* \times L$ which is a series in $(x, t, s_a, r_b, \dots)$. Here x stands for the coordinate on H dual to Δ_0, t for the coordinate on K^* as in (6.37), s_a (resp. r_b) are coordinates on L dual to $\kappa(a)$ (resp. $\mu(a)$), and dots stand for the unexplored part of L. More precisely, potential of the individual theory (6.37) is

$$\Phi(x, t) = \sum_g \Phi_g(x, t) = \sum_g \langle \exp(x\Delta_0) \rangle_g$$

$$= \sum_{g,n} \frac{\langle I_{g,n}(x^n \Delta_0^{\otimes n}) \rangle}{n!} = \sum_{g,n} \frac{x^n}{n!} \int_{\overline{M}_{g,n}} t^{2g-2+n} \exp(l_{g,n}).$$

Now make $l_{g,n}$ generic; that is, put

$$l_{g,n} = \sum_{a,b,\dots} (s_a \kappa_{g,n}(a) + r_b \mu_{g,n}(b) + \dots).$$

Collecting the terms of the right dimension, we finally get the series
(6.44)
$$\Phi(x, t, s_1, \dots, r_1, \dots) = \sum_g \Phi_g(x, t, s_1, \dots, r_1, \dots)$$

$$= \sum_{g,n} \frac{x^n}{n!} t^{2g-2+n} \sum \langle \kappa(1)^{m_1} \kappa(2)^{m_2} \dots \mu(1)^{p_1} \mu(2)^{p_2} \dots \rangle \prod_{a=1}^{\infty} \frac{s_a^{m_a}}{m_a!} \prod_{b=1}^{\infty} \frac{r_b^{p_b}}{p_b!} \dots,$$

where the inner summation is taken over $\mathbf{m}, \mathbf{p}, \dots$ with $|\mathbf{m}| + |\mathbf{p}| + \dots = 3g - 3 + n$.

Putting $t = 1, r_b = \dots = 0$ here, we obtain (6.42). The following theorem shows that by making an invertible change of variables in (6.40) restricted to $t_1 = 0$ we can again get (6.42). Thus (6.44) can be considered as a natural infinite-dimensional extension of (6.40).

Some information about the coefficients of (6.44) with non-vanishing $\mathbf{p}$ is obtained in [FabP]. If L is not generated by the κ- and μ-classes, the total family of these coefficients provide the natural generalization of Hodge integrals from [FabP].

§6. FORMAL FROBENIUS MANIFOLDS OF RANK ONE

6.12. Theorem. *For every* $g = 0, 1, \ldots$ *we have*

$$K_g(x, s_1, \ldots) = F_g(t_0, t_1, \ldots)|_{t_0=x, t_1=0, t_{k+1}=p_k(s_1,\ldots,s_k)}.$$

Here p_k *are Schur polynomials defined by*

$$1 - \exp\left(-\sum_{i=1}^{\infty} \lambda^i s_i\right) = \sum_{k=1}^{\infty} \lambda^k p_k(s_1, \ldots, s_k).$$

Proof. The proof given in [MaZo] is obtained by combining two identities. The explicit formula for the Schur polynomials is

$$p_k(s_1, \ldots, s_k) = \sum_{|\mathbf{m}|=k} \prod_{j=1}^{k} (-1)^{m_j+1} \frac{s_j^{m_j}}{m_j!},$$

and κ–numbers are expressed via τ–numbers by the extension of (6.28):

$$\frac{\langle \kappa(1)^{m_1} \ldots \kappa(3g-3+n)^{m_{3g-3+n}} \rangle}{m_1! \ldots m_{3g-3+n}!}$$

$$= \sum_{k=1}^{\|\mathbf{m}\|} \frac{(-1)^{\|\mathbf{m}\|-k}}{k!} \sum_{\mathbf{m}=\mathbf{m}^{(1)}+\cdots+\mathbf{m}^{(k)}} \frac{\langle \tau_0^n \tau_{\mathbf{m}^{(1)}+1} \ldots \tau_{\mathbf{m}^{(k)}+k} \rangle}{\mathbf{m}_1^{(1)}! \ldots \mathbf{m}_k^{(k)}!}.$$

Here $\mathbf{m}^{(i)} = (m_1^{(i)}, \ldots, m_{3g-3+n}^{(i)}) \neq 0$, $\mathbf{m}^{(i)}! = m_1^{(i)}! \ldots m_{3g-3+n}^{(i)}!$.

In fact, the total formalism of 6.4 holds for every genus. The only exception is the explicit formula (6.14).

6.12.1. Corollaries. It follows from this result that all known properties of the function F can be automatically translated to K. First, F satisfies the Virasoro constraints (see VI.2) and the higher KdV equations. The same is true for K in different coordinates.

Second, we can express each K_g with $g \geq 1$ in terms of K_0 because this was performed in [IZu] for F. Let us recall the genus expansion of F obtained in [IZu]. Put $u_0 = F_0''$ where a prime denotes the derivative with respect to t_0, and define a sequence of formal functions $I_1, I_2, \ldots$ by

$$I_1 = 1 - \frac{1}{u_0'}, \quad I_{k+1} = \frac{I_k'}{u_0'}, \quad k = 1, 2, \ldots.$$

Then $F_1 = \frac{1}{24} \log u_0'$, and for $g \geq 2$

$$(6.45) \quad F_g = \sum_{\sum(i-1)m_i = 3g-3} \langle \tau_2^{m_2} \tau_3^{m_3} \ldots \tau_{3g-2}^{m_{3g-2}} \rangle u_0'^{2g-2+\|\mathbf{m}\|} \prod_{k=2}^{3g-2} \frac{I_k^{m_k}}{m_k!}.$$

Substituting $t_0 = x, t_1 = 0$ and $t_{i+1} = p_i(s_1, \ldots, s_i), i \geq 1$, we get the identical formula for K_g.

6.13. The path semigroup in quantum cohomology. The definition of L reminds us of the definition of the space of primitive elements of a coalgebra (or bialgebra). In fact, there exists a closely related problem in the context of Gromov–Witten invariants, where such a bialgebra appears naturally.

The first term of any operad is a groupoid. Since H_*M is the operad of coalgebras (with comultiplication induced by the diagonal; cf. the next chapter), we get a bialgebra $B_* = \bigoplus_{g\geq 1} H_*(\overline{M}_{g,2})$ and the dual bialgebra $B^* = \prod_{g\geq 1} H^*(\overline{M}_{g,2})$. More precisely, denote here by b the family of boundary morphisms

$$b_{g_1,g_2}: \overline{M}_{g_1,2} \times \overline{M}_{g_1,2} \to \overline{M}_{g_1+g_2,2}$$

gluing the second point of the first curve to the first point of the second curve. It induces a multiplication on the homology coalgebra and a comultiplication on the cohomology algebra

$$b_*: B_* \otimes B_* \to B_*, \quad b^*: B^* \to B^* \otimes B^*$$

making each space a bialgebra. Moreover, renumbering the structure sections $x_1 \leftrightarrow x_2$ we get an involution s of $\overline{M}_{g,2}$ inducing an involution of B_*, resp. B^*. The latter is an automorphism of the comultiplication on homology, resp. multiplication on cohomology, but an antiautomorphism of the remaining two structures, because from the definition one sees that

$$b \circ \sigma_{1,2} = s \circ b \circ (s \times s),$$

where $\sigma_{1,2}$ is the permutation of factors.

6.13.1. Proposition. *Let V be a smooth projective manifold. Then the family of Gromov-Witten correspondences $I_{g,2} \in A_*(V^2 \times \overline{M}_{g,2})$ defines the action of the algebra B_* and coaction of the coalgebra B^* on $H^*(V)$.*

This is part of the general statement that $H^*(V)$ is a cyclic algebra over the modular operad H_*M.

The geometry of this (co)action shows that morally B_* plays the role of the groupoid of paths on V: a stable map of a curve with two marked points to V should be considered as a complex path in V from one marked point to another. This agrees with the composition by gluing the endpoint of one path to the starting point of another, and the smoothing of the resulting singularity plays the role of homotopy. Notice that there may well exist non-constant stable maps of genus zero with only two marked points, whereas there are no stable curves with this property. Nevertheless, one may hope that B_* (and possibly B^*) can be extended by the genus zero component, perhaps by taking an appropriate (co)homology of the respective Artin stack $\mathcal{M}_{0,2}$ which might be close to the (co)homology of BG_m.

It would be highly desirable to introduce a quantum fundamental group of V, with appropriate action of our path groupoid. In any case, we are interested in the primitive elements h of B^* satisfying $b^*(h) = h \otimes 1 + 1 \otimes h$. Clearly, restrictng L to its $n=2$ part we get a supply of such elements. Their full classification may be easier than that of L.

§7. Tensor product of analytic Frobenius manifolds

7.1. Notation. Assume that we are given two germs of pointed Frobenius manifolds with split flat identities in the complex analytic category. Taking their completions at the base points, we can form their tensor product which a priori will be only formal. In this section we show that the resulting formal potential is actually convergent, admits the flat identity, and moreover, shifting the initial base points leads to the canonically isomorphic tensor products.

§7. TENSOR PRODUCT OF ANALYTIC FROBENIUS MANIFOLDS

We start with the easier case of semisimple manifolds, where these statements can be deduced from the explicit expressions for the special initial conditions of the tensor product. These formulas have an independent interest because they are very simple and do not deal with infinitely many quantities as in the flat coordinates description. Let us recall the relevant definitions from Chapter II, §3.

Consider a pointed germ M of the analytic Frobenius manifold over $\mathbf{C}$ (or a formal manifold with zero as the base point) as usual, with flat e, Euler E and $d_0 = 1$, and having pure even dimension. It will be called *tame semisimple* if the operator $E\circ$ has simple spectrum $(u_0^1, \ldots, u_0^n)$ on the tangent space to the base point. We have the following general facts:

a) In a neighborhood of the base point, eigenvalues $(u^1, \ldots, u^n)$ of $E\circ$ on $\mathcal{T}_M$ form a local coordinate system (Dubrovin's canonical coordinates), taking the values $(u_0^1, \ldots, u_0^n)$ at the base point. The potential Φ is an analytic function of these coordinates. If the initial manifold was only assumed to be formal, from tame semisimplicity it follows that it is in fact the completion of a pointed analytic germ.

b) Put $e_i = \partial/\partial u^i$. Then $e_i \circ e_j = \delta_{ij}$. In particular, $e = \sum_i e_i$. It follows that the $\circ$ multiplication on the tangent spaces is semisimple.

c) We have $g(e_i, e_j) = 0$ for $i \neq j$. Furthermore, there exists a function η defined up to addition of a constant such that $g(e_i, e_i) = e_i \eta := \eta_i$. Moreover, we have $eg = $ const, $Eg = (D-1)\eta + $ const. Finally, $E = \sum_i u^i e_i$.

A very important feature of canonical coordinates is that a given tame semisimple germ can be uniquely extended to the Frobenius structure on the universal covering of the total (u^i)–space with deleted partial diagonals. This follows from the Painlevé property of the solutions of Schlesinger's equations: cf. Ch. II, §§1–3. We call this extension *the maximal tame continuation* of the initial germ. The qualification "tame" is essential. It may well happen that a further extension containing non–tame semisimple points or even points with non–semisimple multiplication on the tangent space is possible: e.g., in the example of I.4.5 points where $p'(z)$ has multiple roots have the latter property.

7.1.1. Definition. *Special coordinates of the tame semisimple pointed germ of the Frobenius manifold consist of the values at the base point of the following functions:*

$$(7.1) \qquad (u^i, \eta_j, v_{ij} := \frac{1}{2}(u^j - u^i)\frac{\eta_{ij}}{\eta_j}).$$

Here $\eta_{ij} := e_i e_j \eta$.

To avoid any misunderstanding, let us stress that the canonical coordinates are *functions on a germ*, whereas special coordinates are *functions on the moduli space of germs*.

For a description of the necessary and generically sufficient conditions for a system of numbers to form special coordinates of a Frobenius germ, see II.3.4.3 and II.3.1.3. The following theorem summarizes the properties of special coordinates that we will use. Its first part was proved in Chapter II.

7.2. Theorem. *(i) Any tame semisimple pointed germ with labeled spectrum of $E\circ$ is uniquely (up to isomorphism) defined by its special coordinates.*

(ii) Let (u'^i, η'_j, v'_{ij}) for $i \in S$ and $(u'''^i, \eta''_j, v''_{ij})$ for $j \in T$ be special coordinates of two pointed germs. If the family $(u'^i + u'''^j)$ consists of pairwise distinct elements, then the tensor product of the two germs defined through their completions is again a tame semisimple pointed germ whose canonical coordinates are naturally labeled by the pairs $I \in S \times T$ and have the following form: for $I = (i, j)$, $K = (k, l)$,

(7.2) $$u^I = u'^i + u'''^j, \quad \eta_I = \eta'_i \eta''_j, \quad v_{IK} = \delta_{jl} v'_{ik} + \delta_{ik} v''_{jl}.$$

Proof. Let $(M', m'_0), (M'', m''_0)$ be two pointed germs of semisimple Frobenius manifolds as above, (M, m_0) their formal tensor product. Let $(x'^{a'})$, $(x''^{a''})$ be flat coordinates on them vanishing at the base points, and e'_i, e''_j the idempotent vector fields. Denote by e'^0_i, e''^0_j flat vector fields coinciding with the respective idempotents at base points. Define the transition coefficients λ by

$$\partial'_{a'} = \sum \lambda^i_{a'} e'^0_i, \quad \partial''_{a''} = \sum \lambda^j_{a''} e''^0_j.$$

Denote by $x^{a'a''}$ the tensor product flat coordinates.

At the base point of M, the tangent space is the tensor product of the respective tangent spaces of the factors as metric space and as Frobenius algebra. The second equality in (7.2) follows from this. It follows also that there exist unique idempotents e_{ij} restricting to $e'^0_i \otimes e''^0_j$.

From (4.24) one sees that at this tangent space the tensor product Euler field acts as $(E' \otimes e'' + e' \otimes E'') \circ$. Therefore $e'^0_i \otimes e''^0_j$ are eigenvectors with eigenvalues $u'^i + u'''^j$. This shows the first formula in (7.2).

The last formula can be established by the following calculations modulo second order terms in flat coordinates.

Put
$$e'_i = e'^0_i + \sum x'^{a'} e_{ia'} + O(x'^2)$$
and similarly for the second factor. Then we have

(7.3) $$e_{ij}(x) = e'^0_i \otimes e''^0_j + \sum_{a',a''} x^{a'a''} (\lambda^j_{a''} e'_{ia'} \otimes e''^0_j + \lambda^i_{a'} e'^0_i \otimes e''_{ja''}) + O(x^2).$$

This is checked by the direct calculation which we omit.

Expanding the metrics in flat coordinates we get

$$\eta'_i(x') = \eta'_i(m'_0) + 2 \sum x'^{a'} g(e'_{ia'}, e'^0) + O(x'^2),$$

$$\eta''_j(x'') = \eta''_j(m''_0) + 2 \sum x''^{a''} g(e''_{ia''}, e''^0) + O(x''^2).$$

Now using (7.3) we obtain

$$\eta_{ij}(x) = g(e_{ij}, e_{ij}) = g(e'^0_i \otimes e''^0_j, e'^0_i \otimes e''^0_j)$$
$$+ 2 \sum_{a',a''} x^{a'a''} g(\lambda^j_{a''} e'_{ia'} \otimes e''^0_j + \lambda^i_{a'} e'^0_i \otimes e''_{ja''}, e'^0_i \otimes e''^0_j) + O(x^2)$$
$$= \eta'_i(m'_0) \eta''_j(m''_0)$$
$$+ \sum_{a',a''} x^{a'a''} (\lambda^j_{a''} (\partial_{a'} \eta'_i)(m'_0) \eta''_j(m''_0) + \lambda^i_{a'} \eta'_i(m'_0) (\partial''_{a''} \eta''_j)(m''_0)) + O(x^2).$$

§7. TENSOR PRODUCT OF ANALYTIC FROBENIUS MANIFOLDS

It remains to take the derivatives of this with respect to e_{kl} which is first expressed via flat vector fields by inverting the matrices λ. We obtain

$$\eta_{ij,kl}(0) = \delta_{j,l}\eta'_{ik}(m''_0)\eta''_j(m''_0) + \delta_{i,k}\eta'_i(m''_0)\eta''_{jl}(m''_0)$$

and then apply this to the calculation of $v_{ij,kl}$ using (7.1).

In the remaining part of this section we drop the semisimplicity condition.

7.3. Theorem. *The tensor product of two convergent potentials is convergent.*

Proof. Assume first that there are no odd coordinates. We will prove that the tensor product potential Φ converges at some diagonal point: $x^{a'a''} = x \neq 0$ for all a', a''. In fact, at such a point it can be written as

$$(7.4) \qquad \sum_{n=1}^{\infty} \frac{1}{n!} x^n \left(Y^{(1)} \otimes Y^{(2)}\right)(\Delta_{\overline{M}_{0n}})((\partial^{(1)} \otimes \partial^{(2)})^{\otimes n})$$

with

$$\partial^{(1)} = \sum_{a'} \partial_{a'}, \quad \partial^{(2)} = \sum_{a''} \partial_{a''}.$$

If we choose flat coordinates vanishing at the origin in such a way that $|g^{a'a''}| \leq 1$ for both theories, then (7.4) is dominated by the tensor product potential of two rank one theories given by coordinates

$$C_n^{(1)} = |Y_n^{(1)}(\partial^{(1)\otimes n})|, \quad C_n^{(2)} = |Y_n^{(2)}(\partial^{(2)\otimes n})|$$

studied in §6. Its convergence is established by Theorem 6.8.

In the general case, convergence of a power series in even and odd coordinates is equivalent to the convergence of its reduced derivatives with respect to the odd coordinates (where the reduced series is obtained by putting all odd coordinates equal to zero). A slight modification of the above reasoning can be used to treat this case. We omit the details; but see [Ka2].

7.4. Theorem. *The tensor product of two germs of pointed analytic Frobenius manifolds with flat identities does not depend on the choice of the base points in the convergence domain of the potentials and their tensor product.*

Proof. To be more precise, let $\Phi^{(i)}, i = 1, 2$, be the potentials of two analytic Frobenius manifolds $M^{(i)}$ with flat coordinates $(x^{a'}), (x^{a''})$ vanishing at the initial base points as in 4.10. We may identify $M^{(i)}$ with some neighborhoods of zero of the tangent spaces at the respective base points. Let $\Phi = \Phi^{(1)} \otimes \Phi^{(2)}$ be the tensor product potential.

Let $p = (x_0^{a'}), q = (x_0^{a''})$ be the shifted base points in the convergence domains of $\Phi^{(1)}, \Phi^{(2)}$. They determine the shifted potentials $\Phi_p^{(1)}, \Phi_q^{(2)}$ which by definition are the Taylor formal series of $\Phi^{(1)}, \Phi^{(2)}$ in $(x^{a'} - x_0^{a'})$ and $(x^{a''} - x_0^{a''})$ respectively. We can form the tensor product of shifted potentials $\Phi_p^{(1)} \otimes \Phi_q^{(2)}$ which is a local analytic function on the tensor product of the tangent spaces at the initial base points.

On the other hand, consider the point $\tau(p, q)$ in this tensor product having coordinates

$$(7.5) \qquad x_0^{a'b''} := \delta_{0b''} x_0^{a'} + \delta_{a'0} x_0^{b''}.$$

We stress that in this formula, as elsewhere, the superscripts $a' = 0$, resp. $a'' = 0$, refer to the local flat coordinates such as described in I.2.1.3. The additive constants are here normalized by the condition that these coordinates vanish at the initial base points. The intrinsic description of the point (p, q) involves shifts in the directions of flat identity vector fields.

Denote by $\Phi_{pq} = (\Phi^{(1)} \otimes \Phi^{(2)})_{\tau(p,q)}$ the Taylor series of $\Phi^{(1)} \otimes \Phi^{(2)}$ at $\tau(p, q)$ and assume that it converges at this point.

The precise meaning of the theorem is

(7.6) $$\Phi_{pq} = \Phi_p^{(1)} \otimes \Phi_q^{(2)}$$

up to terms of order ≤ 2.

In order to prove (7.6) we calculate the action of shift on the tree correlation functions of the theories. In Lemma 7.4.1, $Y_n, Y(\tau)$, resp. $Y_{n,p}, Y_p(\tau)$, denote the correlation functions corresponding to some convergent potential Φ, resp. its shifted version Φ_p, flat coordinates are denoted x^a, and p is (x_0^a).

7.4.1. Lemma. *For any stable n-tree τ we have*

(7.7)
$$Y_p(\tau)(\bigotimes_{i=1}^{n} \partial_{a_i}) = \sum_{N \geq 0} \frac{1}{N!} \sum_{b_1,\ldots,b_N} \epsilon(b|a) x_0^{b_N} \ldots x_0^{b_1} Y(\pi_N^*(\tau))(\bigotimes_{i=1}^{n} \partial_{a_i} \otimes \bigotimes_{j=1}^{N} \partial_{b_j}).$$

Here $\epsilon(b|a)$ is the sign defined by

$$\bigotimes_{i=1}^{n} \partial_{a_i} \otimes \bigotimes_{j=1}^{N} \partial_{b_j} = \epsilon(b|a) \bigotimes_{j=1}^{N} \partial_{b_j} \otimes \bigotimes_{i=1}^{n} \partial_{a_i},$$

and $\pi_N^*(\tau)$ is the sum of trees obtained from τ by attaching tails $\{n+1, \ldots, n+N\}$ distributed among vertices in all possible ways.

Proof. We have

$$Y_{n,p}(\bigotimes_{i=1}^{n} \partial_{a_i}) = (\partial_{a_1} \ldots \partial_{a_n} \Phi)(p)$$

$$= \sum_{N \geq 0} \frac{1}{N!} \sum_{b_1,\ldots,b_N} x_0^{b_N} \ldots x_0^{b_1} Y_{n+N}(\bigotimes_{j=1}^{N} \partial_{b_j} \otimes \bigotimes_{i=1}^{n} \partial_{a_i})$$

(7.8) $$= \sum_{N \geq 0} \frac{1}{N!} \sum_{b_1,\ldots,b_N} \epsilon(b|a) x_0^{b_N} \ldots x_0^{b_1} Y_{n+N}(\bigotimes_{i=1}^{n} \partial_{a_i} \otimes \bigotimes_{j=1}^{N} \partial_{b_j}).$$

Clearly, this is (7.7) for one–vertex trees. Inserting these formulas into the rhs of (4.18) and regrouping terms, we get the general identity.

As above, π_N^* is the combinatorial counterpart of the pullback with respect to the forgetful map. In order to use this remark, we will need the following generalization of the second identity (4.34). Let S, T be two disjoint subsets of

$\{1, \ldots, n\}$ with $n - |S| - |T| \geq 3$. We have the following commutative diagram of forgetful maps:

$$\begin{array}{ccc} \overline{M}_{0n} & \xrightarrow{\pi_T} & \overline{M}_{0,n\setminus T} \\ \pi_S \downarrow & & \pi_S \downarrow \\ \overline{M}_{0,n\setminus S} & \xrightarrow{\pi_T} & \overline{M}_{0,n\setminus S\cup T} \end{array}$$

(Notice that the notation for the forgetful maps used here does not show the source.)

Then (4.34) and induction show that

(7.9) $\qquad (\pi_S, \pi_T)^*(\Delta_{\overline{M}_{0,n\setminus S\cup T}}) = (\pi_T, \pi_S)_*(\Delta_{\overline{M}_{0n}}).$

7.4.2. End of proof of Theorem 7.4. We now return to the notation explained right after the statement of the theorem and apply Lemma 7.4.1 to the three series of correlation functions involved in (7.6).

In view of (7.8), (7.5), and (4.21) we have

$$Y_{n,pq}(\bigotimes_{i=1}^{n} \partial_{a'_i a''_i}) = \sum_{N \geq 0} \frac{1}{N!} \sum_{l=0}^{N} \sum_{(b'_1, \ldots, b'_l, b''_1, \ldots, b''_{N-l})} \binom{N}{l} \epsilon(b'0|a'a'')\epsilon(0b''|a'a'')$$

$$\times \prod_{i=N-l}^{1} x_0^{(2)b''_i} \prod_{j=l}^{1} x_0^{(1)b'_j} (Y_p^{(1)} \otimes Y_q^{(2)})(\Delta_{\overline{M}_{0,n+N}})(\bigotimes_{i=1}^{n} \partial_{a'_i a''_i} \otimes \bigotimes_{j=1}^{l} \partial_{b'_j 0} \otimes \bigotimes_{k=1}^{N-l} \partial_{0 b''_k})$$

$$= \sum_{N \geq 0} \frac{1}{N!} \sum_{l=0}^{N} \sum_{(b'_1, \ldots, b'_l, b''_1, \ldots, b''_{N-l})} \binom{N}{l} \epsilon(b'|a')\epsilon(b''|a'')\epsilon(b'|a'')\epsilon(b''|a')$$

$$\times \prod_{i=N-l}^{1} x_0^{(2)b''_i} \prod_{j=l}^{1} x_0^{(1)b'_j} (Y_p^{(1)} \otimes Y_q^{(2)})((\pi_S, \pi_T)_*\Delta_{\overline{M}_{0,n+N}})(\bigotimes_{i=1}^{n} \partial_{a'_i a''_i} \otimes \bigotimes_{j=1}^{l} \partial_{b'_j} \otimes \bigotimes_{k=1}^{N-l} \partial_{b''_k}),$$

where $S = \{n+l+1, \ldots, n+N\}, T = \{n+1, \ldots, n+l\}$. The last equality follows from (4.33).

On the other hand, the respective value of the correlation function of $\Phi_p \otimes \Phi_q$ is

$$(Y_p^{(1)} \otimes Y_q^{(2)})(\Delta_{\overline{M}_{0,n}})(\bigotimes_{i=1}^{n} \partial_{a'_i a''_i})$$

$$= \sum_{N \geq 0} \sum_{l=0}^{N} \sum_{(b'_1, \ldots, b'_l, b''_1, \ldots, b''_{N-l})} \frac{1}{N!(N-l)!} \epsilon(b'|a')\epsilon(b''|a'')\epsilon(b'|a'')\epsilon(b''|a')$$

$$\times \prod_{i=N-l}^{1} x_0^{(2)b''_i} \prod_{j=l}^{1} x_0^{(1)b'_j} (Y_p^{(1)} \otimes Y_q^{(2)})((\pi_T, \pi_S)^*\Delta_{\overline{M}_{0n}})(\bigotimes_{i=1}^{n} \partial_{a'_i a''_i} \otimes \bigotimes_{j=1}^{l} \partial_{b'_j} \otimes \bigotimes_{k=1}^{N-l} \partial_{b''_k}).$$

It remains to apply (7.9).

7.5. The global tensor product. Theorem 7.4 shows the local independence of the tensor product on the choice of base points. Assuming that factors are global Frobenius manifolds admitting globally split identities (see 7.5.8 below), we will prove here existence and uniqueness of the global tensor product, following R. Kaufmann.

The key step in globalizing the results of the last section is to replace any analytic Frobenius manifold by the collection of its analytic germs in such a way that the union of these germs constitutes a tubular neighborhood of the zero section of the tangent bundle. More precisely, at each point the analytic pointed Frobenius germ defines an analytic Frobenius manifold in a neighborhood of zero on each fiber of the tangent bundle. If two fibers of the tangent bundle are sufficiently close to each other, the germs of the analytic Frobenius manifolds near the zero section are identified by the canonical flat affine connection on the tangent bundle associated to the flat structure. This is a version of the exponential map.

Now given two analytic Frobenius manifolds, the germs of all possible tensor products at variable base points can likewise be realized as a tubular neighborhood in the exterior tensor product of the two tangent bundles over the Cartesian product of the two manifolds.

The last step is to identify the above structure with the (pullback of) a structure coming from a Frobenius manifold. The idea is to again use a sort of exponential map in order to construct a Frobenius manifold. To obtain the necessary gluing data one introduces an additional structure, that of a flat affine connection on the exterior product of the two tangent bundles which identifies germs. Note that this bundle a priori only carries a linear connection coming from the flat structures of the factors, but has no canonical affine connection which would allow for gluing the various patches along the base. The affine connection is required to extend the linear one. The role of the linear connection is to transport the flat coordinates whereas the affine extension transports the base points.

In the case of Frobenius manifolds with flat identities such an affine connection is given locally by the formula (7.5) and these local expressions indeed give an affine connection. In order to use this affine connection as gluing data the assumption of split identities is used, since this allows us to take a global quotient of the Cartesian product and thus to avoid potential difficulties coming from identifications induced by parallel displacement with respect to the affine connection. These arise since the associated linear map (7.5) to the affine connection has a kernel.

7.5.1. Notation. For any Frobenius manifold M we denote by Φ^M its potential and for any $m \in M$ after fixing a local basis (∂_a) of T_M^f we write Φ_m^M for the power series expansion of Φ^M in the coordinates x^a dual to the (∂_a) and centered at m. Furthermore, again after fixing a local basis we denote the convergent germ around a point m given by the power series Φ_m^M by (M, m). This germ can naturally be regarded as living in the infinitesimal neighborhood of $0 \in T_m M$, and it defines a structure of Frobenius manifold on the domain of convergence $U_m \subset T_m M$ of Φ_m^M.

7.5.2. Connections and forms. Let $k = \mathbf{R}$ or $\mathbf{C}$ and E be a vector bundle over a manifold M with typical fiber k^n. Denote the group of affine transformations of k^n by A_n. We have the split sequence $0 \to k^n \to A_n \to GL_n \to 1$. Denote by γ be the standard splitting. Consider the pullback of an affine connection $\tilde{\omega}$ on $\tilde{E}$ wrt γ where $\tilde{E}$ is the bundle with affine structure group and typical fiber k^n induced by β. Then the decomposition coming from the split exact sequence of Lie algebras provides the one–to–one correspondence between

(i) {affine connections $\tilde{\omega}$ on $\tilde{E}$};

§7. TENSOR PRODUCT OF ANALYTIC FROBENIUS MANIFOLDS   141

(ii) {pairs (ω, ϕ) where ω is a linear connection form on E and ϕ is a tensorial one–form on E of type (GL_n, k^n)}.

Furthermore, vector–valued 1–forms of type (GL_n, k^n) can be identified with maps $T_x M \to k^n$, $x \in M$.

We say that $\tilde{\omega}$ is compatible with the linear connection ω, and call ϕ the affine part of $\tilde{\omega}$.

On the tangent bundle of any manifold there exists a canonical affine connection given by the canonical one–form θ_{can}, which corresponds to the identity transformation on each $T_x M$.

Notice that the parallel displacement of affine frames splits into the parallel displacement of the frames and the parallel displacement of the base points of these frames. In the case of compatible connections the parallel transport of the frames is governed by the given linear connection and the path of (parallelly displaced) base–points of these frames is given by the affine part.

7.5.3. Remark. Let N be a Frobenius manifold. Consider a tubular neighborhood of the zero section $V \subset TN$ whose restriction V_n to the fiber over any $n \in N$ is inside the domain of convergence of Φ_n^N:

Notice that V is a bundle of Frobenius manifolds. It encodes all germs (N, n) of the Frobenius manifold N: $(N, n) = (V_n, 0)$.

TN has the following canonical structures.

(i) The canonical flat connection which is part of the data for Frobenius manifolds.

(ii) The canonical affine flat connection obtained by extending the Levi–Civita connection from above via the canonical one–form θ_{can}.

The affine connection locally identifies germs of Frobenius manifolds at nearby fibers. More precisely, let x_t be a path in N and $\tilde{x}_t$ be the path of base–points of the path lifted wrt the affine connection satisfying $\tilde{x}_0 = 0$. Then locally

$$(N, x_0) = (V_{x_0}, 0) \cong (V_{x_t}, \tilde{x}_t)$$

where the last isomorphism is given by choosing any local basis of flat fields (∂_a) around x_0 which then defines the flat coordinates on $T_{x_t} N$ centered at $\tilde{x}_t$ by parallel translation. Due to the compatibility of the affine and the linear connection and their flatness, this just amounts to shifting the coordinates in the power series expression of Φ_{x_t} by the coordinates of x_0 in the coordinates centered at x_t. This automatically yields the series Φ_{x_0}.

7.5.4. Remark. Over the Cartesian product of two Frobenius manifolds $M^{(1)}$ and $M^{(2)}$ we have a similar picture:

$$V \hookrightarrow TM^{(1)} \boxtimes TM^{(2)}$$
$$\downarrow$$
$$M^{(1)} \times M^{(2)}$$

(with section s)

Here V is again some neighborhood of the zero section whose fiber over a point (p,q) is this time required to lie inside the domain of convergence of $\Phi_{pq} := \Phi_p \otimes \Phi_q$.

This is again a bundle of Frobenius manifolds and it encodes all possible tensor products between germs from $M^{(1)}$ and $M^{(2)}$,

$$(M^{(1)}, p) \otimes (M^{(2)}, q) = (V|_{(p,q)}, 0).$$

Notice that this diagram cannot serve as a tensor product by itself, since the base has the wrong dimension $\dim M^{(1)} + \dim M^{(2)}$ instead of $\dim M^{(1)} \cdot \dim M^{(2)}$.

Actually it must be made a part of a larger diagram defining the tensor product which includes a third Frobenius manifold, the tensor product itself, and a map to it from the Cartesian product, such that the pull-back of the tangent bundle of the tensor product manifold together with its structure of the bundle of Frobenius manifolds is isomorphic to the above bundle.

On $TM^{(1)} \boxtimes TM^{(2)}$ there is again a canonical flat connection given by the subspace of flat vector fields:

$$(TM^{(1)} \boxtimes TM^{(2)})^f = T^f M^{(1)} \otimes T^f M^{(2)}.$$

However, there is no canonically associated affine connection, since there is no canonical one-form. This leads us to introduce the following extra datum. Its existence is a necessary condition for the existence of the global tensor product.

7.5.5. Definition. *An affine connection on the exterior product bundle over the Cartesian product of two Frobenius manifolds $M^{(1)}$ and $M^{(2)}$ is called the tensor product connection if it satisfies the following conditions:*

(i) It is compatible with the natural linear connection mentioned above.
(ii) It is flat.
(iii) It locally identifies the germs of Frobenius manifolds of V.

Condition (ii) can be rephrased in the following way. Let $\theta_\tau : TM^{(1)} \boxplus TM^{(2)} \to TM^{(1)} \boxtimes TM^{(2)}$ be the map corresponding to the one-form defined by the affine connection. Then θ_τ induces a linear map between the spaces of flat fields:

$$\theta_\tau^f : T^f M^{(1)} \oplus T^f M^{(2)} \to T^f M^{(1)} \otimes T^f M^{(2)}.$$

Here is a more detailed explanation of condition (iii). Let x_t be a path in $M^{(1)} \times M^{(2)}$, and let $\tilde{x}_t$ be the path of base-points of the lifted path wrt the affine connection satisfying $\tilde{x}_0 = 0$. Then locally

$$(TM^{(1)} \boxtimes TM^{(2)}|_{(p,q)}, 0) \stackrel{\theta_\tau}{\cong} (TM^{(1)} \boxtimes TM^{(2)}|_{x_t}, \tilde{x}_t),$$

where the isomorphism is to be understood in the same way as in Remark 7.5.4 above. Notice that the flatness of the affine connection guarantees the triviality of monodromy.

In the present context, Theorem 7.4 implies the following.

7.5.6. Proposition. *If the Frobenius manifolds $M^{(1)}$ and $M^{(2)}$ both carry flat identities, then the affine connection defined by the one-form corresponding to the maps locally given by:*

$$\theta_\tau^U : TM^{(1)} \boxplus TM^{(2)}|_U \to TM^{(1)} \boxtimes TM^{(2)}|_U,$$

$$\theta_\tau(\partial_a^{(1)}) = \partial_{a0}, \quad \theta_\tau(\partial_b^{(2)}) = \partial_{0b}$$

is a tensor product connection.

Here $(\partial_a^{(1)}, \partial_b^{(2)})$ is a local flat basis of $TM^{(1)} \boxplus TM^{(2)}$ and $(\partial_{ab} = \partial_a^{(1)} \otimes \partial_b^{(2)})$ is the corresponding flat local tensor basis of $TM^{(1)} \boxtimes TM^{(2)}$.

The existence of tensor product connections is not obvious or guaranteed in the general case. In the above situation it exists due to the presence of flat identities. That this requirement is essential shows the following fact proved in [Ka2]:

7.5.7. Claim. *The Cartesian product of two one-dimensional Frobenius manifolds admits a tensor product connection iff one of the factors has constant multiplication or equivalently a flat identity.*

In order to ensure the existence and uniqueness (up to equivalence defined below) of the tensor product, we will impose a global restriction on the factors. The restriction is not the most general possible, but it is easy to check and is not too severe. In fact, it is met in all of our main examples such as quantum cohomology, unfolding of singularities, Frobenius manifolds arising from dGBV algebras, and extensions of tame semisimple germs onto **B**, the universal cover of $\mathbf{C}^n \setminus \{\text{partial diagonals}\}$.

7.5.8. Definition. *Let M be a Frobenius manifold with flat identity. The structure of globally split identity is given by the decomposition $M = \bar{M} \times \mathbf{A}^1$ which locally induces the structure of split identity in the sense of I.2.1.3.*

7.5.9. Tensor product diagrams. All the structures involved in the definition of the global tensor product are most conveniently grouped into commutative diagrams. Let $\langle M^{(1)}, T^f M^{(1)}, g_{M^{(1)}}, \Phi^{M^{(1)}} \rangle$, $\langle M^{(2)}, T^f M^{(2)}, g_{M^{(2)}}, \Phi^{M^{(2)}} \rangle$ and $\langle N, T^f N, g_N, \Phi^N \rangle$ be Frobenius manifolds with globally split flat identities. Consider the diagram:

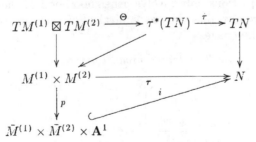

Here i is an embedding of affine flat manifolds; p is the map which is given in some fixed choice of coordinates on the $\mathbf{A}^1$-factors by $p(m_1, x, m_2, y) = (m_1, m_2, x + y)$; τ is an affine map, which factors through p: $\tau = i \circ p$; and Θ is an isomorphism of metric bundles with affine flat structure between the pulled back tangent bundle of N and the exterior product bundle over $M^{(1)} \times M^{(2)}$. In other words, Θ is an isomorphism of metric bundles and $\tau^* T^f N = \Theta(T^f M^{(1)} \otimes T^f M^{(2)})$.

7.5.10. Definition. *We will call such a diagram a tensor product diagram and N a tensor product manifold for $M^{(1)}$ and $M^{(2)}$ if τ additionally preserves the structure of bundles of Frobenius manifolds, that is, if it satisfies the following conditions:*

(i) At all points in the image of τ, $\hat{\tau} \circ \Theta$ gives an isomorphism of pointed germs of Frobenius manifolds:

$$(T_p M^{(1)} \otimes T_q M^{(2)}, 0) \stackrel{\hat{\tau} \circ \Theta}{\cong} (T_{\tau((p,q))} N, 0) = (N, \tau((p,q))).$$

(ii) The affine connection defined on $TM^{(1)} \boxtimes TM^{(2)}$ by the pullback of the canonical affine connection on TN (defined by the flat structure on TN and the canonical one-form θ_{can}) is the tensor product connection θ_τ of Proposition 7.5.6. In other words,

$$\tau^*(\theta_{can}) = \Theta \circ \theta_\tau.$$

The isomorphism in (i) is to be understood as in Remark 7.5.4: for all points $(p, q) \in M^{(1)} \times M^{(2)}$:

$$\left(\bigoplus \mathbf{C} \, \partial_a \otimes \partial_b, g_{M^{(1)}} \otimes g_{M^{(2)}}, \Phi_p^{M^{(1)}} \otimes \Phi_q^{M^{(2)}} \right)$$
$$\stackrel{\hat{\tau} \circ \Theta}{\cong} \left(\bigoplus \mathbf{C} \, (\hat{\tau} \circ \Theta(\partial_a \otimes \partial_b)), g_N, \Phi_{\tau((p,q))}^N \right),$$

where $\hat{\tau} \circ \Theta(\partial_a \otimes \partial_b) \in T_{\tau((p,q)),N}$.

Since the map τ is affine, the image of $M^{(1)} \times M^{(2)}$ in N is a flat submanifold. The condition (i) then says that this submanifold indeed parametrizes all tensor product germs. Notice that the dimension of this submanifold is $\dim M^{(1)} + \dim M^{(2)} - 1$. The lower part of the diagram is introduced in order to rigidify the situation globally and thus to ensure the existence and uniqueness up to equivalence (see below).

For a more general treatment of the tensor product without the restriction of globally split flat identities, see the sections on general tensor product diagrams in [Ka2].

Since all constructions only involve a neighborhood of the image under τ, it is natural to regard two diagrams as equivalent if they are compatible in some neighborhoods of the images.

7.5.11. Definition. *Two tensor product diagrams*

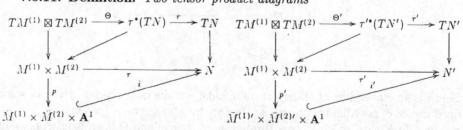

are called equivalent, if there exist open neighborhoods $U_N, U_{N'}$ of the images of τ, τ' and an isomorphism ϕ of Frobenius manifolds $U_N \to U_{N'}$ such that the following induced diagram satisfying the conditions of a tensor product diagram is commutative:

§7. TENSOR PRODUCT OF ANALYTIC FROBENIUS MANIFOLDS

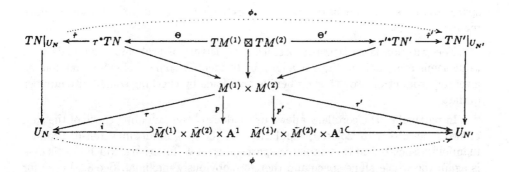

The main theorem on the tensor products of Frobenius manifolds with globally split flat identities proved in [Ka2] can be stated as follows.

7.5.12. Theorem. *For any two Frobenius manifolds with split flat identities there exists a tensor product diagram which is unique up to equivalence.*

Moreover, if both Frobenius manifolds carry Euler fields, then the local Euler fields given by Theorem 4.10.1 patch together to give an Euler field on the tensor product manifold.

After all the preparations, the proof is rather straightforward, but tedious, and we omit it.

7.5.13. Example. If, as is often the case, both $M^{(1)}$ and $M^{(2)}$ are domains in affine spaces or even vector spaces $V^{(1)}$ and $V^{(2)}$ (in the case of affine spaces choose some base–points), then the following data gives a tensor product diagram.

Let $e^{(i)}$ be the flat identities of $V^{(i)}$ and choose some complements $\bar{V}^{(i)}$ of $\langle e^{(i)} \rangle$, i.e. $V^{(i)} = \bar{V}^{(i)} \oplus \langle e^{(i)} \rangle$. Furthermore let $\tau : V^{(1)} \otimes V^{(2)} \to V^{(1)} \otimes V^{(2)}$ be the map $(v, w) \mapsto v \otimes e^{(2)} + e^{(1)} \otimes w$. Consider an extension of the tensor product germ of the two origins again placed at the origin. The theorem above then implies that this germ can be extended to a neighborhood $N \subset V^{(1)} \otimes V^{(2)}$ of $\text{Im}(\tau)$.

Any other choice of origins, extension N, and complements $\bar{V}^{(i)}$ will yield equivalent tensor product diagrams.

This implies that if there exists any Frobenius structure on $V^{(1)} \otimes V^{(2)}$ whose germ at one point of $V^{(1)} \otimes \langle e^{(2)} \rangle + \langle e^{(1)} \rangle \otimes V^{(2)}$ is the germ of a tensor product of germs from $V^{(1)}$ and $V^{(2)}$, then it is unique.

In the case of Frobenius manifolds which are embedded into affine spaces of the same dimension one can always arrive at the situation of split flat identity, by simply enlarging the domain of definition of the potential, since the dependence on the coordinate of a flat identity is only cubic. Therefore, slightly generalizing the above argument we get the following fact.

7.5.14. Corollary. *If one restricts to connected Frobenius manifolds which are embedded into affine spaces of the same dimension, there is a unique maximal connected tensor product Frobenius manifold up to isomorphism.*

Thus, if a connected Frobenius manifold which is embedded into an affine space of the same dimension contains a germ which is the tensor product of two germs stemming from connected Frobenius manifolds which are, too, embedded into affine spaces of the same dimension, it can be extended to a maximal tensor product manifold.

The criterion is particularly useful in the setting of Frobenius manifolds with tame semisimple points, since it allows us to identify some Frobenius manifold as a tensor product of two other Frobenius manifolds by checking some finite number of data.

In particular, the corollary allows us to identify the unfolding space of the sum of two functions with isolated singularities as the unique maximal tensor product manifold associated to the unfolding spaces of the two functions, which in this case is again the whole affine space and therefore obviously maximal. See 8.5 below for details.

§8. K. Saito's frameworks and singularities

8.1. Spectral cover. Let M be a Frobenius manifold in an analytic category. Consider $(\mathcal{T}_M, \circ)$ as a (super)commutative $\mathcal{O}_M$-algebra and denote by C its relative affine spectrum over M. This is an analytic space endowed with the structure maps $\pi : C \to M$ and $s : \mathcal{T}_M \xrightarrow{\sim} \pi_*(\mathcal{O}_C)$. Clearly, π is a finite flat morphism of degree $\dim M$ and s is an isomorphism of $\mathcal{O}_M$-modules.

Moreover, π is étale at a point $x \in M$ iff $(T_x M, \circ)$ is semisimple, and over such points C is a manifold.

We will call (C, π, s) *the spectral cover of M*. In the case of a semisimple manifold with an Euler field E, C actually parametrizes the spectrum of $E \circ$. In this case s also admits the following explicit description. Denote by $\mathcal{K}$ the sheaf of constant functions on C.

8.1.1. Proposition. *If M is semisimple, there exists a section $u \in \Gamma(C, \mathcal{O}_C/\mathcal{K})$ such that for any two local vector fields X, Y on M we have*

(8.1) $$s(X) = \overline{X}u, \quad s(X \circ Y) = \overline{X}u\overline{Y}u$$

or equivalently

(8.2) $$X \circ Y = s^{-1}(\overline{X}u\overline{Y}u).$$

Here $\overline{X}$ is the unique lift of X to C and the product in the right hand side is the product of functions on C.

Proof. Call a connected open subset U in M small if $\pi^{-1}(U)$ is the disjoint union of connected components U_i isomorphically projecting onto U. Let I be the set of these connected components. Then we have natural ring isomorphisms

$$\Gamma(U, p_{C*}(\mathcal{O}_C)) = \bigoplus_{i \in I} \Gamma(U_i, \mathcal{O}_C) = \Gamma(U, \mathcal{O}_M)^I.$$

This $\Gamma(U, \mathcal{O}_M)$-algebra has a basis of idempotents $f_i := \delta_{ij}$ on U_j. Defining $e_i \in \Gamma(U, \mathcal{T}_M)$ by $s(e_i) = f_i$, we get a local $\mathcal{O}_M$-basis of $\mathcal{T}_M$ satisfying $e_i \circ e_j = \delta_{ij} e_j$ and $e = \sum_i e_i$. Let (u^i) be a conjugate set of canonical coordinates on U. Each of them is defined modulo an additive constant. Since small sets cover M and connected components of their inverse images cover C, there exists a global section u of $\mathcal{O}_C/\mathcal{K}$ equal to u^i modulo constants on U_i. The first formula (8.1) is local, linear in X, and is valid for $X = e_i$ by definition. The rest is clear.

Notice in conclusion that it might be more convenient to replace u by the closed 1–form du globally defined on C.

In this section we describe a method for constructing Frobenius manifolds starting with a flat finite morphism $\pi: C \to M$ which will become the spectral cover. The main trick is *to define* the multiplication $\circ$ with the help of the formulas of the type (8.1), but in a changed geometric context.

Namely, we enhance $\pi: C \to M$ to a closed embedding of C into an auxiliary analytic space $p: N \to M$ and postulate that u on C extends to a function F on N in such a way that (8.1) can be taken as a definition of s if one interprets $\overline{X}$ as an arbitrary local lift of X. The latter is unavoidable if we do not want to assume M semisimple and π étale. To make s independent of the choice of lift, we must assume that the relative differential $d_p F$ vanishes on C. In other words, fiberwise over M, C is realized as a (sub)set of critical points of a family of functions, and these critical points must be isolated, for C to be finite over M. In this setting, a residue construction involving F can produce a metric as well.

This situation in fact arises in the theory of unfolding of hypersurface singularities, and the structures relevant for us (most notably, flat metric and split identity) were stressed in K. Saito's work. This is why we call the respective axiomatized setting described below *K. Saito's framework*.

Notice in conclusion that there is a natural closed embedding of the spectral cover C into the cotangent space of M: it is defined by the homomorphism of sheaves of $\mathcal{O}_M$–algebras $S_{\mathcal{O}_M}(\mathcal{T}_M) \to (\mathcal{T}_M, \circ)$ identical on $\mathcal{T}_M$. At least over semisimple points, C becomes in this way a lagrangean submanifold of $T^*(M)$. This symplectic nature of spectral covers is however of secondary importance here.

8.2. Setup. Let $p: N \to M$ be a submersion of complex analytic manifolds, generally non–compact, and F a holomorphic function on N. We consider F as a family of functions on the fibers of p parametrized by points of M. In local coordinates $z = (z_a), t = (t_b)$, where t_b are constant along the fibers of p, we write $F = F(z,t)$.

Let $d_p: \mathcal{O}_N \to \Omega^1_{N/M}$ be the relative differential. Denote by C the closed analytic subspace (or subscheme) of the critical points of the restrictions of F to the fibers given by the equation $d_p F = 0$. Its ideal J_F is locally generated by the partial derivatives XF where X are vertical vector fields on N. Derivatives $\partial F/\partial z_a$ of course suffice. Let $i_C: C \to N$ be the natural embedding, and p_C (former π) the restriction of p to C.

Denote by $\Omega^{\max}_{N/M}$ the invertible sheaf of holomorphic relative volume forms on N, $L := i_C^*(\Omega^{\max}_{N/M})$. The Hessian $\operatorname{Hess}(F) \in \Gamma(C, L^2)$ is a well–defined section of L^2 which in local coordinates as above can be written as

$$\operatorname{Hess}(F) = i_C^* \left[\det\left(\frac{\partial^2 F}{\partial z_a \partial z_b} \right) (dz_1 \wedge \cdots \wedge dz_n)^2 \right].$$

We denote by $G_C \subset C$ the subspace $\operatorname{Hess}(F) = 0$. Let $\mathcal{T}_M$ be the tangent sheaf of M. Finally, let ω be a nowhere vanishing global section of $\Omega^{\max}_{N/M}$.

8.2.1. Definition. *The family $(p: N \to M, F, \omega)$ is called Saito's framework, if the following conditions are satisfied:*

a) *Let the map* $s : \mathcal{T}_M \to p_{C*}(\mathcal{O}_C)$ *be defined by* $X \mapsto \overline{X}F \bmod J_F$, *where* $\overline{X}$ *is any local (in N) lift of X. Then s is an isomorphism of $\mathcal{O}_M$-modules. In particular, C is finite and flat over M. Assume moreover that G_C is a divisor, and $p_C : C \to M$ is étale on the complement to the divisor $G = i_{C*}(G_C) \subset M$.*

Denote by $\circ$ the multiplication in $\mathcal{T}_M$ induced by the one in $p_{C*}(\mathcal{O}_C)$:

(8.3) $\qquad X \circ Y := s^{-1}(\overline{X}F \cdot \overline{Y}F) \bmod J_F, \quad \overline{X \circ Y}F \equiv \overline{X}F \cdot \overline{Y}F \bmod J_F.$

b) *Define the following 1-form ε on $M \setminus G$. Its value on the vector field $X = s^{-1}(f)$ corresponding to the local section f of $\pi_*(\mathcal{O}_C)$ equals*

(8.4) $\qquad i_X(\varepsilon) := \mathrm{Tr}_{C/M}\left(\dfrac{f\, i_C^*(\omega^2)}{\mathrm{Hess}\,(F)}\right) = \sum_{i=1}^{\mu} \dfrac{f(\rho_i)}{\det\left((\partial^2 F/\partial z_a \partial z_b)(\rho_i)\right)},$

where ρ_i are the local branches of the critical locus C over M, and (z_a) is any vertical local coordinate system unimodular with respect to ω.

Then the scalar product $g : S^2(\mathcal{T}_{M\setminus G}) \to \mathcal{O}_{M\setminus G}$ defined by

(8.5) $\qquad\qquad\qquad g(X,Y) := i_{X \circ Y}(\varepsilon)$

is a flat metric. Both ε and g (as flat metrics) extend regularly to M.

8.3. The (pre-)Frobenius structure associated to the Saito framework. Let $(p : N \to M, F, \omega)$ be a Saito framework. As explained above, it induces a multiplication $\circ$ and a flat metric g on M.

Clearly, the vector field $e := s^{-1}(1 \bmod J_F)$ is the identity for $\circ$. Let $\mathcal{T}_M^f$ be the sheaf of vector fields flat with respect to g. Finally, put

$$A(X,Y,Z) = g(X \circ Y, Z) = g(X, Y \circ Z).$$

The last equality follows from (8.5) and the associativity of $\circ$. The tensor A is symmetric because $\circ$ is commutative. Therefore we have:

8.3.1. Claim. *The data $(\mathcal{T}_M^f, g, A)$ define on M the structure of an associative pre-Frobenius manifold in the sense of I.1.3.*

The discussion in the proof of the Proposition 8.1.1 can be repeated almost verbatim over $M \setminus G$ in reverse order. We keep the notation f_i, and $s(e_i) = f_i$ means that $\overline{e}_i F \bmod J_F = f_i$. We define $u^i = F(\rho_i)$ where ρ_i are branches U_i of C over a small open subset U. By a slight abuse of language, we treat u^i also as functions on U. Summarizing, we have:

8.3.2. Proposition. *The data $(\mathcal{T}_M^f, g, A)$ define on $M \setminus G$ the structure of a semisimple pre-Frobenius manifold in the sense of I.3.1 and I.3.2. Moreover, we have $e_i u^j = \delta_{ij}$ so that (u^j) form a local coordinate system (Dubrovin's canonical coordinates) and $[e_i, e_j] = 0$ because $e_i = \partial/\partial u^i$.*

Proof. Only the last statement might need some argumentation. We have $p^*(e_i u^j) = \overline{e}_i p^*(u^j)$ for any lift $\overline{e}_i$ of e_i. To calculate the right hand side we can restrict it to any local section of p since it is constant along the fibers. We choose $\overline{e}_i$ tangent to U_j and restrict the right hand side to U_j where $p^*(u^j)$ coincides with F. The result is δ_{ij} by the definition of e_i.

For future use, we can reformulate this as follows. Dualizing s we get the isomorphism $s^t : \Omega^1_M \to \mathcal{H}om_{\mathcal{O}_M}(p_{C*}(\mathcal{O}_C), \mathcal{O}_M)$. Then $s^t(du^i) : p_{C*}(\mathcal{O}_C) \to \mathcal{O}_M$ is the map which annihilates j–components for $j \neq i$ and coincides with the pushforward on the i–th component.

8.3.3. Theorem. *The structure $(M, \mathcal{T}^f_M, g, A)$ associated to the Saito framework is Frobenius iff $d\varepsilon = 0$.*

Proof. To check the Frobenius property on $M \setminus G$ we appeal to Chapter I, Th. 3.3 (Dubrovin's criterium), both conditions of which, $[e_i, e_j] = 0$ and $d\varepsilon = 0$, are satisfied. In order to pass from $M \setminus G$ to M one can use a continuity argument, e.g., in the following form. Let ∇_0 be the Levi–Civita connection of g, and ∇_λ the pencil of connections on $\mathcal{T}_M$ determined by its covariant derivatives $\nabla_{\lambda, X}(Y) := \nabla_{0, X}(Y) + X \circ Y$. Then M is Frobenius iff ∇_λ is flat for some $\lambda \neq 0$, and so automatically for all λ. Clearly, this is the closed property.

We will now discuss the case when e is flat.

On a small U, we can define functions η_j by $\eta_j = i_{e_j}(\varepsilon) = g(e_j, e_j)$. When $e_j = \partial/\partial u^j$, the closedness of $\varepsilon = \sum_i \eta_i du^i$ means that $\eta_j = e_j \eta$ for a local function η well defined up to addition of a constant, or else $\varepsilon = d\eta$. In the notations of (8.4)

$$(8.6) \qquad \eta_i = \frac{1}{\det((\partial^2 F/\partial z_a \partial z_b)(\rho_i))}.$$

8.3.4. Theorem. *Assume that the conditions of Theorem 8.3.3 are satisfied.*

The identity e is flat, iff for all i, $e\eta_i = 0$, or equivalently, $e\eta = g(e, e) = $ const. This holds automatically in the presence of an Euler field E with $D \neq 2d_0$.

This is Proposition I.3.5.

One more remark about the identity is in order. From the discussion in I.2.1 we know that a locally Frobenius manifold with split identity comes endowed with an open embedding into $M_0 \times \mathbf{A}^1$ compatible with the flat structure and such that e is tangent to the fibers of the projection onto M_0.

This split identity structure appears right from the start and extends to N in the Saito theory of singularities. Namely, there exists a lift $\bar{e}$ of e to N such that $\bar{e}F = 1$ identically, so that in the appropriate coordinate system $(t_0, t' = (t_a))$ on M we have $F = F_0(t', z) + t_0$ where F_0 does not depend on t_0, and $\bar{e} = \partial/\partial t_0$. The coordinate t_0 here corresponds to x^0 of I.2.1.2. This structure can be included in Definition 8.2.1.

It remains to clarify what Euler fields M can have.

8.3.5. Theorem. *Assume that the conditions of Theorem 8.3.3 hold. Let E be a vector field on a small subset U in M.*

a) We have $\mathrm{Lie}_E(\circ) = d_0 \circ$ iff

$$(8.7) \qquad E = d_0 \sum_i (u^i + c^i) e_i$$

for some constants c^i, where (du^i) are 1–forms dual to (e_j).

In particular, for non-zero E we have $d_0 \neq 0$ so that we may normalize E by $d_0 = 1$. Furthermore, if the monodromy representation of the fundamental group of $M \setminus G$ on H_0 of the fibers of $C \to M$ has only one-dimensional trivial subrepresentation, the global vector field E of this form with fixed d_0 is defined uniquely up to addition of a multiple of e.

b) For a field E of the form (8.7) and a constant D, we have $\mathrm{Lie}_E(g) = Dg$ iff

(8.8) $$E\eta = (D - d_0)\eta + \mathrm{const}.$$

In particular, if e is flat, adding a multiple of e does not change the validity of this property.

This follows from I.3.6.

When M comes from the Saito framework, we have a natural candidate for the global Euler field with $d_0 = 1$ suggested by our identification of local coordinates u^i. Namely, put on any small U

(8.9) $$E_F := \sum_{i=1}^{\mu} F(\rho_i) e_i = \sum_{i=1}^{\mu} u^i e_i.$$

Assume that it is in fact an Euler field and that we are in the conditions when it is defined uniquely up to a shift by a multiple of e. Assume furthermore that there exists a point 0 in M to which E_F extends and at which it vanishes (0 may lie in G, and in the theory of singularities it does so). Since e cannot vanish, the choice of such 0 fixes E_F completely.

8.3.6. Definition. *Saito's framework* $(p : N \to M, F, \omega)$ *is called the strong Saito framework, if the structure* $(M, \mathcal{T}_M^f, g, A)$ *described above is Frobenius, with flat identity e and Euler field E_F.*

8.3.7. Remark. Since the definitions of the pre-Frobenius and Frobenius structures, and also of the identity and Euler fields, are local, we can lift all these structures from $M \setminus G$ to $C \setminus G_C$.

8.4. Unfolding singularities. K. Saito's theory (cf. [S1], [S2], [Od] and the references therein) produces (a germ of) a strong Saito's framework starting with a germ of the holomorphic function $f(z_1, \ldots, z_n)$ with isolated singularity at zero.

Namely, one can choose holomorphic germs $\phi_0 = 1, \phi_1, \ldots, \phi_{\mu-1}$ whose classes constitute a basis of the Milnor ring $\mathbf{C}\{\{z\}\}/(\partial f/\partial z_a)$ in such a way that $F := f + \sum t_i \phi_i$ is the miniversal unfolding of f. Then $N = N_f$, resp. $M = M_f$, is a neighborhood of zero in the (z,t)-, resp. (t)-, space and F is defined above.

The crucial piece of the structure is the choice of ω encoded in the Saito notion of a good primitive form. Generally its existence is established in an indirect way. However, if f is of the type A, D or E, one can take $\omega = dz_1 \wedge \cdots \wedge dz_n$. For general quasi-homogeneous singularities most of the data constituting the Saito framework are algebraic varieties, rational maps, and rational differential forms so that the whole setup has considerably more global character.

Unfolding of the A_n-singularity $f(z) = z^{n+1}$ produces the Frobenius manifold discussed in I.4.5: N has coordinates $(z; a_1, \ldots, a_n)$, p is the projection onto $M = (a_i)$-space, and $F = z^{n+1} + a_a z^{n-1} + \cdots + a_n$ was denoted $p(z)$ in I.4.5.

We briefly discuss one interesting example motivated by physics.

8.4.1. Example: Gepner's manifolds $V_{n,k}$.
Let $n \geq 2, k \geq 1, h = n+k$. We will call *Gepner's Frobenius manifold* $V_{n,k}$ the manifold which is produced from the Saito's framework obtained by unfolding the polynomial

$$f_{n,k}(z_1,\ldots,z_{n-1}) := \frac{1}{h} \sum_{i=1}^{n-1} y_i^h,$$

where y and z are related by

$$\prod_{j=1}^{n-1}(1+y_j T) = 1 + \sum_{l=1}^{n-1} z_l T^l.$$

In particular, if one assigns to z_l weight l, $f_{n,k}$ becomes quasi-homogeneous of weight h. Its unfolding space is spanned by the classes of appropriate monomials, and a Zariski open dense subset $V_{n,k}$ of this space carries the structure of the Frobenius manifold as above. This subspace contains the point m corresponding to the *fusion potential*

$$g_{n,k}(z_1,\ldots,z_{n-1}) := f_{n+1,k-1}(z_1,\ldots,z_{n-1},1).$$

As D. Gepner ([Ge2]) proved, the tangent space $\mathcal{T}_m V_{n,k}$ with o-multiplication, that is, the Milnor algebra of $g_{n,k}$, is isomorphic to the Verlinde algebra (fusion ring) of the $su(n)_k$ WZW model of the conformal field theory. Zuber in [Zu] conjectured, and Varchenko and Gusein–Zade in [G–ZV] proved, that the lattice of the Verlinde algebra and the respective bilinear form can be interpreted in terms of vanishing cycles of $f_{n,k}$.

The total Frobenius manifold $V_{n,k}$ is thus a deformation of this fusion ring, in much the same way as quantum cohomology is the deformation of the usual cohomology ring.

8.5. Direct sum diagram.
We will consider now three Saito frameworks $(p : N \to M, F, \omega)$ and $(p_i : N_i \to M_i, F_i, \omega_i)$, $i = 1, 2$. We will call *the direct sum diagram* any cartesian square

(8.10)
$$\begin{array}{ccc} N_1 \times N_2 & \xrightarrow{\nu} & N \\ {\scriptstyle (p_1,p_2)}\downarrow & & \downarrow{\scriptstyle p} \\ M_1 \times M_2 & \xrightarrow{\nu_M} & M \end{array}$$

with the following properties:

(i) $\nu^*(F) = F_1 \boxplus F_2$.

(ii) $\nu^*(\omega) = \omega_1 \boxtimes \omega_2$.

Thus in a neighborhood of any point of N lying over the image of ν_M there exist local coordinates $(z_a^{(1)}, z_b^{(2)}; t_e)$ such that t_e are lifted from M, and (i) can be written as

(8.11) $$F(z_a^{(1)}, z_b^{(2)}; \nu_M^*(t_e)) = F_1(z_a^{(1)}; t_c^{(1)}) + F_2(z_b^{(2)}; t_d^{(2)})$$

and similarly (ii) can be written as

(8.12) $$\omega(z_a^{(1)}, z_b^{(2)}; \nu_M^*(t_e)) = \omega_1(z_a^{(1)}; t_c^{(1)}) \wedge \omega_2(z_b^{(2)}; t_d^{(2)}).$$

8.5.1. Properties of the direct sum diagrams. Clearly, $\nu^{-1}(C)$ is defined by the equations $d_{p_1}(F_1) \boxplus d_{p_2}(F_2) = 0$. Both summands then must vanish so that $\nu^{-1}(C) = C_1 \times C_2$. Denote by $\nu_C : C_1 \times C_2 \to C$ the restriction of ν. From (8.11) one then sees that

$$(8.13) \qquad \nu_C^*(\operatorname{Hess}(F)) = \operatorname{Hess}(F_1) \boxtimes \operatorname{Hess}(F_2)$$

and hence $\nu_M^{-1}(G) = G_1 \times M_2 \cup M_1 \times G_2$. Now let $m = \nu_M(m_1, m_2)$, $m_i \in M_i$. Choose small neighborhoods $m \in U$ in M, $m_i \in U_i$ in M_i such that $\nu_M(U_1 \times U_2) \subset U$. Number the connected components $U_i^{(1)}$ of $p_{C_1}^{-1}(U_1)$, resp. $U_j^{(2)}$ of $p_{C_2}^{-1}(U_2)$, by some indices i, resp. j. Then the connected components of $p_C^{-1}(U)$ are naturally numbered by the ordered pairs $I = (ij)$ in such a way that

$$\nu_C(U_i^{(1)} \times U_j^{(2)}) \subset U_I.$$

From now on we will assume that all the frameworks we are considering are strong ones. Then one can define e_I, u^I, η_I, etc., as above, and from (8.3), (8.6), (8.11)–(8.13) one immediately sees that

$$(8.14) \qquad u^I(m) = u_1^i(m_1) + u_2^j(m_2), \quad \eta_I(m) = \eta_i^{(1)}(m_1)\eta_j^{(2)}(m_2),$$

where in the right hand side we have the respective local functions on M_1, M_2.

The following restriction formula is slightly less evident.

8.5.2. Proposition. *Let $I = (ij), K = (kl), \eta_{IK} = e_I \eta_K = e_K \eta_I$, and similarly $\eta_{ik}^{(1)} = e_i \eta_k^{(1)}$, etc. Then we have in the same notations as in (8.14):*

$$(8.15) \qquad \eta_{IK}(m) = \delta_{jl}\eta_{ik}^{(1)}(m_1)\,\eta_l^{(2)}(m_2) + \delta_{ik}\eta_k^{(1)}(m_1)\,\eta_{jl}^{(2)}(m_2).$$

Proof. Calculate $\nu_M^*(d\eta_I)$ in two ways. On the one hand, we have

$$(8.16) \qquad \nu_M^*(d\eta_I) = \sum_K \nu_M^*(\eta_{IK})\nu_M^*(du^K).$$

As at the end of the proof of Proposition 8.3.2, we can identify du^K with a map from $p_{C*}(\mathcal{O}_C)$ to $\mathcal{O}_M$ vanishing on all components except for the K-th one where it is the canonical pushforward. After restriction to $M_1 \times M_2$ it may therefore be non-vanishing only on $U_k^{(1)} \times U_l^{(2)}$ so that we can calculate $\nu_M^*(\eta_{IK})$ by restricting $\nu_M^*(d\eta_I)$ to this product.

On the other hand, in view of (8.14),

$$\nu_M^*(d\eta_I) = d\nu_M^*(\eta_I) = d(\eta_i^{(1)} \boxtimes \eta_j^{(2)}) = d\eta_i^{(1)} \boxtimes \eta_j^{(2)} + \eta_i^{(1)} \boxtimes d\eta_j^{(2)}$$

$$(8.17) \qquad = \sum_r \eta_{ir}^{(1)} du_1^r \boxtimes \eta_j^{(2)} + \sum_s \eta_i^{(1)} \boxtimes \eta_{js}^{(2)} du_2^s.$$

Only the k-th summand in the first sum restricted to $U_k^{(1)} \times U_l^{(2)}$ may be non-vanishing, and considered as a map (cf. above) it equals $\delta_{jl}\eta_{ik}^{(1)} \boxtimes \eta_j^{(2)}$ times the pushforward map. We have the similar expression for the l-th summand of the second sum. Comparison with (8.16) furnishes (8.15) because $\nu_M^*(du^K) = du_1^k \boxplus du_2^l$.

Comparing this with Theorem 7.2, we immediately obtain

8.5.3. Corollary. *If we have a direct sum diagram (8.10), then the Frobenius manifold M is (canonically isomorphic to) the tensor product of Frobenius manifolds M_1 and M_2.*

As an example, consider the "adding squares" operation. The unfolding space A_1 of z^2 is the flat one-dimensional Frobenius manifold which is identity with respect to the tensor product. This agrees with the fact that adding free extra z_j-coordinates to N and the sum of their squares to F, and multiplying the primitive form ω by $\wedge dz_j$, we do not change the resulting Frobenius structure of M.

8.6. Direct sums of singularities. In the theory of singularities, one can compare the versal unfolding spaces M_f, M_g, and $M_{f \oplus g}$ of the germs f, g, and $f \oplus g$ ($\oplus$ means that the sets of arguments of f and g are disjoint). It so happens that they fit into the direct sum diagram (8.10) (the only choice that remains is that of the volume form ω on the space of $f \oplus g$ which is natural to take decomposable as in 8.5 above).

By iteration, we can consider an arbitrary number of summands. In particular, the Frobenius manifold $A_{n_1,\ldots,n_k}$ which is obtained by unfolding the quasi-homogeneous singularity at zero, $f(z) := z_1^{n_1} + \cdots + z_k^{n_k}$, is the tensor product of A_{n_i}.

We now give some calculations for A_n.

We return to the notation of I.4.5, with p replaced by F.

8.6.1. Proposition. *Consider the points of A_n where $a_1, \ldots, a_{n-2} = 0$, a_{n-1}, a_n arbitrary. Choose a primitive root ζ of $\zeta^n = 1$ and a root b of $b^n = -\dfrac{a_{n-1}}{n+1}$. At these points we have:*

$$(8.18) \qquad u^i = a_n + \frac{n}{n+1} \zeta^i a_{n-1} b,$$

$$(8.19) \qquad \eta_i = \frac{\zeta^i}{n(n+1) b^{n-1}},$$

$$(8.20) \qquad v_{jk} = \frac{1}{(n+1)(1 - \zeta^{k-j})}.$$

Remark. It is suggestive to compare these coordinates with those for the quantum cohomology of $\mathbf{P}^{n-1}$ (cf. II.4.9.1) on the plane spanned by the identity (coordinate x_0) and the dual hyperplane section (coordinate x_1):

$$(8.18a) \qquad u^i = x_0 + n \zeta^i e^{\frac{x_1}{n}},$$

$$(8.19a) \qquad \eta_i = \frac{\zeta^i}{n} e^{-x_1 \frac{n-1}{n}},$$

$$(8.20a) \qquad v_{jk} = \frac{1}{1 - \zeta^{k-j}}.$$

Proof. At our subspace, $F(z) = z^{n+1} + a_{n-1} z + a_n$. Hence

$$F'(z) = (n+1)\left(z^n + \frac{a_{n-1}}{n+1}\right)$$

has roots $\rho_i = \zeta^i b$. But for A_n–manifolds we have universally $u^i = F(\rho_i)$, $\eta_i = \dfrac{1}{F''(\rho_i)}$. This furnishes (8.18) and (8.19).

The proof of (8.20) is longer. We have to calculate the values of functions

$$\text{(8.21)} \qquad \frac{1}{2}(u^k - u^j)\frac{\eta_{jk}}{\eta_k},$$

restricted to the plane of our base points.

At a generic point of A_n, we can calculate η_{jk} in the following three-step way:

$$\text{(8.22)} \qquad \eta_{jk} = \frac{\partial \eta_j}{\partial u^k} = \sum_{l,m=1}^{n} \frac{\partial \eta_j}{\partial \rho_m}\frac{\partial \rho_m}{\partial a_l}\frac{\partial a_l}{\partial u^k}.$$

Since

$$\eta_j = \frac{1}{F''(z)} = \frac{1}{(n+1)\prod_{i:\,i\neq j}(\rho_i - \rho_j)},$$

we have

$$\text{(8.23)} \qquad \frac{\partial \eta_j}{\partial \rho_m} = -\frac{\eta_j}{\rho_m - \rho_j}$$

if $m \neq j$, and

$$\text{(8.24)} \qquad \frac{\partial \eta_j}{\partial \rho_j} = \eta_j \sum_{i:\,i\neq j}\frac{1}{\rho_i - \rho_j}.$$

Moreover,

$$\text{(8.25)} \qquad \frac{\partial \rho_m}{\partial a_l} = -(n-l)\rho_m^{n-l-1}\eta_m.$$

This can be checked by derivating the identity $F'(\rho_m) = 0$.

Finally, from Chapter I, (4.24), we have

$$\text{(8.26)} \qquad \sum_{l=1}^{n}\frac{\partial a_l}{\partial u^k}\rho_i^{n-l} = \delta_{ik}.$$

We will now restrict (8.23)–(8.26) to our plane.

Using (8.18) and (8.19), we get consecutively:

$$\text{(8.27)} \qquad \frac{\partial \eta_j}{\partial \rho_m} = \frac{1}{n\,a_{n-1}}\frac{1}{\zeta^{m-j}-1}$$

if $m \neq j$, and

$$\text{(8.28)} \qquad \frac{\partial \eta_j}{\partial \rho_j} = \frac{n-1}{2n\,a_{n-1}},$$

$$\text{(8.29)} \qquad \frac{\partial \rho_m}{\partial a_l} = -\frac{n-l}{n(n+1)}b^{-l}\zeta^{-ml}.$$

Solving (3.11) for partial derivatives, we also find

$$\text{(8.30)} \qquad \frac{\partial a_l}{\partial u^k} = \frac{1}{n}b^{l-n}\zeta^{kl}.$$

It remains to substitute (3.12)–(3.14) into (2.7) to get after some calculation

$$(8.31) \qquad \eta_{jk} = \frac{2\zeta^{k-j}}{(\zeta^{k-j}-1)^2} \frac{1}{n\, a_{n-1}^2}.$$

Finally, substituting (8.18), (8.19), and (8.31) in (8.21), we obtain (8.18).

8.7. Spectrum. The d-spectrum of A_n is $d = \frac{n-1}{n+1}, q_i = \frac{i}{n+1}$ for $i = 0, \ldots, n-1$: see 4.10.4 and I.4.5.1. The additivity with respect to the tensor product (4.38) allows us to calculate the spectrum of $A_{n_1,\ldots,n_k}$. The integral part of this spectrum sometimes looks like the spectrum of quantum cohomology.

We will now show that in some important cases a Frobenius manifold M contains a closed Frobenius submanifold corresponding to the integral part of its spectrum.

8.7.1. Proposition. *Assume that we have an analytic or formal Frobenius manifold M with an Euler field E, $d_0 = 1$, $D \in \mathbf{Z}$, and flat identity. Let $-\mathrm{ad}\, E$ be semisimple on flat vector fields with spectrum d_a, (x_a) a flat coordinate system with*

$$E = \sum_{a:\, d_a \neq 0} d_a x^a \partial_a + \sum_{b:\, d_b = 0} r_b \partial_b,$$

and $e = \partial_0$. Define the submanifold $HM \subset M$ by the equations

$$x_c = 0 \text{ for all } c \text{ such that } d_c \notin \mathbf{Z}.$$

Finally, assume that at least one of the following conditions is satisfied:

(i) $r_b = 0$ for all b with $|\Delta_b| = 2$.

(ii) M is of qc-type, and $\sum_{b:\, d_b = 0} r_b \Delta_b$ takes only integral values on B (for definition, see 5.4.1).

Then HM with induced metric, $\circ$-multiplication, E and e is a Frobenius manifold.

Remark. From the proof it will be clear that one can replace integers in this statement by any arithmetic progression containing 0 to which D and d_0 belong.

Proof of Proposition 8.7.1. If d_c is not integral, the functions $Ex_c = d_c x_c$, $ex_c = 0$ vanish on HM. Hence E and e are tangent to HM and can be restricted to it. From the equation $(d_a + d_b - D)g_{ab} = 0$ (see I.(2.17)) one sees that if $d_a, D \in \mathbf{Z}$, $d_b \notin \mathbf{Z}$, we have $g_{ab} = 0$. Therefore the restriction of g to HM is non-degenerate (it is obviously flat), and x_a for $d_a \in \mathbf{Z}$ restrict to a flat coordinate system on HM. The $\circ$-product of two vector fields tangent to HM at the points of M does not contain the transverse components. In fact, we have $E\Phi_{ab}{}^c = (d_0 - d_a - d_b + d_c)\Phi_{ab}{}^c$ (see I.(2.18)). Hence if $d_a, d_b \in \mathbf{Z}$, $d_c \notin \mathbf{Z}$, then in the case (i) every monomial in the series Φ_{ab}^c must be an eigenvector of E with non-integral eigenvalue, and therefore it must contain some x_e with $d_e \notin \mathbf{Z}$ so that it vanishes on HM. In the case (ii) we apply the same reasoning separately to the generalized (involving exponentials) monomials contributing to the third derivatives of Ψ in (5.23) and to the third derivatives of c.

The same reasoning shows that the induced multiplication of vector fields on $\mathcal{T}_{HM}$ is defined by the third derivatives of the induced potential.

8.7.2. Example. Consider the tensor products of A_n's. Proposition 8.7.1 is applicable because the Euler field vanishes at the origin (corresponding to $f = z^{n+1}$). We want to describe $(n_1, \ldots, n_N)$ with non-trivial $H(A_{n_1} \otimes \cdots \otimes A_{n_N})$. We can assume $n_i \geq 2$ because A_1 is identity with respect to the tensor multiplication. The first necessary condition is

$$\mathbf{d} := \sum_{i=1}^{N} \frac{n_i - 1}{n_i + 1} \in \mathbf{Z}. \tag{8.32}$$

If it is satisfied, the full $\mathbf{d}$–spectrum of the tensor product consists of certain rational points between 0 and $\mathbf{d}$. Multiplicity of 0 and $\mathbf{d}$ is one. Generally, the multiplicity of some $m \leq \mathbf{d}$ is

$$h^{2m}\left(H\left(\bigotimes_i A_{n_i}\right)\right) := \text{the number of } (i_1, \ldots, i_N) \in \mathbf{Z}_{\geq 0}^N \text{ satisfying}$$

$$\sum_{k=1}^{N} \frac{i_k}{n_k + 1} = m, \ 0 \leq i_k \leq n_k - 1. \tag{8.33}$$

The $\mathbf{d}$–spectrum of $H(\bigotimes A_{n_k})$ consists of the part of (8.33) for all integer m.

In particular, the manifold $A_n^{\otimes n+1}$ contains a canonically defined pointed Frobenius submanifold $HA_n^{\otimes n+1}$ whose spectrum looks formally like that of the even-dimensional part of quantum cohomology of an $(n-1)$–dimensional algebraic (or symplectic) manifold V.

More precisely, V must have Betti numbers

$$h^{2m}(V) := \text{the number of } (i_1, \ldots, i_{n+1}) \in \mathbf{Z}_{\geq 0}^{n+1}, \text{ satisfying}$$

$$\frac{1}{m} \sum_{k=1}^{n+1} i_k = m(n+1), \ 0 \leq i_k \leq n - 1, \tag{8.34}$$

and vanishing (modulo torsion) $c_1(V)$.

For example, even Betti numbers must be $(1,19,1)$ for $n = 3$, and $(1,101,101,1)$ for $n = 4$.

The Poincaré symmetry of them is generally established by the involution $(i_k) \mapsto (n-1-i_k)$, $m \mapsto n-1-m$.

Problem. Is there actually a manifold V_n whose Quantum Cohomology contains $HA_n^{\otimes n+1}$? Is it at least true that $HA_n^{\otimes n+1}$ is a Frobenius manifold of qc–type? (Notice that A_n itself is not of qc–type.)

As was explained above, $A_n^{\otimes n+1}$ is the unfolding space at zero of the singularity of $x_1^{n+1} + \ldots + x_{n+1}^{n+1}$. A geometric argument shows that $HA_n^{\otimes n+1}$ carries the variation of Hodge structure corresponding to the middle cohomology of the hypersurface $x_1^{n+1} + \ldots + x_{n+1}^{n+1} = 0$. More precisely, the volume form periods constitute the horizontal sections of one of the structure connections of the Frobenius manifold in question.

§9. Maurer–Cartan equations and Gerstenhaber–Batalin–Vilkovyski algebras

9.1. Maurer–Cartan equations. Fix a supercommutative $\mathbf{Q}$–algebra k. All our structures are $\mathbf{Z}_2$–graded, and notation like $\tilde{x}$ means the parity of a homogeneous element x. Let $g = g_0 \oplus g_1$ be a Lie superalgebra over k, supplied with an odd differential d satisfying $d[a,b] = [da, b] + (-1)^{\tilde{a}}[a, db]$.

Put $Z = Z(g,d) := \operatorname{Ker} d, B = B(g,d) := \operatorname{Im} d, H := H(g,d) = Z/B$. Clearly, Z is a Lie subalgebra, and B its ideal, so that H with induced bracket product is a Lie superalgebra.

The differential d can be shifted.

For $\gamma \in g_1$, put $d_\gamma(a) := da + [\gamma, a]$. Clearly, $d_\gamma[a,b] = [d_\gamma a, b] + (-1)^{\tilde{a}}[a, d_\gamma b]$.

9.1.1. Claim. *a) We have $d_\gamma^2 = 0$ if*

$$(9.1) \qquad d\gamma + \frac{1}{2}[\gamma, \gamma] = 0.$$

b) Let $\gamma' = \gamma + \epsilon\beta$, $\beta \in g$, ϵ an even or odd constant with $\epsilon^2 = 0$ such that $\epsilon\beta$ is odd. Assume that γ satisfies (9.1). Then γ' satisfies (9.1) as well iff

$$(9.2) \qquad d_\gamma(\beta) = d\beta + [\gamma, \beta] = 0.$$

This is straightforward.

If K is another supercommutative k–algebra, we define $g_K = K \otimes_k g, d_K = 1 \otimes d$. We will always work with K flat over k so that $Z_K := \operatorname{Ker} d_K = K \otimes_k Z$, and similarly for B and H. Claim 9.1.1 is of course applicable to (g_K, d_K) as well.

We want to produce from this setting a non–linear version of the homology $H(g,d)$ or rather of the diagram $g \supset Z \to H$.

The most straightforward is the case when, say, g is free of finite rank over k. We then replace g by the linear superspace $\mathcal{G} := \operatorname{Spec} k[\Pi g^t]$, where Π is the parity inversion functor, and replace Z by the closed subspace $\mathcal{Z} \subset \mathcal{G}$ defined by the equations (9.1). In order to understand what should be the non–linear version of B, we interpret Claim 9.1.1 b) as saying that d_γ–cycles form the Zariski tangent space to the point γ of the Maurer–Cartan space (9.1). It then contains the subspace of d_γ–boundaries, and we can construct the distribution $\mathcal{B}$ generated by the boundaries. If the quotient space $\mathcal{H} = \mathcal{Z}/\mathcal{B}$ in some sense exists, it can be regarded as the non–linear cohomology of (g,d).

In more down–to–earth terms, choose a (homogeneous) basis $\{\gamma_i\}$ of g and a family of independent (super)commuting variables t^i such that $\tilde{t}^i = \tilde{\gamma}_i + 1$. Then $\Gamma := \sum_i t^i \gamma_i$ is a generic odd element of g (or rather of $k[t^i] \otimes g$), and the equation $d\Gamma + \frac{1}{2}[\Gamma, \Gamma] = 0$ is equivalent to the system of equations

$$(9.3) \qquad \forall k: \quad \sum_i (-1)^i t^i D_i^k + \frac{1}{2} \sum_{i,j} t^i t^j (-1)^{i(j+1)} L_{ij}{}^k = 0.$$

Here we define the structure constants by $d\gamma_i = \sum_k D_i^k \gamma_k$ and $[\gamma_i, \gamma_j] = \sum_k L_{ij}{}^k \gamma_k$ and use the following shorthand for the signs: $(-1)^{i(j+1)}$ means $(-1)^{\tilde{\gamma}_i(\tilde{\gamma}_j+1)}$, etc.

These equations define the coordinate ring R of the affine scheme which we called $\mathcal{Z}$. Obviously, $\mathcal{Z}$ represents the following functor on the category of super-commutative k–algebras K:

(9.4) $$K \mapsto \{\text{solutions to (9.1) in } (K \otimes g)_1\}.$$

Similarly, if we have any odd d_Γ–cycle $\epsilon\beta = \sum_a \epsilon s^a \gamma_a$ with coefficients in $K \otimes R$, then statement 9.1.1 b) means that the map $X_\beta : t^a \to s^a$ descends to the derivation of $K \otimes R$ over K, that is, to a vector field on $\mathcal{G}_K$ of parity $\tilde{\epsilon}$. Of course, the adequate functorial language for derivations is that of the first order infinitesimal deformations of points, because generally the vector fields implied by 9.1.1 b) are defined only in the infinitesimal neighborhood of γ.

We will now stop discussing the case of finite rank g because in most interesting examples this does not hold, and only $H(g,d)$ is of finite rank.

So we step back and try to produce a formal section of $\mathcal{Z}$ passing through $\gamma = 0$ and transversal to the distribution $\mathcal{B}$. We want it to be of the same size as H, or rather ΠH, and we will assume henceforth that H is free of finite rank. Before passing to the odd Lie superalgebras in 9.3 below, we denote $K := k[[\Pi H^t]] = k[[x_i]]$ where x_i are coordinate functions on ΠH dual to a basis of ΠH. Any element $\Gamma \in g_K$ can be uniquely written as $\sum_{n \geq 0} \Gamma_n$ where Γ_n is homogeneous of degree n in x_i. Such an element can be naturally called a formal section of $\mathcal{Z}$, or *a generic (formal) solution to (9.1)*, if it has the following properties:

a) $\Gamma \in (g_K)_1, \Gamma_0 = 0, \Gamma_1 = \sum_i x_i c_i$ where $dc_i = 0$ and classes of c_i form a basis of H odd dual to $\{x_i\}$.

b) $d_K \Gamma + \dfrac{1}{2}[\Gamma, \Gamma] = 0$.

The necessary condition for the existence of Γ is the identical vanishing of the Lie bracket induced on $H(g,d)$. In fact, the equation $d\Gamma + \dfrac{1}{2}[\Gamma,\Gamma] = 0$ implies (assuming a) above) $d\Gamma_2 + \dfrac{1}{2}[\Gamma_1,\Gamma_1] = 0$. Hence $[c_i, c_j] \in \mathcal{B}$. However, generally it is not sufficient. In fact, the next equation reads $d\Gamma_3 + [\Gamma_1, \Gamma_2] = 0$, but since Γ_2 may be non–closed, we cannot conclude that $[\Gamma_1, \Gamma_2]$ is a boundary. The manageable sufficient condition is stronger: (g, d) must be quasi–isomorphic to the differential Lie algebra $H(g,d)$ with zero bracket and zero differential. For a considerably more general treatment see [GoM]. Our direct and elementary approach is self–contained and produces slightly more detailed information in the cases essential for the theory of Frobenius manifolds.

9.2. Theorem. *(i) Assume that there exists a surjective morphism of differential Lie superalgebras $\phi : (g, [,], d) \to (H, 0, 0)$ inducing an isomorphism on the homology. Then there exists a generic formal solution Γ to (9.1).*

Moreover, Γ can be chosen in such a way that for any $n \geq 2$, $\Gamma_n \in K \otimes \text{Ker}\,\phi$. In other words, $(\text{id} \otimes \phi)(\Gamma) = \sum_i x_i[c_i]$. Such a solution will be called normalized.

(ii) If (i) is satisfied, then for any generic solution Γ, non–necessarily normalized, the map $\phi_K = \text{id} \otimes \phi : g_K \to H_K$ is the surjective morphism of differential Lie superalgebras $(g_K, [,]_K, d_{K,\Gamma}) \to (H_K, 0, 0)$ inducing an isomorphism on the homology.

§9. MAURER–CARTAN EQUATIONS

Proof. (i) Let $n \geq 1$. Assuming that Γ_i for $i \leq n$ are already constructed, and writing d instead of d_K, we must find Γ_{n+1} from the equation

$$(9.5) \qquad d\Gamma_{n+1} = -\frac{1}{2} \sum_{i,j:\, i+j=n+1} [\Gamma_i, \Gamma_j].$$

First of all we check that the right hand side of (9.5) is closed in g_K. In fact, since the components $\Gamma_1, \ldots, \Gamma_n$ satisfy similar equations by the inductive assumption, the differential of the rhs equals

$$\frac{1}{2} \sum_{i+j+k=n+1} [[\Gamma_i, \Gamma_j], \Gamma_k].$$

This expression vanishes because the Jacobi identity for odd elements reads

$$[[\Gamma_i, \Gamma_j], \Gamma_k] + [[\Gamma_k, \Gamma_i], \Gamma_j] + [[\Gamma_j, \Gamma_k], \Gamma_i] = 0.$$

Hence the coefficients of the rhs of (9.5) (as polynomials in x_i) belong to $Z \cap [g, g]$. But $[g, g] \in \operatorname{Ker} \phi$ and $Z \cap \operatorname{Ker} \phi = B$ because ϕ is a quasi–isomorphism. Thus we can solve (9.5).

We can add to any solution elements of Z_K of degree $n+1$. But $Z + \operatorname{Ker} \phi = g$ because ϕ induces surjection on homology. Hence we can normalize Γ_{n+1} by the requirement $\Gamma_{n+1} \in K \otimes \operatorname{Ker} \phi$.

(ii) Now fix Γ satisfying (9.5) for all n. We will write d_Γ instead of $d_{K,\Gamma}$ and put $Z_\Gamma := \operatorname{Ker} d_\Gamma \subset g_K$, $B_\Gamma = d_\Gamma(g_K)$. We have $B \subset \operatorname{Ker} \phi_K$ and $[g_K, g_K] \subset \operatorname{Ker} \phi_K$, hence $B_\Gamma \subset \operatorname{Ker} \phi_K$. Therefore, ϕ_K is compatible with the zero bracket and zero differential on H_K. The natural inclusion $Z_\Gamma + \operatorname{Ker} \phi_K \to g_K$ becomes surjection after the reduction modulo the ideal (x_i) of K, because ϕ is surjective. Hence this inclusion is surjective, and ϕ_K is surjective as well. It remains to show that ϕ_K induces injection on homology, that is,

$$(9.6) \qquad Z_\Gamma \cap \operatorname{Ker} \phi_K \subset B_\Gamma.$$

Let $c = \sum_{n \geq 0} c_n \in Z_\Gamma$. This means that $dc_0 = 0$, and in general

$$(9.7) \qquad dc_n = -\sum_{i+j=n} [\Gamma_i, c_j]$$

(we keep writing d for d_K). Assuming that $\phi_K(c) = 0$ we want to deduce the existence of homogeneous elements a_n of degree n in g_K such that

$$(9.8) \qquad c_{n+1} = da_{n+1} + \sum_{i+j=n+1} [\Gamma_i, a_j].$$

We have $dc_0 = 0$ and $\phi(c_0) = 0$, hence c_0 is a boundary because ϕ is the quasi-isomorphism. Assuming that $a_0, \ldots, a_n$ are found, we will establish the existence of a_{n+1} satisfying (9.8), if we manage to prove that $c_{n+1} - \sum_{i+j=n+1} [\Gamma_i, a_j]$ is d–closed. In fact, this element also belongs to $\operatorname{Ker} \phi_K$ and so must be a boundary.

The differential of this element is

$$(9.9) \qquad dc_{n+1} + \sum_{i+j=n+1} [\Gamma_i, da_j] - \sum_{i+j=n+1} [d\Gamma_i, a_j].$$

Replace da_j in the first sum by $c_j - \sum_{k+l=j}[\Gamma_k, a_l]$ for $j \leq n$ (this holds by induction). Replace $d\Gamma_i$ in the second sum by the sum of commutators from (9.5). The terms containing c_j will cancel thanks to (9.7). The remaining terms can be written as
$$-\sum_{i+j+k=n+1}[\Gamma_i,[\Gamma_j,a_k]] + \frac{1}{2}\sum_{i+j+k=n+1}[[\Gamma_i,\Gamma_j],a_k].$$
This expression vanishes because of the Jacobi identity.

9.2.1. Corollary. *Define the map $\psi = \psi_\Gamma : H_K \to g_K$ as the K-linear extension of*
$$H \to g_K : X \mapsto \overline{X}\Gamma,$$
where $\overline{X}$ acts on $K \otimes g$ as the right g-linear extension of the derivation on K acting as $(\Pi X, *)$ on ΠH^t.

Then ψ is a section of ϕ_K if Γ is normalized.

Proof. First of all, we have $\overline{X}(d\Gamma + \frac{1}{2}[\Gamma, \Gamma]) = 0$ from which it follows that $d_\Gamma(\overline{X}\Gamma) = 0$, that is, $\overline{X}\Gamma$ is a d_Γ–cycle. Its image in H_K is $(\overline{X}\Gamma_1 + \sum_{n\geq 2}\overline{X}\Gamma_n)$ mod B_Γ. The first term is clearly X. The remaining ones are in $K \otimes \operatorname{Ker}\phi_K$, if Γ is normalized.

9.3. Odd Lie (super)algebras. As in 9.1, now let $g = g_0 \oplus g_1$ be a k-module endowed with a bilinear operation *odd bracket* $(a, b) \mapsto [a \bullet b]$ which satisfies the following conditions:

a) *parity of $[a \bullet b]$ equals $\tilde{a} + \tilde{b} + 1$,*

b) *odd anticommutativity:*

(9.10) $$[a \bullet b] = -(-1)^{(\tilde{a}+1)(\tilde{b}+1)}[b \bullet a],$$

c) *odd Jacobi identity:*

(9.11) $$[a \bullet [b \bullet c]] = [[a \bullet b] \bullet c] + (-1)^{(\tilde{a}+1)(\tilde{b}+1)}[b \bullet [a \bullet c]].$$

Such a structure will be called an odd Lie (super)algebra. We consider such algebras endowed with an odd differential satisfying

(9.12) $$d[a \bullet b] = [da \bullet b] + (-1)^{\tilde{a}+1}[a \bullet db].$$

Physicists sometimes denote such multiplication $\{,\}$ (see e.g. [LZ]). Our choice of notation allows one to use consistently the standard sign mnemonics of superalgebra, if $\bullet$ counts as an element of parity one.

If (g, d) is the usual differential Lie superalgebra, the parity change functor $g \mapsto \Pi g$ turns the usual bracket product $[,]$ into the odd bracket product, and defines an equivalence of the two categories (the differential changes sign). It seems therefore that there is not much point in considering odd brackets. However, in the context of GBV-algebras they come together with usual supercommutative multiplication, and parity change then turns this multiplication into odd one (see the next section). This is, of course, a particular case of the general operadic formalism over the category of superspaces, where any operation can be inherently even or odd.

In what follows we choose to work with even multiplication and odd bracket product. But we will use the results established for the usual Lie superalgebras, with appropriately modified parities and signs, for odd Lie superalgebras. In particular, the odd Maurer–Cartan equation in the physical literature is called *the master equation*:

$$(9.13) \qquad d\Gamma + \frac{1}{2}[\Gamma \bullet \Gamma] = 0.$$

Theorem 9.2 provides conditions of its solvability in $k[[H^t]] \otimes g$ rather than $k[[\Pi H^t]] \otimes g$. Notice also that Γ in (9.13) must be even.

9.4. Gerstenhaber–Batalin–Vilkovyski algebras. Let $\mathcal{A}$ be a supercommutative algebra with identity over another supercommutative algebra k (constants). Consider an odd k–linear operator $\Delta : \mathcal{A} \to \mathcal{A}, \Delta(1) = 0$, with the following property:

$$\forall a \in \mathcal{A}, \qquad \partial_a := (-1)^{\tilde{a}}([\Delta, l_a] - l_{\Delta a})$$

(9.14) \qquad is the derivation of parity $\tilde{a} + 1$ over k.

Here l_a denotes the operator of left multiplication by a, and brackets denote the supercommutator. Explicitly,

$$\partial_a b = (-1)^{\tilde{a}} \Delta(ab) - (-1)^{\tilde{a}} (\Delta a) b - a \Delta b.$$

The sign ensures the identity $\partial_{ca} = c \partial_a$ for any constant c. By definition of derivation,

$$(9.15) \qquad [\partial_a, l_b] = l_{\partial_a b}.$$

The pair $(\mathcal{A}, \Delta)$ is called a *GBV-algebra* if, in addition, $\Delta^2 = 0$. There is an obvious operation of scalar extension.

9.4.1. Lemma. *In any GBV-algebra we have*

$$(9.16) \qquad [\Delta, \partial_a] = \partial_{\Delta a},$$

$$(9.17) \qquad [\partial_a, \partial_b] = \partial_{\partial_a b}.$$

Proof. From (9.14) we have

$$[\Delta, \partial_a] = (-1)^{\tilde{a}}([\Delta, [\Delta, l_a]] - [\Delta, l_{\Delta a}]).$$

From the Jacobi identity for operators and $[\Delta, \Delta] = 0$ we find $[\Delta, [\Delta, l_a]] = 0$ because

$$[\Delta, [\Delta, l_a]] = [[\Delta, \Delta], l_a] - [\Delta, [\Delta, l_a]].$$

From (9.14) with Δa replacing a we have $[\Delta, l_{\Delta a}] = (-1)^{\tilde{a}+1} \partial_{\Delta a}$. Hence

$$[\Delta, \partial_a] = (-1)^{\tilde{a}+1}[\Delta, l_{\Delta a}] = \partial_{\Delta a}.$$

To prove (9.17), we notice that since $[\partial_a, \partial_b]$ must be a derivation, in the intermediate calculations we are allowed not to register all the summands which are left multiplications: they will cancel anyway. So we have, denoting such summands by dots and using consecutively (9.14), (9.15), Jacobi and (9.16), and again (9.14) with $\partial_a b$ replacing a:

$$[\partial_a, \partial_b] = (-1)^{\tilde{b}}[\partial_a, [\Delta, l_b] - l_{\Delta b}] = (-1)^{\tilde{b}}[\partial_a, [\Delta, l_b]] + \ldots$$
$$= (-1)^{\tilde{b}+\tilde{a}+1}[\Delta, [\partial_a, l_b]] + \ldots = (-1)^{\tilde{b}+\tilde{a}+1}[\Delta, l_{\partial_a b}] + \ldots = \partial_{\partial_a b}.$$

Define now the odd bracket operation on $\mathcal{A}$ by the formula

(9.18) $$[a \bullet b] := \partial_a b.$$

9.4.2. Proposition. *The pair of bilinear operations (multiplication and odd bracket) defines on $\mathcal{A}$ the structure of the odd Poisson algebra in the following sense:*

(i) The odd bracket satisfies the odd anticommutativity, the odd Jacobi, and the odd Poisson identities:

$$[a \bullet b] = -(-1)^{(\tilde{a}+1)(\tilde{b}+1)}[b \bullet a], \quad [a \bullet [b \bullet c]] = [[a \bullet b] \bullet c] + (-1)^{(\tilde{a}+1)(\tilde{b}+1)}[b \bullet [a \bullet c]],$$

(9.19) $$[a \bullet bc] = [a \bullet b]c + (-1)^{\tilde{b}(\tilde{a}+1)} b [a \bullet c].$$

(ii) Δ is the derivation with respect to the odd brackets so that $(\mathcal{A}, \bullet, \Delta)$ is the differential odd Lie algebra.

Proof. The anticommutativity can be checked directly. The Jacobi identity follows from (9.17) written as $[\partial_a, \partial_b] = \partial_{[a \bullet b]}$. The Poisson identity means that ∂_a is a derivation. The last statement follows from (9.14).

Notice that with respect to the usual multiplication Δ is the differential operator of order ≤ 2 and not necessarily derivation.

9.5. Additional differential. Assume now that we have an additional k-linear odd map $\delta : \mathcal{A} \to \mathcal{A}$ which is the derivation with respect to the multiplicative structure of $\mathcal{A}$ satisfying

(9.20) $$\delta^2 = [\delta, \Delta] = \delta\Delta + \Delta\delta = 0.$$

We will say that $(\mathcal{A}, \Delta, \delta)$ is *a differential GBV-algebra (dGBV)*.

9.5.1. Lemma. *We have*

(9.21) $$[\delta, \partial_a] = \partial_{\delta a}.$$

Therefore δ is the derivation with respect to the odd bracket as well.

Proof. Since $[\delta, \partial_a]$ is a derivation of $\mathcal{A}$, we can calculate omitting the multiplication operators as above:

$$[\delta, \partial_a] = (-1)^{\tilde{a}}[\delta, [\Delta, l_a]] + \ldots = (-1)^{\tilde{a}}([[\delta, \Delta], l_a] - [\Delta, [\delta, l_a]]) + \ldots$$
$$= -(-1)^{\tilde{a}}[\Delta, l_{\delta a}] + \ldots = \partial_{\delta a}.$$

Furthermore,

$$\delta[a \bullet b] = \delta \partial_a b = [\delta, \partial_a]b + (-1)^{\tilde{a}+1}\partial_a \delta b$$

(9.22) $$= \partial_{\delta a} b + (-1)^{\tilde{a}+1}\partial_a \delta b = [\delta a \bullet b] + (-1)^{\tilde{a}+1}[a \bullet \delta b].$$

9.5.2. Shifted differential. Let δ be a differential satisfying (9.20). For an even $a \in \mathcal{A}$ put

(9.23) $$\delta_a := \delta + \partial_a, \quad \delta_a(b) = \delta b + [a \bullet b].$$

Then we have $\delta_a^2 = 0$ if the odd Maurer–Cartan equation is satisfied:

(9.24) $$\delta a + \frac{1}{2}[a \bullet a] = 0.$$

Furthermore,

(9.25) $$[\delta_a, \Delta] = 0 \quad \text{if} \quad \Delta a = 0.$$

Therefore, from (9.24), (9.25) it follows that $(\mathcal{A}, \Delta, \delta_a)$ is a differential GBV–algebra (dGBV). In particular,

(9.26) $$[\delta_a, \partial_b] = \partial_{\delta_a b}.$$

We can in the same way shift Δ. The essential difference is that, as Δ itself, the shifted differential generally will not be the derivation with respect to the associative multiplication.

9.6. Homology of $(\mathcal{A}, \delta)$. Since δ is the derivation with respect to both multiplications in $\mathcal{A}$ (associative one and the bracket), $\mathrm{Ker}\,\delta$ is the subalgebra with respect to both of them, and $\mathrm{Im}\,\delta$ is the ideal in this subalgebra with respect to both structures. Therefore the homology group $H(\mathcal{A}, \delta)$ inherits both multiplications, satisfying the identities (9.19).

This reasoning holds for $H(\mathcal{A}, \delta_a)$ as well, if a satisfies the Maurer–Cartan equation (9.24).

9.7. Homology of $(\mathcal{A}, \Delta)$. The same reasoning furnishes only the structure of the odd Lie algebra on $H(\mathcal{A}, \Delta)$, because Δ is not a derivation with respect to the associative multiplication. However, if δ and Δ satisfy conditions (A) and (B) below, we will have the natural isomorphism $H(\mathcal{A}, \Delta) = H(\mathcal{A}, \delta)$.

The lemma below is well known; see e.g. [GoM].

9.7.1. Lemma. *Let $\mathcal{A}$ be an additive group supplied with two endomorphisms δ and Δ satisfying $\delta^2 = \Delta^2 = 0$ and $\delta \Delta = \alpha \Delta \delta$ where α is an automorphism of $\mathcal{A}$ such that $\alpha(\mathrm{Im}\,\Delta\delta) = \mathrm{Im}\,\Delta\delta$. Then clearly, $\mathrm{Im}\,\delta\Delta = \mathrm{Im}\,\Delta\delta \subset \mathrm{Im}\,\delta \cap \mathrm{Ker}\,\Delta$ and similarly with δ and Δ permuted.*

The following statements are equivalent:

(i) The inclusions of the differential subgroups $i : (\mathrm{Ker}\,\Delta, \delta) \subset (\mathcal{A}, \delta)$ and $j : (\mathrm{Ker}\,\delta, \Delta) \subset (\mathcal{A}, \Delta)$ are quasi-isomorphisms (that is, induce isomorphisms of homology).

(ii) We have actually equalities:

(A) $$\mathrm{Im}\,\delta\Delta = \mathrm{Im}\,\Delta\delta = \mathrm{Im}\,\delta \cap \mathrm{Ker}\,\Delta,$$

(B) $$\mathrm{Im}\,\delta\Delta = \mathrm{Im}\,\Delta\delta = \mathrm{Im}\,\Delta \cap \mathrm{Ker}\,\delta.$$

Assume that these conditions are satisfied. Then both homology groups in (i) are naturally isomorphic to

$$(\mathrm{Ker}\,\Delta \cap \mathrm{Ker}\,\delta)/\mathrm{Im}\,\delta\Delta.$$

Moreover, the natural map $\mathrm{Ker}\,\Delta \to H(\mathcal{A}, \Delta)$ induces the surjection of the differential groups $(\mathrm{Ker}\,\Delta, \delta) \to (H(\mathcal{A}, \Delta), 0)$ which is a quasi-isomorphism, and similarly with δ and Δ interchanged. Hence both the differential groups $(\mathcal{A}, \Delta)$ and $(\mathcal{A}, \delta)$ are formal.

Proof. We have:

(9.27) $\qquad H(i)$ is injective $\iff \operatorname{Ker} \Delta \cap \operatorname{Im} \delta = \delta (\operatorname{Ker} \Delta)$.

(9.28) $\qquad H(i)$ is surjective $\iff \operatorname{Ker} \delta \subset \operatorname{Ker} \Delta + \operatorname{Im} \delta \Rightarrow \Delta (\operatorname{Ker} \delta) = \operatorname{Im} \delta \Delta$.

Here and below all kernel and images are taken in $\mathcal{A}$. In the right hand side of (9.27), the inclusion $\supset$ is evident, and the injectivity of $H(i)$ supplies the reverse inclusion. The last arrow in (9.28) is obtained by applying Δ to the previous inclusion: this gives $\Delta(\operatorname{Ker}\delta) \subset \Delta(\operatorname{Im}\delta) = \operatorname{Im}\delta\Delta$ whereas the reverse inclusion is obvious.

Interchanging δ and Δ we find

(9.29) $\qquad H(j)$ is injective $\iff \operatorname{Ker}\delta \cap \operatorname{Im}\Delta = \Delta(\operatorname{Ker}\delta)$.

(9.30) $\qquad H(j)$ is surjective $\iff \operatorname{Ker}\Delta \subset \operatorname{Ker}\delta + \operatorname{Im}\Delta \Rightarrow \delta(\operatorname{Ker}\Delta) = \operatorname{Im}\delta\Delta$.

Taken together, (9.27) and (9.30) prove (A), and (9.28) and (9.29) prove (B), so that we have established the implication (i) $\Rightarrow$ (ii).

Conversely, assume that (A) and (B) hold.

Then $H(i)$ induces surjection on the homology, because if $\delta a = 0$, we have $\Delta a \in \operatorname{Ker}\delta \cap \operatorname{Im}\Delta$ so that by (B), $\Delta a = \Delta \delta b$, and then $a - \delta b \in \operatorname{Ker}\Delta$ represents the same homology class as a.

Moreover, $H(i)$ induces injection on the homology, because if $a \in \operatorname{Ker}\Delta$, $a = \delta b$ for some $b \in \mathcal{A}$, then $a \in \operatorname{Ker}\Delta \cap \operatorname{Im}\delta$ so that by (A), $a = \delta c$ for some $c \in \operatorname{Im}\Delta \subset \operatorname{Ker}\Delta$.

By symmetry, the same holds for $H(j)$.

The cycle subgroup for both differential groups $(\operatorname{Ker}\delta, \Delta)$ and $(\operatorname{Ker}\Delta, \delta)$ is $(\operatorname{Ker}\delta \cap \operatorname{Ker}\Delta)$, and if (i) and (ii) hold, the boundaries can be identified with $\operatorname{Im}\delta\Delta$; cf. (9.28) and (9.30). It remains to deduce formality, say, from (A) and (B).

The natural map $\operatorname{Ker}\Delta \to H(\mathcal{A}, \Delta)$ is compatible with differentials, because if $a \in \operatorname{Ker}\Delta$, then $\delta a \in \operatorname{Im}\delta \cap \operatorname{Ker}\Delta$ so that by (A), $\delta a = \Delta \delta b$ for some b, and hence the map is compatible with the zero differential on $H(\mathcal{A}, \Delta)$.

This map is surjective on the homology. In fact, consider the class of a, $\Delta a = 0$ in $H(\mathcal{A}, \Delta)$. Then $\delta a \in \operatorname{Im}\delta \cap \operatorname{Ker}\Delta$ so that in view of (A), $\delta a = \delta \Delta b$, and the δ–cycle $a - \Delta b$ represents the same class as a.

Finally, the map is injective on the homology. In fact, if $a \in \operatorname{Im}\Delta$ and $\delta a = 0$, then in view of (B), $a \in \operatorname{Im}\delta\Delta \subset \delta(\operatorname{Ker}\Delta)$.

Thus we have established the two–step quasi–isomorphism of $(\mathcal{A}, \delta)$ with $(H(\mathcal{A}, \Delta), 0)$ and by symmetry of $(\mathcal{A}, \Delta)$ with $(H(\mathcal{A}, \delta), 0)$. But the first two groups are also naturally quasi–isomorphic. So they are formal.

9.7.2. Remarks. In the context of dGBV–algebras, we will apply this identification to $(\mathcal{A}, \Delta, \delta_a)$ with variable or formal generic a. Then we will be able to interpret the "constant" space $H = H(\mathcal{A}, \Delta)$ as the flat structure on the family of algebras $H(\mathcal{A}, \delta_a)$ parametrized by the points of the generic formal section of the

Maurer–Cartan manifold. The important technical problem will then be deriving the conditions (A) and (B) for the variable a.

Notice that taken together, (A) and (B) are equivalent to

(C) $$\operatorname{Im} \delta\Delta = \operatorname{Im} \Delta\delta = (\operatorname{Ker} \delta \cap \operatorname{Ker} \Delta) \cap (\operatorname{Im} \delta + \operatorname{Im} \Delta).$$

To deduce, say, (A) from (C), one omits the last term in (C) and gets $\operatorname{Im} \delta\Delta \supset \operatorname{Im} \delta \cap \operatorname{Ker} \Delta$ whereas the inverse inclusion is obvious. Similarly, (C) follows from (A) and (B) together.

Assume that $\mathcal{A}$ is finite dimensional over a field and δ varies in a family, say $\{\delta_a\}$. After a generalization, dimension of $\operatorname{Im} \delta\Delta$ can only jump, and that of $\operatorname{Ker} \delta$ only drop. Hence if (B) holds at a point, it holds in an open neighborhood of it. In the case of the Dolbeault complex (cf. [BK]), only the cohomology will be finite–dimensional. The validity of (C) for a particular $\delta = \delta_0$ follows from the Kähler formalism. Theorem 9.2 (ii) furnishes the same result for the generic formal deformation.

9.8. Integral. Let $(\mathcal{A}, \Delta, \delta)$ be a dGBV–algebra. An even k–linear functional $\int : \mathcal{A} \to k$ is called *an integral* if the following two conditions are satisfied:

(9.31) $$\forall a, b \in \mathcal{A}, \quad \int (\delta a) b = (-1)^{\tilde{a}+1} \int a \delta b,$$

(9.32) $$\forall a, b \in \mathcal{A}, \quad \int (\Delta a) b = (-1)^{\tilde{a}} \int a \Delta b.$$

Notice that (9.31) is equivalent to $\forall a \in \mathcal{A}, \int \delta a = 0$ because δ is a k–derivation. Applying (9.32) to $b = 1$, we see that $\forall a \in \mathcal{A}, \int \Delta a = 0$ as well.

9.8.1. Proposition. *Let $\int$ be an integral for $(\mathcal{A}, \Delta, \delta)$.*

(i) If a or b belongs to $\operatorname{Ker} \Delta$, we have

(9.33) $$\int \partial_a b = \int [a \bullet b] = 0.$$

Hence if a satisfies (9.24) and (9.25), $\int$ is an integral for $(\mathcal{A}, \Delta, \delta_a)$ as well.

(ii) $\int$ induces a linear functional on $H(\mathcal{A}, \Delta)$ and $H(\mathcal{A}, \delta_a)$ for all a as above. These functionals are compatible with the identifications following from the condition (C).

Proof. If, say, $\Delta a = 0$, we have

$$\int \partial_a b = \int \left((-1)^{\tilde{a}} \Delta(ab) - (-1)^{\tilde{a}} (\Delta a) b - a \Delta b \right) = - \int a \Delta b = -(-1)^{\tilde{a}} \int \Delta a \, b = 0.$$

The rest is straightforward.

9.9. Metric. If $\int$ is an integral on $(\mathcal{A}, \Delta, \delta)$, we can define the scalar products on $H(\mathcal{A}, \delta_a)$ induced by the symmetric scalar product $(a, b) \mapsto \int ab$ on $\mathcal{A}$. For the construction of Frobenius manifolds, it is necessary to ensure that these scalar products are non–degenerate.

Integral and metric are compatible with base extensions.

9.10. Additional grading. Assume now that $\mathcal{A}$ as a commutative k–super-algebra is graded by an additive subgroup of k. Thus $\mathcal{A} = \bigoplus_n \mathcal{A}^n$, $k \in \mathcal{A}^0$, $\mathcal{A}^m \mathcal{A}^n \subset \mathcal{A}^{m+n}$, and each $\mathcal{A}^i$ is graded by parity. We write $|a| = i$ if $a \in \mathcal{A}^i$. Various induced gradings and degrees of homogeneous operations are denoted in the same way. (In the main example of [BK], $\mathcal{A}$ is $\mathbf{Z}$–graded, and each $\mathcal{A}^i$ is either even, or odd, but this plays no role in general.)

All base extensions must then be furnished by the similar grading or its topological completion.

We will assume also that $|\Delta| = -1$. It follows that $|[a \bullet b]| = |a| + |b| - 1$ which we interpret as $|\bullet| = -1$. Moreover, we postulate that $|\delta| = 1$. This means that the shifted differential δ_γ can be homogeneous only for $|\gamma| = 2$, and similarly for extended base.

Homology space H in all its incarnations (cf. Lemma 9.7.1) inherits the grading from $\mathcal{A}$. The dual space H^t is graded in such a way that the pairing $H^t \otimes H \to k$ has degree zero. This induces the additional grading (or more precisely, the notion of homogeneity) on $K = k[[H^t]]$ (which might be the product rather than the sum of its homogeneous components).

Integral is supposed to have a definite degree, not necessarily zero (and usually non–zero).

9.11. Tensor product of GBV–algebras. Let $(\mathcal{A}_i, \Delta_i)$, $i = 1, 2$, be two GBV–algebras over k. Put $\mathcal{A} := \mathcal{A}_1 \otimes \mathcal{A}_2$, and $\Delta := \Delta_1 \otimes 1 + 1 \otimes \Delta_2 : \mathcal{A} \to \mathcal{A}$.

9.11.1. Proposition. $(\mathcal{A}, \Delta)$ *is a GBV–algebra. We have for* $a_i, b_i \in \mathcal{A}_i$

$$(9.34) \qquad \partial_{a_1 \otimes a_2} = \partial_{a_1} \otimes (-1)^{\tilde{a}_2} l_{a_2} + l_{a_1} \otimes \partial_{a_2},$$

or equivalently

$$(9.35) \quad [a_1 \otimes a_2 \bullet b_1 \otimes b_2] = (-1)^{\tilde{a}_2(\tilde{b}_1+1)}[a_1 \bullet b_1] \otimes a_2 b_2 + (-1)^{\tilde{b}_1(\tilde{a}_2+1)} a_1 b_1 \otimes [a_2 \bullet b_2].$$

Proof. (9.34) is established by a straightforward calculation which we omit. From (9.34) it follows that $\partial_{a_1 \otimes a_2}$ are derivations. Hence ∂_a are derivations for all $a \in \mathcal{A}$ so that $(\mathcal{A}, \Delta)$ is a GBV–algebra. (9.35) is a rewriting of (9.34).

Clearly, tensor product is commutative and associative with respect to the standard isomorphisms.

If $\delta_i : \mathcal{A}_i \to \mathcal{A}_i$ are odd derivations of $(\mathcal{A}_i, \Delta_i)$ satisfying (9.20), then $\delta := \delta_1 \otimes 1 + 1 \otimes \delta_2$ is an odd derivation of $\mathcal{A}_1 \otimes \mathcal{A}_2$ satisfying (9.20).

If $\mathcal{A}_i$ are furnished with additional gradings having the properties postulated above, then the total grading on $\mathcal{A}_1 \otimes \mathcal{A}_2$ satisfies the same conditions.

9.11.2. Decomposable solutions to the Maurer–Cartan equation. In the notation of the previous subsection, let $(\mathcal{A}, \Delta, \delta)$ be the tensor product of $(\mathcal{A}_i, \Delta_i, \delta_i), i = 1, 2$. Assume that $a_i \in \mathcal{A}_i$ satisfy the Maurer–Cartan equation (9.24). Then from (9.35) it follows that $a := a_1 \otimes 1 + 1 \otimes a_2$ satisfies (9.24) as well. Moreover, if $\Delta_i a_i = 0$, then $\Delta a = 0$, so that $(\mathcal{A}, \Delta, \delta_a)$ is the differential GBV–algebra. Such structures will be called decomposable ones.

9.12. Example: dGBV algebras related to the Calabi–Yau manifolds ([BarK]). Let W be a compact complex Kähler manifold with the property $\Omega_W^{\max} \cong$

$\mathcal{O}_W$ ("weak Calabi–Yau"). Choose once and for all a non–zero holomorphic volume form Ω on W. Consider the **C**-algebra

$$(9.36) \qquad \mathcal{A}_W := \bigoplus_{p,q \geq 0} \Gamma_{C^\infty}(W, \wedge^q(\overline{T}_M^*) \otimes \wedge^p(T_W))$$

with $\mathbf{Z}_2$-grading $(p+q) \bmod 2$.

The map $\gamma \mapsto \gamma \vdash \Omega$ identifies $\mathcal{A}_W$ with the complexified de Rham complex of W. Let Δ correspond to ∂ with respect to this identification. One can directly check that Δ satisfies conditions of the first paragraph of 9.4 (Tian–Todorov lemma), hence claims 9.4.1 and 9.4.2 as well. Furthermore, since $\wedge^*(T_W)$ is a holomorphic vector bundle, $\mathcal{A}_W$ can be endowed with the differential $\overline{\partial}$ which we identify with δ. Again, (9.20) can be checked directly so that $(\mathcal{A}_W, \Delta, \delta)$ is a dGBV.

The key property is the validity of Lemma 9.7.1: this is essentially the $\partial\overline{\partial}$-lemma from [DGMS]. Notice that only the existence of Kähler structure on W is needed for its validity, concrete choice does not matter.

The homology space is

$$(9.37) \qquad H(\mathcal{A}_W, \delta) = H^*(W, \wedge^*(T_W)).$$

Clearly, using Ω, one can identify it with $H^*(W, \Omega^*_W)$ as well.

Define the integral by

$$(9.38) \qquad \int \gamma := \int_W (\gamma \vdash \Omega) \wedge \Omega.$$

It does not vanish only on the component $q = p = \dim W$. Properties (9.31) and (9.32) follow from the Stokes formula.

The algebra $\mathcal{A}_W$ possesses the additional **Z**-grading by $q + p$ satisfying all the conditions of sec. 9.10.

9.13. Example: dGBV algebras related to the symplectic manifolds satisfying the strong Lefschetz condition ([Me1]). Now let (U, ω) be a real manifold of dimension $2m$ endowed with a symplectic form ω. Denote by $\langle,\rangle$ the pairing on $\Omega^*(U)$ induced by the symplectic form. Put

$$(9.39) \qquad (\mathcal{B}_U, \Delta, \delta) := (\Omega^*(U), (-1)^{*+1} \star d\star, d),$$

where $\star : \Omega^k(U) \to \Omega^{2m-k}(U)$ is the symplectic star operator defined by

$$(9.40) \qquad \beta \wedge (\star \alpha) = \langle \beta, \alpha \rangle \frac{\omega^m}{m!}.$$

$\mathbf{Z}_2$-grading is the degree of the differential form $\bmod 2$. Calculating Δ in local coordinates, one sees that it is the differential operator of second order satisfying (9.14), whereas (9.20) follows from (9.40). Thus $(\mathcal{B}_U, \Delta, \delta)$ is a dGBV–algebra.

9.13.1. Proposition. *Assume that (U, ω) satisfies the strong Lefschetz condition, that is, the cup product*

$$[\omega^k] \cup : H^{m-k}(U) \to H^{m+k}(U)$$

is an isomorphism for each $k \leq m$. Then $(\mathcal{B}_U, \Delta, \delta)$ satisfies Lemma 9.7.1.

S. Merkulov [Me1] proves this, completing some earlier results from [Kos], [Br], and [Mat].

From now on, we will assume that the strong Lefschetz condition holds, so that U is compact. Then we can define the integral on $\mathcal{B}_U$:

$$(9.41) \qquad \int \gamma := \int_U \gamma.$$

Properties (9.31) and (9.32) follow from the Stokes formula combined with the identities $\star(\star\alpha) = \alpha$ and $\beta \wedge (\star\alpha) = (\star\beta) \wedge \alpha$.

The standard $\mathbf{Z}$-grading of $\Omega^*(W)$ then satisfies all conditions of sec. 9.10.

§10. From dGBV–algebras to Frobenius manifolds

10.1. Normalized formal solution to the master equation. In this section, we fix a dGBV k–algebra $(\mathcal{A}, \Delta, \delta)$ and the derived odd bracket $[\bullet]$ on it. We will assume that this algebra satisfies a series of assumptions which will be introduced and numbered consecutively.

ASSUMPTION 1. $(\mathcal{A}, \Delta, \delta)$ *satisfies the conditions of Lemma 9.7.1. Moreover, the homology group $H = H(\mathcal{A}, \delta)$ (and any group naturally isomorphic to it) is a free k–module of finite rank.*

Choosing an indexed basis $[c_i], c_i \in \mathcal{A}$, of H and the dual basis (x_i) of H^t we will always assume that $c_0 = 1$. As in 9.1, but now conserving parity, we put $K := k[[H^t]] = k[[x_i]]$. We will denote by $X_i = \partial/\partial x_i$ the respective partial derivatives acting on K and on $K \otimes \mathcal{A}, K \otimes H$, etc., via the first factor.

10.1.1. Proposition. *If $(\mathcal{A}, \Delta, \delta)$ satisfies Assumption 1 above, then there exists a generic even formal solution $\Gamma = \sum_i \Gamma_i \in K \otimes \mathrm{Ker}\,\Delta$ to the master equation*

$$\delta\Gamma + \frac{1}{2}[\Gamma \bullet \Gamma] = 0$$

with the following properties:

(i) $\Gamma_0 = 0, \Gamma_1 = \sum x_i c_i, \Gamma_n \in K \otimes \mathrm{Im}\,\Delta$ for all $n \geq 2$. Here $c_i \in \mathrm{Ker}\,\Delta \cap \mathrm{Ker}\,\delta$, and Γ_n is the homogeneous component of Γ of degree n in (x_i).

(ii) Moreover, this Γ can be chosen in such a way that $X_0 \Gamma = 1$.

Such a solution will be called normalized.

Proof. The first statement follows from Theorem 9.2 (i) applied to the odd differential Lie superalgebra $(\mathrm{Ker}\,\Delta, [\bullet], \delta)$.

We must only check that the conditions of the applicability of this theorem are satisfied. To facilitate the bookkeeping for the reader, we register the correspondences between the old and the new notation: $\mathrm{Ker}\,d$ becomes $\mathrm{Ker}\,\Delta \cap \mathrm{Ker}\,\delta$, $\mathrm{Ker}\,\phi$ turns into $\mathrm{Im}\,\Delta$, and $\mathrm{Im}\,d$ corresponds to $\mathrm{Im}\,\delta\Delta$. All of this forms a part of Lemma 9.7.1. From the second formula in 9.4 it follows that $\mathrm{Ker}\,\Delta$ is closed with respect to $[\bullet]$: $[a \bullet b] = (-1)^{\tilde{a}}\Delta(ab)$. This formula shows as well that $[\bullet]$ induces zero operation on $H(\mathrm{Ker}\,\Delta, \delta)$: if $a, b \in \mathrm{Ker}\,\Delta \cap \mathrm{Ker}\,\delta$, then $[a \bullet b] \in \mathrm{Im}\,\Delta \cap \mathrm{Ker}\,\delta = \mathrm{Im}\,\delta\Delta$.

It remains to check the assertion (ii). Clearly, our choice $c_0 = 1$ assures that $X_0\Gamma_1 = 1$. Assume by induction that $\Gamma_2, \ldots, \Gamma_n$ do not depend on x_0. Clearly,

$[c_0 \bullet a]=0$ for any a, so that in the (odd version of the) equation (9.5) the right hand side is independent of x_0 as well. Since the argument showing the existence of Γ_{n+1} in the proof of 9.2 can be applied to each coefficient of the monomials in x_i separately, we may find Γ_{n+1} independent of x_0. The final normalization argument can also be applied coefficientwise.

10.2. The (pre)–Frobenius manifold associated to $(\mathcal{A}, \Delta, \delta)$. Consider the formal manifold M, the formal spectrum of K over k. The flat coordinates will be by definition (x_i) so that the space of flat vector fields can be canonically identified with H. We fix a normalized Γ as above.

10.2.1. Lemma. *The bi-differential group $(\mathcal{A}_K, \Delta_K, \delta_\Gamma)$ satisfies the conditions and conclusions of Lemma 9.7.1.*

Proof. Clearly, $\delta_\Gamma^2 = \Delta_K^2 = [\Delta_K, \delta_\Gamma] = 0$ (the latter follows from (9.16)). From Assumption 1 and the proof of 10.1.1 above we see that we can apply Theorem 9.2 (ii) to $(K \otimes \text{Ker } \Delta, \delta_\Gamma)$ instead of $(g_K, d_{K, \Gamma})$. The inclusion (9.6) reads in this context:

$$\text{Im } \Delta_K \cap \text{Ker } \delta_\Gamma \subset \text{Im } \delta_\Gamma \Delta_K$$

which implies the condition (B).

To check (A), consider the inclusion map

$$\text{Im } \delta_\Gamma \Delta_K \to \text{Im } \delta_\Gamma \cap \text{Ker } \Delta_K.$$

It becomes an isomorphism after reduction modulo (x_i) in view of Assumption 1. Hence it is an isomorphism.

10.2.2. $\circ$–multiplication on tangent fields. We will define now the K-linear $\circ$–multiplication on the K–module of all vector fields $\mathcal{T}_M = K \otimes H = H_K$.

To this end we first apply Theorem 9.2 (ii) to the odd differential Lie algebra $(\text{Ker } \Delta, \delta_\Gamma)$. It shows that the homology of this algebra is naturally identified with H_K.

In view of Lemmas 9.7.1 and 10.2.1 we know that the injection $(\text{Ker } \Delta_K, \delta_\Gamma) \to (\mathcal{A}_K, \delta_\Gamma)$ induces the isomorphism of homology $H_K = \text{Ker } \delta_\Gamma / \text{Im } \delta_\Gamma$. But $\text{Ker } \delta_\Gamma$ is a commutative K–subalgebra of $\mathcal{A}_K$ and $\text{Im } \delta_\Gamma$ is an ideal in it. Hence H_K inherits the multiplication which we denote $\circ$. We record the following "explicit" formula for it. Interpreting any $X \in H_K$ as the derivation $\overline{X}$ of $K \otimes \mathcal{A}$ acting through the first factor (cf. Corollary 9.2.1), we have:

(10.1) $$\overline{X \circ Y} \Gamma \equiv \overline{X} \Gamma \cdot \overline{Y} \Gamma \bmod \text{Im } \delta_\Gamma$$

(dot here means the associative multiplication in $\mathcal{A}_K$). This follows directly from Corollary 9.2.1 applied to our situation. Notice that whereas $\overline{X}\Gamma$ and $\overline{Y}\Gamma$ lie in $\text{Ker } \Delta_K \cap \text{Ker } \delta_\Gamma$, their product generally lies only in the larger group $\text{Ker } \delta_\Gamma$.

Directly from the initial definition one sees that $e := X_0$ is the flat identity for $\circ$.

In order to complete the description of the pre–Frobenius structure, it remains to choose a flat metric on M.

ASSUMPTION 2. *There exists an integral $\int$ for $(\mathcal{A}, \Delta, \delta)$ such that the bilinear form on $H = H(\mathcal{A}, \delta)$ induced by $(X, Y) \mapsto \int \overline{X}\Gamma \cdot \overline{Y}\Gamma$ is non–degenerate.*

Denoting this form g we clearly have the invariance property defining the symmetric multiplication tensor A:

(10.2) $$g(X, Y \circ Z) = g(X \circ Y, Z) := A(X, Y, Z).$$

We will check now that this structure is actually Frobenius. Since the $\circ$–multiplication is associative, we have only to establish its potentiality.

To this end we will check Dubrovin's criterium: the structure connection ∇_λ on $\mathcal{T}_M$ is flat (cf. I.1.5).

To be more precise, let ∇_0 be the flat connection on $\mathcal{T}_M$ whose horizontal sections are H. Clearly, $\nabla_{0,Y}(Z) = \overline{Y}(Z)$ where this time $\overline{Y}$ means Y acting on $K \otimes H$ via K. By definition,

(10.3) $$\nabla_{\lambda,Y}(Z) = \overline{Y}(Z) + \lambda Y \circ Z,$$

where λ is an even parameter.

We have the canonical surjection $\operatorname{Ker} \Delta_K \to H_K$ and the two lifts of X both denoted by $\overline{X}$ are compatible with this surjection, and also with the embedding $\operatorname{Ker} \Delta_K \subset \mathcal{A}_K$. Therefore the section of $\operatorname{Ker} \Delta_K \to H_K$ denoted ψ in Corollary 9.2.1 sends $\overline{Y}(Z)$ to $\overline{Y}(\overline{Z}\Gamma)$, and $\nabla_{\lambda,Y}(Z)$ lifts to $\overline{Y}(\overline{Z}\Gamma) + \lambda \overline{Y}\Gamma \cdot \overline{Z}\Gamma$ in view of (10.1) and (10.3).

Since our preparations are now complete, we can prove

10.2.3. Theorem. *The connection ∇_λ is flat. Hence the pre-Frobenius structure defined above is potential.*

Proof. Applying (10.3) twice, we find

(10.4) $$\nabla_{\lambda,X}\nabla_{\lambda,Y}(Z) = \overline{X}(\overline{Y}(Z)) + \lambda \overline{X}(Y \circ Z) + \lambda X \circ \overline{Y}(Z) + \lambda^2 X \circ Y \circ Z.$$

We may and will consider only the case when X, Y supercommute (e.g. $X, Y \in H$). In order to establish flatness, it suffices to check that

(10.5) $$\overline{X}(Y \circ Z) + X \circ \overline{Y}(Z) = (-1)^{\widetilde{X}\widetilde{Y}}(\overline{Y}(X \circ Z) + Y \circ \overline{X}(Z)).$$

We will see that the ψ–lifts of both sides of (10.5) already coincide up to $\operatorname{Im} \delta_\Gamma$. In fact, $\overline{X}(Y \circ Z)$ lifts to $\overline{X}(\overline{Y}\Gamma \cdot \overline{Z}\Gamma)$ and $X \circ \overline{Y}(Z)$ lifts to $\overline{X}\Gamma \cdot \overline{Y}(\overline{Z}\Gamma)$ so that (10.5) becomes

$$\overline{X}(\overline{Y}\Gamma) \cdot \overline{Z}\Gamma + (-1)^{\widetilde{X}\widetilde{Y}}\overline{Y}\Gamma \cdot \overline{X}(\overline{Z}\Gamma) + \overline{X}\Gamma \cdot \overline{Y}(\overline{Z}\Gamma)$$
$$= (-1)^{\widetilde{X}\widetilde{Y}}\overline{Y}(\overline{X}\Gamma) \cdot \overline{Z}\Gamma + \overline{X}\Gamma \cdot \overline{Y}(\overline{Z}\Gamma) + (-1)^{\widetilde{X}\widetilde{Y}}\overline{Y}\Gamma \cdot \overline{X}(\overline{Z}\Gamma).$$

This finishes the proof.

10.3. Euler field. Assume now that $\mathcal{A}$ is endowed with a grading satisfying the conditions of 9.10. All of the previous discussion makes sense, and the results hold true if we add appropriate grading conditions at certain places, the most important of which is $|\Gamma| = 2$ implying $|\circ| = 2$ in view of (10.1).

Denote by E the derivation of K defined by the following Euler condition:

(10.6) $$\forall f \in K, \ Ef = \frac{1}{2}|f|f,$$

where $|\ |$ is the grading induced on H^t from $\mathcal{A}$ via H. In coordinates as in 10.1 we have

(10.7) $$E = \frac{1}{2}\sum_i |x_i|\, x_i X_i.$$

ASSUMPTION 3. *Assume that the integral is homogeneous and denote its degree by $2D - 4$.*

10.3.1. Proposition. *E is an Euler field on the formal Frobenius manifold described in 10.2. Its spectrum is $(D;\ d_i$ with multiplicity $\dim H^{-2d_i})$, and $d_0 = 1$.*

Proof. Comparing (10.7) with the notation of I.2.4, we see that the spectrum of $-\mathrm{ad}\, E$ on $H = T_M^f$ is d_i with multiplicity $\dim H^{-2d_i}$ where

(10.8) $$d_i = \frac{1}{2}|x_i| = -\frac{1}{2}|X_i|.$$

Since $X_0 \circ X = X$ and $|\circ| = 2$, we have $d_0 = 1$.
We must now check the formula

(10.9) $$E(g(X,Y)) - g([E,X],Y) - g(X,[E,Y]) = D g(X,Y).$$

It suffices to do this for the case when X, Y are flat vector fields having definite degrees. Then $[E, X] = \dfrac{|X|}{2} X$. Since $g(X, Y) \in k$, (10.9) becomes

$$(|X| + |Y| + 2D) g(X, Y) = 0.$$

But $g(X, Y) = \int \overline{X\Gamma} \cdot \overline{Y\Gamma}$ vanishes unless $2D - 4 + |X| + 2 + |Y| + 2 = 0$ which proves (10.9).

Furthermore, from (10.1) we infer that $|X \circ Y| = |X| + |Y| + 2$. Hence if $X_i \circ X_j = \sum_k A_{ij}{}^k X_k$, we have

$$E A_{ij}{}^k = \frac{1}{2}|A_{ij}{}^k| \cdot A_{ij}{}^k = \frac{1}{2}(|X_i| + |X_j| - |X_k| + 2)\, A_{ij}{}^k.$$

Comparing this with formula (2.18) of Chapter I and taking into account (10.8), we see that E satisfies Definition I.2.2.1. This finishes the proof.

Notice that the Euler field (10.7) contains no flat summand: X_i with $d_i = 0$ do not contribute. Hence if this construction furnishes a Frobenius manifold which is quantum cohomology of some V, then $c_1(V)$ must vanish (modulo torsion).

10.3.2. Remark. Comparing the Frobenius manifold produced from a dGBV-algebra with grading as above to a quantum cohomology Frobenius manifold, one must first shift the dGBV-grading by two. Then X_0 and $\circ$ acquire degree zero.

10.4. Explicit potential. The direct way to establish potentiality is to find an even series $\Phi \in K$ such that for all $X, Y, Z \in H$ we have $A(X, Y, Z) = XYZ\Phi$ (from now on, we write X instead of $\overline{X}$ in order to denote derivations on various K-modules acting through K). Moreover, it suffices to check this for $X = Y = Z$. We will give here the beautiful formula of Chern–Simons type for Φ discovered in [BarK].

Extend the integral to the K-linear map $\int : \mathcal{A}_K \to K$. For a fixed normalized Γ put $\Gamma = \Gamma_1 + \Delta B$ where $B_0 = B_1 = 0$ and Δ means Δ_K.

10.4.1. Theorem. *The formal function*

$$\Phi := \int \left(\frac{1}{6} \Gamma^3 - \frac{1}{2} \delta B \, \Delta B \right) \tag{10.10}$$

is a potential for the Frobenius manifold defined above.

Proof. We have to prove that for any $X \in H$

$$A(X, X, X) = \int (X\Gamma)^3 = X^3 \Phi. \tag{10.11}$$

We supply below the detailed calculation consisting of the series of elementary steps, each being an application of one of the identities (9.31), (9.32), Leibniz' rule for (super)derivations, and the fact that δ, Δ, X pairwise supercommute. Moreover, we use the master equation in the form $\Delta \Gamma^2 = -2\delta\Gamma$ following from $\Delta\Gamma = 0$. Finally, $\delta\Gamma_1 = X^n \Gamma_1 = 0$ for $n \geq 2$ so that $\delta\Gamma = \delta\Delta B$, $X^n \Gamma = X^n \Delta B$.

We start with treating the first summand of the right hand side of (10.10). The derivation X is interchangeable with integration, so we have by the Leibniz rule

$$X^3 \left(\frac{1}{6} \int \Gamma^3 \right) = \int (X\Gamma)^3 + \int \left((2 + (-1)^{\widetilde{X}})\Gamma \cdot X\Gamma \cdot X^2\Gamma + \frac{1}{2} \Gamma^2 \cdot X^3\Gamma \right). \tag{10.12}$$

The second summand of (10.10) is added in order to cancel the extra terms in (10.12).

First, we rewrite it:

$$\frac{1}{2} \int \delta B \, \Delta B = \frac{1}{2} \int B \delta \Delta B = \frac{1}{2} \int B \delta\Gamma = -\frac{1}{4} \int B \Delta(\Gamma^2) = \frac{1}{4} \int \Delta B \cdot \Gamma^2. \tag{10.13}$$

(We could have chosen the last expression in (10.13) from the start.)

Now, again by the Leibniz rule,

$$\frac{1}{4} X^3 \int \Delta B \cdot \Gamma^2 = \frac{1}{4} \int \Big(X^3(\Delta B)\, \Gamma^2 + (2 + (-1)^{\widetilde{X}}) X^2(\Delta B) \cdot X(\Gamma^2)$$
$$+ (2 + (-1)^{\widetilde{X}}) X(\Delta B) \cdot X^2(\Gamma^2) + \Delta B \cdot X^3(\Gamma^2) \Big). \tag{10.14}$$

The first two summands in (10.14) can be directly rewritten in the same form as in (10.12):

$$X^3(\Delta B) \cdot \Gamma^2 + (2 + (-1)^{\widetilde{X}}) X^2(\Delta B) \cdot X(\Gamma^2)$$
$$= X^3 \Gamma \cdot \Gamma^2 + (2 + (-1)^{\widetilde{X}}) X^2 \Gamma \cdot X(\Gamma^2). \tag{10.15}$$

The third summand takes somewhat more work:

$$\int X(\Delta B) \cdot X^2(\Gamma^2) = -\int XB \cdot X^2(\Delta\Gamma^2) = 2 \int XB \cdot X^2(\delta\Gamma)$$
$$= 2 \int XB \cdot \delta(X^2\Gamma) = 2 \int XB \cdot \delta\Delta(X^2 B) = -2 \int \delta\Delta XB \cdot X^2 B$$
$$= -2 \int X\delta\Gamma \cdot X^2 B = \int X\Delta(\Gamma^2) \cdot X^2 B = \int X(\Gamma^2) \cdot X^2 \Delta B$$
$$= \int X(\Gamma^2) \cdot X^2 \Gamma. \tag{10.16}$$

Finally, the fourth summand is calculated similarly, but in two steps. We start with an expression of the second order in X:

$$\int \Delta B \cdot X^2(\Gamma^2) = -\int B \cdot X^2(\Delta \Gamma^2) = 2\int B \cdot X^2(\delta \Gamma)$$

$$= 2\int B \cdot \delta(X^2 \Gamma) = 2\int B \cdot \delta\Delta(X^2 B) = -2\int \delta\Delta B \cdot X^2 B$$

$$= -2\int \delta\Gamma \cdot X^2 B = \int \Delta(\Gamma^2) \cdot X^2 B = \int \Gamma^2 \cdot X^2 \Delta B$$

(10.17)
$$= \int \Gamma^2 \cdot X^2 \Gamma.$$

Now apply X to the first and the last expressions of (10.17). We get

(10.18) $$\int X(\Delta B) \cdot X^2(\Gamma^2) + \int \Delta B \cdot X^3(\Gamma^2) = \int X(\Gamma^2) \cdot X^2 \Gamma + \int \Gamma^2 \cdot X^3 \Gamma.$$

Comparing this with (10.16), one gets

(10.19) $$\int \Delta B \cdot X^3(\Gamma^2) = \int \Gamma^2 \cdot X^3 \Gamma.$$

Putting all of this together, one obtains finally (10.11).

10.5. Example: Frobenius manifolds of B–type, related to the Calabi–Yau manifolds. Returning now to the examples of 9.12, one sees that all assumptions of this section hold so that we get a class of Frobenius manifolds, which we may call BK–models of Calabi–Yau manifolds W. In particular, we can easily calculate the d–spectrum which is:

(10.20) $$(w; d \text{ with multiplicity } \sum_{q+p=2d} h^{p,w-q}(W)), \ w := \dim_\mathbb{C}(W).$$

10.6. Example: Frobenius manifolds related to the symplectic manifolds satisfying the strong Lefschetz condition. Similarly, in the situation of 9.13 we obtain the Frobenius manifold with the d–spectrum

(10.21) $$(m; d \text{ with multiplicity } \dim H^{2d}(W)).$$

Notice that in this case as well the anticanonical component of the Euler field vanishes.

It would be interesting to establish isomorphisms between these examples and to understand when they furnish Frobenius manifolds of qc–type. Notice that if W, U are mirror dual Calabi–Yau manifolds, then the spectra of $\mathcal{A}_W$ and $\mathcal{B}_U$ coincide.

For a statement on the mirror isomorphism of this type, see [CaoZh1] and [CaoZh2].

CHAPTER IV

Operads, Graphs, and Perturbation Series

§1. Classical linear operads

1.1. Setup. In this section we fix a ground field k of characteristic zero and denote by $VECT$ the category of linear spaces over k. All tensor products are taken over k unless it is explicitly stated otherwise. The symmetric group $\mathbf{S}_n$ is defined as the group of the bijections $\underline{n} \to \underline{n}$ where $\underline{n} = \{1, \ldots, n\}$.

Classical linear algebra deals with a linear space V endowed with a family $\mathcal{O}$ of linear operators $V \to V$. Usually it is convenient to close $\mathcal{O}$ by adding all operator compositions and their linear combinations to $\mathcal{O}$. In this way linear algebra becomes the study of associative k-algebras and their linear representations.

Classical linear operads arise in the same way when we start with a linear space V endowed with a family $\mathcal{P}$ of *poly*linear operators $V^{\otimes m} \to V$, $m = 1, 2, 3, \ldots$ (for example, an associative algebra is such a space endowed with a multiplication map $V^{\otimes 2} \to V$). Closing $\mathcal{P}$ with respect to compositions (of functions with many variables) and linear combinations we get a (concrete) classical linear operad together with its linear representation in V. Axiomatizing the universal properties of such an object, we get the following notion.

1.1.1. Definition. *A classical linear operad $\mathcal{P}$ consists of the data a) – d) satisfying the axioms A) – C) below.*

a) A family of linear spaces $\mathcal{P}(j)$, for all $j \geq 1$.

b) A left/right linear action of $\mathbf{S}_j$ on $\mathcal{P}(j)$, for all $j \geq 1$: $s \in \mathbf{S}_j$ maps $f \in \mathcal{P}(j)$ to $fs = s^{-1}f$.

c) A family of composition maps $\gamma(k_1, \ldots, k_j)$, for all $j \geq 1$, $k_a \geq 1$:

$$(1.1) \quad \gamma(k_1, \ldots, k_j): \mathcal{P}(j) \otimes \mathcal{P}(k_1) \otimes \cdots \otimes \mathcal{P}(k_j) \to \mathcal{P}(k_1 + \cdots + k_j).$$

d) (Optional). An identity element $I \in \mathcal{P}(1)$.

We will state the axioms for these data in two forms: directly in terms of γ and in functional notation. For the latter, put $\underline{\mathcal{P}} = \bigoplus_{k=1}^{\infty} \mathcal{P}(k)$ and notice that (1.1) allows us to consider each $f \in \mathcal{P}(j)$ as a polylinear function $\underline{\mathcal{P}}^j \to \underline{\mathcal{P}}$:

$$(1.2) \quad f(g_1, \ldots, g_j) := \gamma(f \otimes g_1 \otimes \cdots \otimes g_j),$$

where $\gamma = \gamma(k_1, \ldots, k_j)$ if $g_a \in \mathcal{P}(k_a)$. We will often write simply γ for such multigraded components of the operadic composition.

A) *The symmetric group $\mathbf{S}_j$ acts on the functions (represented by) $\mathcal{P}(j)$ by permutation of arguments:*

$$(1.3) \quad (fs)(g_1, \ldots, g_j) = f(s(g_1, \ldots, g_j)).$$

In γ-notation:

(1.3') $$\gamma(fs \otimes g_1 \otimes \cdots \otimes g_j) = \gamma(f \otimes s(g_1 \otimes \cdots \otimes g_j)).$$

In addition, for $s_1 \in S_{k_1}, \ldots, s_j \in S_{k_j}$, denote by $s_1 \times \cdots \times s_j \in S_{k_1 + \cdots + k_j}$ the image of $(s_1, \ldots, s_j)$ acting blockwise upon

$$(1, \ldots, k_1 | k_1 + 1, \ldots, k_1 + k_2 | \ldots | k_1 + \cdots + k_{j-1} + 1, \ldots, k_1 + \cdots + k_j).$$

Then

(1.4) $$f(g_1 s_1, \ldots, g_j s_j) = (f(g_1, \ldots, g_j))(s_1 \times \cdots \times s_j).$$

In γ-notation:

(1.4') $$\gamma(f \otimes g_1 s_1 \otimes \cdots \otimes g_j s_j) = (\gamma(f \otimes g_1 \otimes \cdots \otimes g_j))(s_1 \times \cdots \times s_j).$$

B) The composition maps are associative with respect to the substitution (in functional notation). That is, for any $f \in \mathcal{P}(j)$, $g_a \in \mathcal{P}(k_a), a = 1, \ldots, j$, *and* $h_{a,b} \in \mathcal{P}(l_{a,b}), b = 1, \ldots, k_a$, *we have*

$$[f(g_1, \ldots, g_j)](h_{1,1}, \ldots, h_{1,k_1}; \ldots; h_{j,1}, \ldots, h_{j,k_j})$$

(1.5) $$= f(g_1(h_{1,1}, \ldots, h_{1,k_1}), \ldots, g_j(h_{j,1}, \ldots, h_{j,k_j})).$$

In γ-notation:

$$\gamma(\gamma(f \otimes g_1 \otimes \cdots \otimes g_j) \otimes h_{1,1} \otimes \cdots \otimes h_{j,k_j})$$

(1.5') $$= \gamma(f \otimes \gamma(g_1 \otimes h_{1,1} \otimes \cdots \otimes h_{1,k_1}) \otimes \cdots \otimes \gamma(g_j \otimes h_{j,2} \otimes \cdots \otimes h_{j,k_j})).$$

C) (Optional). If $\mathcal{P}$ is endowed with identity $I \in \mathcal{P}(1)$, then I (resp. $I^{\otimes n}$) become left (resp. right) identical functions:

(1.6) $$I(g) = g; \quad f(I, \ldots, I) = f,$$

(1.6') $$\gamma(I \otimes g) = g; \quad \gamma(f \otimes I \otimes \cdots \otimes I) = f.$$

An operad endowed with identity which is considered as a part of its structure will be called a unital operad.

We will often call the classical linear operads simply operads until the introduction of other versions of this notion.

1.2. Example. Let (E, μ) be an associative algebra with multiplication $\mu : E \otimes E \to E$. Define an operad $\mathcal{P}_E$ by $\mathcal{P}_E(1) = E$, $\mathcal{P}_E(j) = \{0\}$ for $j \geq 2$, $\gamma(1) = \mu$, the rest of the data being self–explanatory. Operadic associativity of γ is clearly equivalent to the associativity of μ.

Conversely, for any operad $\mathcal{P}$, $\mathcal{P}(1)$ with multiplication $\gamma(1)$ is an associative algebra. Operadic identity becomes algebra identity and vice versa.

1.3. Example. Let V be a linear space. Define the operad $OpEnd(V)$ by the following data:

(1.7) $$OpEnd(V)(j) = Hom_{VECT}(V^{\otimes j}, V),$$

S_j acts by permuting arguments as in (1.3), the composition γ is defined by substitution as in the lhs of (1.2), and $I = id_V$.

1.4. Definition. *A morphism of operads $\varphi : \mathcal{P} \to \mathcal{Q}$ is a family of linear maps $\varphi(j) : \mathcal{P}(j) \to \mathcal{Q}(j)$, $j \geq 1$, compatible with the action of symmetric groups, composition, and optionally, mapping $I_\mathcal{P}$ to $I_\mathcal{Q}$.*

Thus we have defined a category of classical linear operads $OPER$. In fact, we allow some ambiguity, because the existence of the identity is optional, and, even if it exists, we may decide not to consider it as a part of the structure when we define morphisms. This extends the common ambiguity in the definition of the category of associative algebras.

1.5. Operads as a generalization of associative algebras. Denote by ASS one of the two categories of associative k-algebras (with or without identity). Constructions of sec. 1.2 extend to the functors $ASS \to OPER$ and $OPER \to ASS$ which are adjoint to each other from both sides so that we have canonical identifications
$$Hom_{OPER}(\mathcal{P}, \mathcal{P}_A) = Hom_{ASS}(\mathcal{P}(1), A),$$
$$Hom_{OPER}(\mathcal{P}_A, \mathcal{P}) = Hom_{ASS}(A, \mathcal{P}(1)).$$
In particular, ASS is a full subcategory of $OPER$.

1.6. Operads as classifiers of algebras of different species. By species we mean here a general notion whose specializations include, e.g., associative, Lie, commutative, and Poisson algebras; cf. 1.6.4 below.

1.6.1. Definition. *Let $\mathcal{P}$ be an operad and V a linear space. A structure of $\mathcal{P}$-algebra on V, or equivalently, a linear representation of $\mathcal{P}$ in V, is a morphism of operads $\rho : \mathcal{P} \to OpEnd(V)$ sending I to id_V if $\mathcal{P}$ is unital.*

As Definition 1.1.1 shows, $\underline{\mathcal{P}} = \bigoplus_{j \geq 1} \mathcal{P}(j)$ has a canonical structure of $\mathcal{P}$-algebra (regular representation).

Generally, to define a structure of $\mathcal{P}$-algebra on V is the same as to define for every element $f \in \mathcal{P}(j)$ a j-ary multiplication map $m_f : V^{\otimes j} \to V$ linearly depending on f, translating γ-composition to substitution and the action of the symmetric groups to the permutation of the arguments.

1.6.2. Definition. *Let V, W be two $\mathcal{P}$-algebras. A morphism between them is a linear map $\varphi : V \to W$ such that for every $f \in \mathcal{P}(j)$ we have*

(1.8) $$\varphi(m_f^V(v_1 \otimes \cdots \otimes v_j)) = m_f^W(\varphi(v_1) \otimes \cdots \otimes \varphi(v_j)).$$

We will show that for certain species C of k-algebras which we may call "operadic" one can find a unital operad COp such that COp-algebras and morphisms between them "are" algebras of the species C and their morphisms.

1.6.3. An informal discussion. Let us start with an example, taking again associative algebras without unit, this time considered as a species. Besides the identity map $V \to V$, any associative algebra is commonly given by one generating bilinear multiplication $m : V \otimes V \to V$, but the transposition of arguments transforms it into another multiplication m^{op}. Therefore we must put $AssOp(1) = \langle I \rangle$ (brackets denoting the linear span), $AssOp(2) = \langle m, m^{op} \rangle$, the regular representation of S_2. In $AssOp(3)$ we have then twelve ternary operations that can be constructed from I, m, m^{op}: in the functional notation they are $m(m, I), m(I, m^{op})$, etc. In plain words, each such operation applied to $v_1 \otimes v_2 \otimes v_3 \in V^{\otimes 3}$ picks two

v_i's, multiplies them in some order, and then multiplies the result by the remaining v_j.

These twelve ternary operations are related by identities expressing the associativity of m and its consequence, that of m^{op}: $m(m, I) = m(I, m)$, etc. As a result, $AssOp(3)$ is isomorphic to the regular representation of S_3 generated by, say, $m(m, I)$.

The general pattern is as follows. Pick an infinite sequence of independent non-commuting but associative variables $x_1, x_2, x_3, \ldots$. Instead of $m, m^{op}, m(I, m^{op})$, etc., write the value of the respective operation applied to the initial segment of this sequence, getting respectively $x_1 x_2, x_2 x_1, x_1 x_3 x_2$, etc. A contemplation shows that one can thus identify $AssOp(n)$ with the linear space generated by all associative monomials $x_{s(1)} \ldots x_{s(n)}$ where $s \in \mathbf{S}_n$, with the evident action of $\mathbf{S}_n$.

Namely, $m(\ldots(m(m, I), I)\ldots)$ produces the monomial $(\ldots((x_1 x_2) x_3) \ldots) x_n = x_1 \ldots x_n$, and the application of $\mathbf{S}_n$ furnishes the rest. It remains to describe the γ-composition of a monomial $x_{s(1)} \ldots x_{s(n)}$ with $g_1 \otimes \cdots \otimes g_n \in \bigotimes_{a=1}^n AssOp(j_a)$. We first replace arguments $x_1, \ldots, x_{j_a}$ in g_a by adding $j_1 + \cdots + j_{a-1}$ to all subscripts thus getting $\widetilde{g}_a$, and then put

$$\gamma(x_{s(1)} \otimes \cdots \otimes x_{s(n)} \otimes g_1 \otimes \cdots \otimes g_n) := \widetilde{g}_{s(1)} \ldots \widetilde{g}_{s(n)}.$$

Now let us try to construct a functor from $AssOp$-algebras to associative algebras.

A structure of an $AssOp$-algebra on V, clearly, is uniquely determined by the restriction of the operadic morphism

$$\rho(2): AssOp(2) \to Hom_{VECT}(V^{\otimes 2}, V).$$

However, the image of $\rho(2)$ is a two-dimensional space of multiplications $\{am + bm^{op}\}$ whereas classically we need just one associative multiplication. Let us write the associativity equation $\mu(\mu, I) = \mu(I, \mu)$ for $\mu = am + am^{op}$ in the functional notation with free arguments x, y, z:

$$\mu(\mu, I) = a[(axy + byx)z] + b[z(axy + byx)],$$

$$\mu(I, \mu) = a[x(ayz + bzy)] + b[(ayz + bzy)x].$$

Comparing coefficients, one sees that the *universal* associativity (in any linear representation) is equivalent to $ab = 0$. Hence the best we can do is to pinpoint in any $AssOp$-algebra V two lines of associative multiplications: $\langle \rho(m) \rangle$ and $\langle \rho(m^{op}) \rangle$. An additional choice of unit would reduce each line to one (non-zero) element, however there is nothing in the structure of $AssOp$ that would help us to do this. In fact, we encounter here a general problem: how to account for eventual structural special elements, i.e. 0-ary operations? In principle, we could have extended the definition of an operad $\mathcal{P}$ by including $\mathcal{P}(0)$ and extending correspondingly (1.1). In particular, we can put $AssOp(0)$ = ground field, $OpEnd(V)(0) = V$, and define the identity in V as the image of 1. In other cases this might not work.

We will now summarize the preceding discussion in a deliberately vague "metatheorem" (for more precise statements, see below).

1.6.4. Species of algebras and operads.
Let C be a category of algebras which is defined by a family of multilinear operations $\{m_i | i \in I\}$ and a family of

universal identities between them constructed of compositions and linear combinations. Morphisms in C are linear maps compatible with m_i's. Examples: associative algebras without identity (multiplication; associativity); Lie algebras (bracket; skew–symmetry; Jacobi identity); Poisson algebras without identity (multiplication, bracket; associativity, commutativity, Jacobi, Leibniz); commutative rings with an m–dimensional linear space of pairwise commuting derivations, etc.

Then one can construct a classical linear operad COp with the following properties.

a) $COp\,(j)$ as a representation space of $\mathbf{S}_j$ is isomorphic to a subspace of the free algebra $F_C(x_1,\ldots,x_j)$ of the species C freely generated by j independent variables $x_1,\ldots,x_j$. This subspace consists of forms of total degree j linear in each x_a, upon which $\mathbf{S}_j$ acts by permuting arguments.

b) Compositions γ are induced by substitution.

c) To give a structure of a COp–algebra on a space V is the same as giving a set of structures of species C on V. Various elements of this set are obtained by choosing in COp various generating families of solutions $\{m'_i\}$ of the universal identities defining C. The group $\text{Aut}(COp)$ acts transitively on this set.

d) The category of COp–algebras is equivalent to the category of algebras of the species C. Every choice of generators $\{m_i\}$ as above fixes one equivalence functor. However, two different choices may lead to non–isomorphic functors.

This happens, e.g., with $AssOp$ and functors corresponding to m and m^{op}.

To give a precise statement and proof of this theorem, we would have to explain in more detail the two different notions of "freeness" and "defining an object by generators and relations": separately for operads and algebras over a given operad. Above we used them on an intuitive level.

Before proceeding further, we want to list the limitations of the operadic approach to species, some of which can be overcome by modifying the notion of the classical operad.

• We cannot account for the structure constants, partly because of the lack of $\mathcal{P}(0)$.

• We cannot account for the use of dual spaces in the definitions of some species, e.g., algebras with invariant scalar products interpreted as $V \to V^*$. (In this case, a remedy is the introduction of the cyclic operads).

• We cannot account for the structure morphisms like comultiplication $V \to V \otimes V$, and generally tensors of various co– and contravariant degrees.

• We cannot account for non–linear and not everywhere defined operations like inversion in the multiplicative group of a field.

1.7. Operads as analogs of associative algebras. In 1.2, 1.5 we have shown that the associative algebras naturally form a part of the classical operads (with $\mathcal{P}(j) = 0$ for $j \geq 2$). We will now demonstrate that the total classical operad $\mathcal{P}$ is in a very definite sense an analog of associative algebra.

To do this convincingly, we must start with a definition of an associative algebra as a couple (V,m) where V is an object of the monoidal category $VECT$, and m is an associative morphism $V \otimes V \to V$, eventually endowed with identity which is a morphism $\mathbf{1} \to V$ where $\mathbf{1}$ is the ground field considered as an identity object in

$VECT$. The two categories $(VECT, \otimes)$ and ASS obtained in this way are connected by the two adjoint functors

$$\text{forget } m: ASS \to VECT,$$

$$\text{free (tensor) algebra}: VECT \to ASS.$$

In order to present the classical linear operads in the same way we have to start with specifying an analog of the functor "forget m". This can be done in several ways because we can choose to forget any subset of the data given in Definition 1.1.1. Here we will decide that m corresponds to all γ's. What is left then is the following category $SMOD$ of **S**-modules:

1.7.1. Definition. *An object of $SMOD$ is a family of linear spaces $V(j)$, $j \geq 1$, endowed with an action of $\mathbf{S}_j$.*

A morphism in $SMOD$ is a family of linear maps $V(j) \to W(j)$ compatible with the $\mathbf{S}_j$-action.

We will sometimes say that $V(j)$ is the part of V of degree j.

1.7.2. Lemma–Definition. *a) The category $SMOD$ possesses a bifunctorial product $*$ which can be defined on the objects by the following formula:*

$$(1.9) \qquad V * W(n) = \bigoplus_{j=1}^{n} V(j) \otimes_{\mathbf{S}_j} \left(\bigoplus_{\pi: \underline{n} \to \underline{j}} \bigotimes_{i=1}^{j} W(|\pi^{-1}(i)|) \right).$$

Here $\underline{n} = \{1, \ldots, n\}$ and π runs over all surjective maps. The action of $\mathbf{S}_j$ must be self-explanatory, and the tensor product is taken over the group ring of $\mathbf{S}_j$.

This product is functorially associative but not commutative so that $(SMOD, *)$ is a monoidal category. It possesses a two-sided identity object $\mathbf{1}$: the ground field placed in degree 1, zero elsewhere.

b) The map $V \mapsto (V, 0, 0, \ldots)$ extends to a functor identifying $(VECT, \otimes, \mathbf{1})$ with a full monoidal subcategory of $(SMOD, *, \mathbf{1})$.

Now consider an associative algebra (V, μ), $\mu: V * V \to V$ in the monoidal category of **S**-modules. From (1.9) we see that μ is a family of maps

$$(1.10) \qquad \mu(n): \bigoplus_{j=1}^{n} V(j) \otimes_{\mathbf{S}_j} \left(\bigoplus_{\pi: \underline{n} \to \underline{j}} \bigotimes_{i=1}^{j} V(|\pi^{-1}(i)|) \right) \to V(n), \ n \geq 1.$$

For given $(j; k_1, \ldots, k_j)$, $k_1 + \cdots + k_j = n$, consider the component of (1.10) corresponding to the $\mathbf{S}_j$-orbit of the map sending $\{1, \ldots, k_1\}$ to 1, $k_1 + 1, \ldots, k_1 + k_2$ to 2, etc. We can identify this part of the source with $V(j) \otimes V(k_1) \otimes \cdots \otimes V(k_j)$ so that μ generates a family of maps

$$(1.11) \qquad \gamma(k_1, \ldots, k_j): V(j) \otimes V(k_1) \otimes \cdots \otimes V(k_j) \to V(k_1 + \cdots + k_j).$$

1.7.3. Proposition. *a) The associativity of μ translates into the associativity of γ's in the sense of (1.5).*

b) The fact that μ is a morphism in $SMOD$ translates into the compatibility axioms (1.3) and (1.4).

c) In this way we get a functor

$$\text{Associative algebras in } (SMOD, *) \to OPER$$

which is an equivalence of categories.

There exists a similar equivalence between associative algebras with identity and unital operads.

Proof. We will now sketch a proof of the main statements in 1.7.2 and 1.7.3. In order to understand the main formula (1.9), we will show that it expresses the substitution law of "formal series in $VECT$".

To be more precise, denote by $FSETS$ the category of finite non–empty sets and bijections. Let $F[\cdot]: FSETS \to VECT$ be a functor.

$FSETS$ is equivalent to its full subcategory whose objects are $\underline{n}$. Restricting F to this subcategory we get an **S**–module $V_F : V_F(n) := F[\underline{n}]$, the action of $\mathbf{S}_n$ being induced by the bijections of $\underline{n}$.

Now consider $V_F(n)$ as coefficients of the formal series defining the functor $F(\cdot): VECT \to VECT$:

$$F(X) := \bigoplus_{n \geq 1} V_F(n) \otimes_{\mathbf{S}_n} X^{\otimes n}.$$

Such functors will be called *analytic ones*.

We will show that these constructions establish an equivalence of the three categories involved: functors $F[\cdot]$ and their morphisms, $SMOD$, analytic functors. Moreover, the composition of analytic functors is again analytic, and it induces on the coefficients exactly the $*$–product:

$$V_{F \circ G}(n) = (V_F * V_G)(n).$$

The equivalence of the category of functors $F[\cdot]$ and $SMOD$ is a part of general nonsense because $FSETS$ is equivalent to its subcategory of natural numbers. The only point deserving explication is the possibility to lift every **S**–module to an $F[\cdot]$ canonically without using the axiom of choice. Namely, for a finite set M with $|M| = m$ put

$$\widetilde{F}[M] := F[\underline{m}] \otimes_{\mathbf{S}_m} \langle Iso(\underline{m}, M) \rangle.$$

Here $\langle Iso(\underline{m}, M) \rangle$ is the linear space freely generated by the bijections $\underline{m} \to M$. Strictly speaking, now $\widetilde{F}[\underline{m}]$ is not $F[\underline{m}]$, but these $\mathbf{S}_m$–modules are canonically isomorphic, and we forget about this subtlety and say, for example, that the **S**–module $F[\underline{m}] = X^{\otimes m}$ extends to the functor $F[M] = X^{\otimes M}$ on the category of finite sets.

The equivalence of $SMOD$ and the category of analytic functors $F(\cdot)$ also becomes a formal fact once we learn how to reconstruct functorially the coefficients $V_F(n)$. Let $F(\cdot)$ be given. Multiplication by any element λ of the ground field is an endomorphism of the identical functor of $VECT$. Hence it acts functorially on each $F(X)$, and the λ^n-eigenspace of $F(X)$ is exactly $F_n(X) := V_F(n) \otimes_{\mathbf{S}_n} X^{\otimes n}$, at least when λ is not 0 or a root of unity. Now consider the space $X_n = \langle \underline{n} \rangle$ freely generated by the vectors $e_1, \ldots, e_n$. Then $e_1 \otimes \cdots \otimes e_n$ generates the regular $\mathbf{S}_n$-submodule R_n which is the image of the projector $p_n : X_n^{\otimes n} \to X_n^{\otimes n}$. Since F_n is a functor, we can define $\mathrm{Im}\,(F_n(p_n)) = V_F(n) \otimes_{\mathbf{S}_n} R_n = V_F(n)$, both equalities denoting canonical isomorphisms.

We will apply this prescription to the calculation of the coefficients of the composition of analytic functors:

$$(F \circ G)(X) = \bigoplus_{j=1}^{\infty} V_F(X) \otimes_{\mathbf{S}_j} \left(\bigoplus_{k=1}^{\infty} V_G(k) \otimes_{S_k} X^{\otimes k} \right)^{\otimes j}$$

$$= \bigoplus_{j=1}^{\infty} V_F(X) \otimes_{\mathbf{S}_j} \left[\bigoplus_{k_1,\ldots,k_j=1}^{\infty} (V_G(k_1) \otimes_{S_{k_1}} X^{\otimes k_1}) \otimes \cdots \otimes (V_G(k_j) \otimes_{S_{k_j}} X^{\otimes k_j}) \right].$$

It follows that

$$(F \circ G)_n(X) = \bigoplus_{j=1}^{\infty} V_F(X) \otimes_{\mathbf{S}_j} \left[\bigoplus_{k_1+\cdots+k_j=n} \bigotimes_{a=1}^{j} (V_G(k_a) \otimes_{S_{k_a}} X^{\otimes k_a}) \right].$$

Now we must put $X = X_n$ as above and look at the image of $(F \circ G)_n(p_n)$ or, more intuitively, at the tensor coefficients of the vectors $e_{s(1)} \otimes \cdots \otimes e_{s(n)}$. Clearly, for a given j, such terms in square brackets correspond to the partitions of $\underline{n}$ into j blocks indexed by $1, \ldots, j$, i.e. to the surjections $\underline{n} \to \underline{j}$ as in (1.9).

To finish the proof, it remains to establish that the functor $F \circ G(\cdot)$ is isomorphic to the sum of $\mathrm{Im}\,(F \circ G)_n(p_n)$. We leave this to the reader.

1.7.4. Tensor algebra in $SMOD$**.** Let V be an object of $SMOD$. Put

$$F(V) := \sum_{n=1}^{\infty} V^{*n}.$$

There is an obvious multiplication map $V^{*m} * V^{*n} \to V^{*m+n}$ which makes $F(V)$ an associative algebra, or an operad. It is called *the free operad generated by* V (without identity). As in the classical linear algebra, F is adjoint to the forgetful functor $OPER \to SMOD$. This completes the analogy sketched at the beginning of 1.7.

1.8. Operads and topology: homology of moduli spaces. We will now introduce the basic operad of the quantum cohomology. Denote by $H_*(\overline{M}_{0,n+1})$ the homology space of the moduli space of stable curves of genus zero (with coefficients in the ground field for $VECT$). We will define the classical linear operad $H_*\overline{M}_0$ by the following data.

a) $H_*\overline{M}_0(n) = H_*(\overline{M}_{0.n+1})$ for $n \geq 2$, the first component being the ground field.

In the following, it will be convenient to assume that the structure sections of $\overline{C}_{n+1} \to \overline{M}_{0.n+1}$ are labeled by $\{0, \ldots, n\}$.

b) $\mathbf{S}_n$ acts upon $H_*\overline{M}_0(n)$ by renumbering the sections $x_1, \ldots, x_n$.

c) The structure map

$$\gamma(k_1,\ldots,k_j): H_*(\overline{M}_{0,j+1}) \otimes H_*(\overline{M}_{0,k_1+1}) \otimes \cdots \otimes H_*(\overline{M}_{0,k_j+1})$$

(1.12)
$$\to H_*(\overline{M}_{0,k_1+\cdots+k_j+1})$$

is induced by the embedding of the boundary stratum

(1.13) $b(k_1,\ldots,k_j): \overline{M}_{0,j+1} \times \overline{M}_{0,k_1+1} \times \cdots \times \overline{M}_{0,k_j+1} \to \overline{M}_{0,k_1+\cdots+k_j+1}.$

On the level of geometric points, given $j+1$ stable labeled curves of genus zero,

$$(C; x_0, x_1, \ldots, x_j); \ (D_a; y_{0,a}, \ldots, y_{k_a,a}), \ a = 1,\ldots,j.$$

$b(k_1,\ldots,k_j)$ produces from them the stable curve

$$\left(C \amalg \left(\coprod_{a=1}^{j} D_a \right) /(\sim); \ z_0,\ldots,z_{k_1+\cdots+k_j} \right),$$

where $(\sim)$ is the equivalence relation gluing x_a and $y_{0,a}$ for all $a = 1,\ldots,j$, and furthermore $z_0 = x_0, (z_1,\ldots,z_{k_1+\cdots+k_j}) = (y_{1,1},\ldots,y_{k_1,1};\ldots;y_{1,a},\ldots,y_{k_a,a}).$

Operadic axioms for $H_*\overline{M}_0$ follow from their evident versions for the spaces $\overline{M}_{0n}$.

What we are actually saying here is that we can define the more general notion of operad by replacing the basic category $VECT$ by any symmetric monoidal category, eventually with the identity object, and that the moduli spaces form such an operad. The homology functor (with respect to the pushforward maps) from the monoidal category of manifolds to $(VECT, \otimes)$ then produces from a geometric operad the classical linear operad. This viewpoint will be discussed in more detail in the following section.

1.8.1. Proposition. *Let (H, g) together with multiplications $\circ_n : H^{\otimes n} \to H$ be a cyclic $Comm_\infty$-algebra in the sense of III.1.2. Then there is a unique structure of $H_*\overline{M}_0$-algebra on H such that the fundamental class of $\overline{M}_{0,n+1}$ induces $\circ_n$.*

This is somewhat hidden in the calculations of III.1 and III.4.

Notice that we did not say that $H_*\overline{M}_0$ actually classifies $Comm_\infty$-algebras: the presence of the metric g on H complicates the situation.

The point is that the action of $\mathbf{S}_n$ on $\overline{M}_{0,n+1}$ and $H_*\overline{M}_{0n}$ actually extends to the action of $\mathbf{S}_{n+1}$ involving the 0-th point as well. The interaction of this action with operadic multiplications satisfies the axioms of *cyclic operads* introduced by E. Getzler and M. Kapranov in [GeK1] (cf. also 2.6 below). One can define the notion of the invariant bilinear form on an algebra over a cyclic linear operad. In this way one can refine Proposition 1.8.1.

Moreover, the families of spaces $\overline{M}_{g,n+1}$ and their homology, for all stable pairs $(g, n+1)$, together with $\mathbf{S}_{n+1}$-actions and boundary maps, form the basic pattern for another extension of the notion of operad, that of *modular operad* introduced and studied in [GeK2]. Complete CohFT's introduced in III.4.5 are cyclic algebras over the modular operad $H_*(\overline{M}_{g,n+1})$.

Notice in conclusion that $H_*\overline{M}_0$ is endowed with important additional structures. Namely, the components of this operad are in fact coalgebras (pushforward with respect to the diagonal map), and compositions (1.12) as well as representations of $\mathbf{S}_n$ are coalgebra morphisms. This is the intrinsic reason for the existence of the operation of the tensor product on the category of $H_*\overline{M}_0$-algebras and $Comm_\infty$-algebras.

§2. Operads and graphs

2.1. Introduction. In this section we sketch in their natural generality several themes which have already emerged in the previous section. Briefly speaking, there are many useful types of operads, and each type is determined by the choice of two categories:

1) Basic symmetric monoidal category $(\mathcal{C}, \boxtimes)$ replacing $(VECT, \otimes)$ which supports the classical linear operads.

2) A category of (labeled) graphs Γ reflecting the combinatorics of the operadic data and axioms.

A concrete operad from this viewpoint is a functor $\Gamma \to \mathcal{C}$.

To clarify the role of $\mathcal{C}$, we first explain how to extend Definition 1.1.1.

2.2. Classical (May) operads in a monoidal category. Let us recall that a symmetric monoidal category $(\mathcal{C}, \boxtimes)$ is a category endowed with the bifunctor $\boxtimes : \mathcal{C} \times \mathcal{C} \to \mathcal{C}$ together with an involutive commutativity constraint and an associativity constraint. Taken together, they define a family of compatible and functorial isomorphisms $s_* : X_1 \boxtimes \cdots \boxtimes X_n \widetilde{\to} X_{s^{-1}(1)} \boxtimes \cdots \boxtimes X_{s^{-1}(n)}$, for any objects $X_1, \ldots, X_n$ of $\mathcal{C}$ and all $s \in \mathbf{S}_n$.

Most of our monoidal categories will have an identity object $1_\mathcal{C} = 1$. The functors $1\boxtimes$ and $\boxtimes 1 : \mathcal{C} \to \mathcal{C}$ are canonically isomorphic to the identity functor.

In order to be able to extend the constructions of 1.7, we will assume that $\mathcal{C}$ has small colimits preserved by any functor $X \boxtimes$. In particular, $\mathcal{C}$ must have an initial object 0.

We can now define *a classical operad $\mathcal{P}$ in $\mathcal{C}$* by closely following Definition 1.1.1. Components $\mathcal{P}(n)$ will be objects of $\mathcal{C}$ endowed with the action of $\mathbf{S}_n$, $\otimes$ will be replaced by $\boxtimes$, and operadic multiplications γ will be morphisms in $\mathcal{C}$. Axioms A)–C) must be written down as commutative diagrams, involving in particular permutation isomorphisms of tensor products in $\mathcal{C}$.

A neater version of the definition is again obtained by passing to the category $S\mathcal{C}$ every object which is a family of $\mathbf{S}_n$–objects $\mathcal{P}(n)$ in $\mathcal{C}$ given for $n \geq 1$. To be able to write it as a sum of its components, we will require that $\mathcal{C}$ has small limits. The category $S\mathcal{C}$ admits a *non-symmetric* monoidal structure $*$, furnished by the formula (1.9). It has the unit object $1_{S\mathcal{C}}$ with 1 as the first component, 0 elsewhere. An associative monoid in $S\mathcal{C}$ is a pair $(\mathcal{P}, \mu)$ where $\mu : \mathcal{P} * \mathcal{P} \to \mathcal{P}$ is an associative multiplication. Giving an additional morphism $1 \to \mathcal{P}$ with the usual properties defines unital monoids. An analog of Proposition 1.7.3 holds true, establishing the equivalence of the category of associative monoids in $(S\mathcal{C}, *)$ and the category of classical operads in $\mathcal{C}$. However, the proof of Proposition 1.7.3 must be changed, because we have used in it not only the monoidal structure of $VECT$ but the linear structure and the language of elements as well. This can be avoided in different ways. Here we will take this fact for granted, and we leave to the reader the transposition of other constructions of §1 to the present context.

§2. OPERADS AND GRAPHS

2.2.1. Main classes of monoidal categories. Sets with direct product and linear spaces with tensor product form two archetypal classes of symmetric monoidal categories.

Variations include imposing additional structure on the objects. Sets more often appear endowed with a topology or manifold structure (in smooth or analytic category). Linear spaces come equipped with grading and/or differential. In this way we get classical topological operads, classical operads in the category of complexes, and so on. Monoidal functors between symmetric monoidal categories extend to the respective categories of operads.

2.3. Oriented trees and ∗–product. We start with describing trees as a natural bookkeeping device.

a) *Trees as substitution schemes.* Let τ be a tree with at least two flags at each vertex. Orient τ by choosing one tail as *root* and declaring that direction to the root is positive. Then every vertex has at least one incoming flag and exactly one outgoing flag. Label each vertex of τ by a symbol of the function whose arguments are labeled by the incoming flags of this vertex, and whose value labels the outgoing flag f and also its j–image, that is, the other half of the edge if f belongs to an edge. Then the whole tree symbolizes a computation, or substitution scheme. The input values are assigned to the incoming tails of τ, and the output value is assigned to the root. For example, one vertex tree with n incoming tails symbolizes $f(x_1, \ldots, x_n)$ and the $(m+1)$–vertex tree with the appropriate distribution of flags symbolizes $f(g_1(x_1^{(1)}, \ldots, x_{n_1}^{(1)}), \ldots, g_m(x_1^{(m)}, \ldots, x_{n_m}^{(m)}))$.

If we label flags by objects of a symmetric monoidal category and label each vertex v by a morphism mapping the $\boxtimes$–product of the labels of incoming flags to the label of the outgoing flag, the tree will describe the respective composite morphism from the $\boxtimes$–product of input objects to the output object.

If $\boxtimes$ is not supposed to be symmetric, we must assume that all sets of incoming flags of each vertex are totally ordered. For a symmetric $\boxtimes$–product, the respective actions of symmetric groups on arguments of various levels can be succinctly described by saying that this construction is functorial on the category of oriented trees with isomorphisms compatible with orientation.

b) *Trees and free operads.* Let $(\mathcal{C}, \boxtimes)$ be a symmetric monoidal category, V an object of non–symmetric monoidal category $(S\mathcal{C}, \ast)$, and $F(V) = \coprod_{n=1}^{\infty} V^{\ast n}$ the free operad generated by V, as in 1.7.4. We can define $\boxtimes$–products indexed by arbitrary finite sets and extend $n \mapsto V(n)$ to a functor $T \mapsto V(T)$ on the category of non–empty finite sets and their bijections. For an oriented tree as above put

(2.1) $$V(\tau) := \boxtimes_{v \in V_\tau} V(F_\tau^{\text{in}}(v)).$$

Then we have functorial isomorphisms

(2.2) $$F(V)(n) = \coprod_{\{n-\text{trees}\}\ \tau/(iso)} V(\tau).$$

Here n–trees are oriented trees with the set $\{1, \ldots, n\}$ of incoming tails.

This statement summarizes in a more conceptual way the bookkeeping scheme described above. It can be deduced with some pain from the formalism of analytic functors as in 1.7, and we leave the derivation to the reader. We will also reproduce the relevant combinatorics in the context of formal series below, in the section dedicated to sums over graphs.

2.4. Classical operads as functors. Denote by Γ_{clas} the category whose objects are finite rooted trees with the following properties: a) the multiplicity of each vertex is at least two; b) at each vertex either all incoming flags are halves of edges, or all incoming flags are tails. Morphisms are generated by the following two classes of maps:

a) Isomorphisms compatible with orientation.

b) Contraction of all edges having a common vertex with some outgoing flag and keeping orientation.

More formally, a morphism $\varphi : \sigma \to \tau$ consists of two maps $\varphi_V : V_\sigma \to V_\tau$ and $\varphi^F : F_\tau \to F_\sigma$ compatible with boundaries and involutions and such that φ^F sends tails to tails. Composition of the morphisms corresponds to the composition of the induced maps on vertices and flags. A morphism contracts an edge e if φ_V glues its vertices, and both flags of this edge do not belong to the image of φ^F.

Contractions of different edges commute in an evident sense.

Let v be a vertex of a rooted tree τ. Its *star* τ_v is a one–vertex tree with vertex v, tails $F_\tau(v)$, and the outcoming flag as a root.

2.4.1. Proposition. *The category of classical operads (without identity) in a symmetric monoidal category $(\mathcal{C}, \boxtimes)$ is equivalent to the category of functors $\mathcal{P} : \Gamma_{clas} \to \mathcal{C}$ isomorphic to a functor satisfying the following condition:*

$$(2.3) \qquad \mathcal{P}(\tau) = \boxtimes_{v \in V_\tau} \mathcal{P}(\tau_v).$$

Sketch of proof. a) *From functors to operads.* Given such a functor $\mathcal{P}$, we construct the data of Definition 1.1.1 in the following way: $\mathcal{P}(j) := \mathcal{P}(\tau_j)$ where τ_j is the one–vertex tree with tails $0, 1, 2, \ldots, j$ and root 0. The action of $\mathbf{S}_j$ corresponds to the automorphisms of τ_j permuting the tails $1, \ldots, j$. The multiplication map $\gamma(k_1, \ldots, k_j)$ corresponds to the morphism contracting all edges $\sigma \to \tau_{k_1 + \cdots + k_j}$, where σ has $j+1$ vertices and j edges and the tails are distributed in an obvious way. The relations A) and B) follow from the functoriality.

b) *From operads to functors.* Given an operad $(\mathcal{P}(n), \gamma)$, we first extend it to the functor from finite sets to $\mathcal{C}$, then define $\mathcal{P}(\tau)$ by (2.1), and finally use γ in order to define $\mathcal{P}$ on morphisms contracting all edges having a common vertex with some outgoing flag.

2.5. Markl's operads. From the graph theoretic viewpoint it would be more natural to allow all rooted trees with $|v| \geq 2$ as objects, and contractions of any subset of edges as morphisms. The functors from this category Γ_M to $\mathcal{C}$ satisfying (2.3) (up to functor isomorphism) are called Markl's operads.

2.6. Cyclic operads. Consider now the category Γ_{cyc} of finite trees with $|v| \geq 2$, with morphisms generated by contraction of edges and isomorphisms. Neither root nor orientation is a part of the structure. Functors $\Gamma_{cyc} \to \mathcal{C}$ satisfying (2.3) are essentially *cyclic operads* in the sense of [GeK1]. The most essential new feature of cyclic operads is the action of $\mathbf{S}_{j+1}$ upon $\mathcal{P}(j)$.

§3. Sums over graphs

In this section we prove basic formal identities for certain infinite sums (partition functions) taken over graphs of various topological types. The simplest "Euler product" identity relates sums over not necessarily connected graphs to those over connected ones. Summation over trees is interpreted as a calculation of the critical value of a formal potential. Finally, summation over graphs of arbitrary topology is interpreted as the perturbation series for a formal Feynman integral.

3.1. Setup. Let $(\mathcal{E}, \circ)$ be a symmetric monoidal category with the identity object **1** satisfying the following conditions.

a) $\mathcal{E}$ has a countable set of isomorphism classes of objects. Every object has a finite automorphism group.

b) Every object of $\mathcal{E}$ is isomorphic to a product $\bigcirc_i \pi_i^{a_i}$ where π_i are indecomposable with respect to $\circ$ ("primes"), $\pi_i^{a_i}$ is the $\circ$–product of a_i copies of π_i, and $\pi_i \neq \pi_j$ for $i \neq j$. This product is defined uniquely up to permutation of factors.

c) We have

$$(3.1) \qquad |\mathrm{Aut}\, \bigcirc_i \pi_i^{a_i}| = \prod_i a_i! |\mathrm{Aut}\, \pi_i|^{a_i},$$

in particular, $|\mathrm{Aut}\,(1)| = 1$.

In addition let R be a commutative topological ring and let $w : \mathrm{Ob}\,\mathcal{E} \to R$ be a weight function depending only on the isomorphism class of the object and multiplicative: $w(\sigma \circ \tau) = w(\sigma)w(\tau)$.

3.1.1. Theorem. *If the sums and products involved absolutely converge, we have*

$$(3.2) \qquad \prod_{\{\pi\}/(iso)} \exp \frac{w(\pi)}{|\mathrm{Aut}\,\pi|} = \sum_{\sigma \in \mathrm{Ob}\,\mathcal{E}/(iso)} \frac{w(\sigma)}{|\mathrm{Aut}\,\sigma|} = \exp\left(\sum_{\{\pi\}/(iso)} \frac{w(\pi)}{|\mathrm{Aut}\,\pi|}\right).$$

Proof. We have

$$\prod_{\{\pi\}/(iso)} \exp \frac{w(\pi)}{|\mathrm{Aut}\,\pi|} = \prod_{\{\pi\}/(iso)} \sum_{a=0}^{\infty} \frac{w(\pi)^a}{a! |\mathrm{Aut}\,\pi|^a},$$

and it remains to apply (3.1).

3.2. Application to sums over graphs. Throughout this section, we will take for $\mathcal{E}$ various categories of finite graphs, $\circ$ will denote the disjoint sum, and "primes" π will be connected graphs. Property (3.1) will be evident from the definition of isomorphisms. The second equality in (3.2) says that a weighted sum taken over all graphs can be obtained by exponentiation from the similar sum taken only over connected graphs.

We will now introduce a family of weights which will be called standard.

3.2.1. Definition. *A standard weight on a category of finite graphs is defined by the following choices:*

a) A set of "colors" A, finite or countable.

b) A family of symmetric tensors $C_{a_1,\ldots,a_k}$, $k = 1, 2, \ldots$, whose subscripts belong to A and coordinates belong to a topological commutative ring R.

c) A symmetric tensor g^{ab} with the same properties. The matrix (g^{ab}) must be invertible, and we put $(g_{ab}) = (g^{ab})^{-1}$.

In other words, we have a free R-module H with metric and a sequence of symmetric polynomials of all degrees on H expressed in terms of a basis indexed by A. (We can generalize this setting considering supercommutative R and $\mathbf{Z}_2$-graded H.)

With these choices made, we put for a graph τ:

(3.3) $$w(\tau) = \sum_{u:\, F_\tau \to A} \prod_{e \in E_\tau} g^{u(\partial e)} \prod_{v \in V_\tau} C_{u(F_\tau(v))}.$$

3.2.2. Remarks. a) The expression ∂e in (3.3) means the set of two flags constituting the edge e. When a marking $u : F_\tau \to A$ is given, $u(\partial e) = \{a, b\}$ consists of two elements of A which produce g^{ab}. We can similarly define $C_{u(F_\tau(v))}$ thanks to the symmetry.

b) If A is finite, the whole sum (3.3) is finite. Otherwise we have to postulate convergence already at this step. In our applications R will be a formal series ring. The multiplicativity of w with respect to disjoint union is evident.

c) Consider now the sum of type (3.2) with a standard weight:

(3.4) $$Z_\mathcal{E}(w) = \sum_{\tau/(iso)} \frac{1}{|\mathrm{Aut}\,\tau|} \sum_{u:\, F_\tau \to A} \prod_{e \in E_\tau} g^{u(\partial e)} \prod_{v \in V_\tau} C_{u(F_\tau(v))}.$$

Such sums occur in some models of statistical and quantum physics. Coloring of flags corresponds to the picture of A types of particles propagating along the edges with amplitudes g^{ab} and interacting at vertices with amplitudes $C_{a_1,\ldots,a_k}$. In this context, graphs are Feynman diagrams, and (3.4) can be called *the partition function*. The same formalism emerges in the general operadic context and in the topology of moduli spaces.

3.3. Summation over trees. In this subsection, we will calculate the partition function (3.3) in which the summation is taken over the set T of isomorphism classes of all (connected) trees *without tails* and having at least one edge.

We will treat here $C_{a_1,\ldots,a_k}$ as independent formal variables over a subring $R_0 \subset R$ containing g_{ab}, g^{ab}, and $\mathbf{Q}$. Then all our sums make sense as formal series.

We will express Z via a simpler formal function of auxiliary variables $\phi = \{\phi^a \mid a \in A\}$ independent over R:

(3.5) $$\Phi(\phi) = -\frac{1}{2} \sum_{a,b} g_{ab} \phi^a \phi^b + \sum_{k=1}^\infty \frac{1}{k!} \sum_{a_1,\ldots,a_k \in A} C_{a_1,\ldots,a_k} \phi^{a_1} \ldots \phi^{a_k}.$$

Put $C^a = \sum_{b \in A} g^{ab} C_b$ and denote by $N \subset R$ the ideal generated by $C_{a_1,\ldots,a_k}$ for all $k \geq 2$.

3.3.1. Theorem. *a) The equations*

(3.6) $$\forall a \in A,\ \frac{\partial \Phi(\phi)}{\partial \phi^a} = 0$$

admit the unique solution $\phi_0 = \{\phi_0^a\} \in R^A$ *satisfying the condition*

(3.7) $$\phi_0^a \equiv C^a \bmod N.$$

b) *The partition function* $Z = Z_T$ *satisfies the differential equations*

$$\frac{\partial Z}{\partial C_a} = \phi_0^a, \ a \in A, \tag{3.8}$$

and is the critical value of $\Phi(\phi)$:

$$Z = \Phi(\phi_0). \tag{3.9}$$

Remark. The assumption that $C_{a_1,\ldots,a_k}$ are independent formal variables is used several times in the statements and proofs: to locate the critical point ϕ_0, to make sense of the lhs of (3.8), etc. However, when the identities (2.5) and (2.6) are proved in the formal context, they can be specialized to other topological rings R.

Proof. a) Rewrite (3.6) as

$$\forall a \in A, \ \sum_{b \in A} g_{ab}\phi^b = C_a + \sum_{k \geq 2} \frac{1}{k!} \sum_{a_1,\ldots,a_k \in A} \frac{\partial}{\partial \phi^a}(C_{a_1,\ldots,a_k}\phi^{a_1}\ldots\phi^{a_k}), \tag{3.10}$$

that is,

$$\forall a \in A, \ \phi^a = C^a + \sum_{k \geq 2} \frac{1}{k!} \sum_{a_1,\ldots,a_k,b \in A} g^{ab}\frac{\partial}{\partial \phi^a}(C_{a_1,\ldots,a_k}\phi^{a_1}\ldots\phi^{a_k}). \tag{3.11}$$

Comparing (3.7) and (3.11) one sees that the critical point in question can be calculated by iterating (3.11). More precisely, consider the formal operator T mapping $\psi = (\psi^a \,|\, a \in A)$ to $(T^a(\psi) \,|\, a \in A)$ where

$$T^a(\psi) = \sum_{k \geq 2} \frac{1}{k!} \sum_{i=1}^{k} \sum_{a_1,\ldots,a_k,b \in A} g^{ab}C_{a_1,\ldots,a_k}\psi^{a_1}\ldots\widehat{\psi^{a_i}}\ldots\psi^{a_k}\,\delta_{a_i,b}. \tag{3.12}$$

The equation (3.11) can be rewritten as $\phi_0 = C + T(\phi_0)$ and solved by means of a version of the geometric progression formula

$$\phi_0 = C + T(C + T(C + T(C + \ldots))). \tag{3.13}$$

The solution is clearly unique.

b) In order to make more transparent the formal structure of (3.13) as a sum over trees, we will consider the case when $A = \{*\}$ is a one–element set.

Put $g^{**} = g$, $g_{**} = g^{-1}$, $C_{*\ldots*}$ (k subscripts) $= C_k, \phi^* = \phi$, and $C^1 = gC_1$. Then (3.12) becomes

$$T(\psi) = \sum_{k=1}^{\infty} \frac{gC_{k+1}}{k!}\psi^k$$

and (3.13) takes the form

$$\phi_0 = \sum_{k=0}^{\infty} \frac{gC_{k+1}}{k!}\left(\sum_{l=0}^{\infty} \frac{gC_{l+1}}{l!}\left(\sum_{m=0}^{\infty} \frac{gC_{m+1}}{m!}(gC_1+\ldots)^m\right)^l\right)^k. \tag{3.14}$$

Opening the brackets we will represent ϕ_0 as a sum of monomials in $\dfrac{gC_{i+1}}{i!}$. We will say that such a monomial has height $\leq N$ if it is a product of terms situated before the N-th opening bracket in (3.14) or directly after it (the terms of the latter type are gC_1).

E.g. the only monomial of height 0 is gC_1. Monomials of height 1 are
$$\frac{gC_{k+1}}{k!}(gC_1)^k, \quad k \geq 1.$$

Monomials of height 2 are indexed by the families of integers $\{k; l_1, \ldots, l_k\}$, $k \geq 1$, $l_i \geq 0$, each such family contributing

(3.15) $$\frac{C_{k+1}}{k!}\frac{C_{l_1+1}}{l_1!}\cdots\frac{C_{l_k+1}}{l_k!}(gC_1)^{l_1+\cdots+l_k}.$$

To establish the general pattern, we need a definition. Consider a tree without tails τ. *The pinning* of τ is given by the choice of the following data:

a) The choice of a vertex $v_0 \in V_\tau$ with $|F_\tau(v_0)| = 1$ called *the root*.

Such choice determines a unique orientation of all flags (or edges) of τ such that the unique flag of v_0 is incoming and every vertex $v \neq v_0$ has exactly one outgoing flag.

b) A total ordering of all sets $V_\tau(k) \subset V_\tau$ where $V_\tau(k)$ denotes the set of all vertices of τ separated by k edges from v_0. This total ordering must satisfy the following condition. Let $f_k : V_\tau(k+1) \to V_\tau(k)$ be the map "going along the outgoing edge to the next vertex". Then f_k must be monotone with respect to the chosen orderings.

A pinned tree is a tree with pinning. An isomorphism of pinned trees is an isomorphism of trees compatible with orientation and pinning. *The height* of a pinned tree is $\max\{k \mid V_\tau(k+1) \neq \emptyset\}$.

A contemplation shows that there is a natural bijection between the isomorphism classes of pinned trees (τ, p) with $|E_\tau| \geq 1$ of height $\leq N$ and monomials of height $\leq N$ which can be directly obtained from (3.14). Moreover, various pinnings of the same τ generate the differently ordered but equal monomials which can be written in the form dependent only on τ:

(3.16) $$\frac{1}{C_1}g^{|E_\tau|}\prod_{v \in V_\tau}C_{|v|}/(|v|-1)!.$$

Now, the number of different pinnings of τ is $|T_\tau|\prod_{v \in V_\tau}(|v|-1)!$ where T_τ is the set of potential roots and factorials count orderings of incoming edges. The automorphism group of τ effectively acts on the set of pinnings. Hence (3.16) appears with the coefficient

$$|T_\tau|\prod_{v \in V_\tau}(|v|-1)!/|\mathrm{Aut}\,\tau|.$$

We now turn to the proof of (3.8) which for the one-element A becomes

(3.17) $$\phi_0 = \frac{\partial}{\partial C_1}\left(\sum_\tau \frac{1}{|\mathrm{Aut}\,\tau|}g^{|E_\tau|}\prod_{v \in V_\tau}C_{|v|}\right).$$

In fact, the discussion above shows that the tree τ with all its pinnings contributes to ϕ_0 the term

$$\frac{|T_\tau|}{C_1}\frac{\prod_{v \in V_\tau}(|v|-1)!}{|\mathrm{Aut}\,\tau|}g^{|E_\tau|}\prod_{v \in V_\tau}\frac{C_{|v|}}{(|v|-1)!}.$$

§3. SUMS OVER GRAPHS

In view of (3.4), this is the same as the contribution of τ to $\dfrac{\partial Z}{\partial C_1}$. This gives (3.8).

We leave to the reader the discussion of the case $|A| > 1$.

To derive (3.9) from (3.8), consider both sides of (3.9) as formal series in C_a, $a \geq 1$. Their constant terms (values at $(C_a) = 0$) vanish. For Z, this follows from the fact that any tree in (3.4) has at least two vertices with $|v| = 1$. For ϕ_0, this follows from (2.4). Hence it suffices to check that $\dfrac{\partial}{\partial C_a} Z = \dfrac{\partial}{\partial C_a} \Phi(\phi_0)$ for all $a \in A$. But we have

$$\frac{\partial}{\partial C_a} \Phi(\phi_0) = \sum_{b \in A} \frac{\partial \Phi}{\partial \phi^b}(\phi_0) \frac{\partial \phi_0^b}{\partial C_a} + \frac{\partial \Phi}{\partial C_a}(\phi_0).$$

The first sum vanishes because $(dS)(\phi_0) = 0$, and the second term equals ϕ_0^a because of (3.5). It remains to apply (3.8).

3.4. Summation over graphs of arbitrary topology. We will now study the partition function (3.4) for more general graphs, keeping the same assumptions about the coefficient ring R and tensors C, g as in 3.2 and 3.3. In order to keep track of the Euler characteristic of the graphs, we extend R to the Laurent formal series ring $R_\lambda = R((\lambda^{-1}))$.

3.4.1. Definition. *An R_λ-linear functional*

$$\langle \cdot \rangle : R_\lambda[[\phi]] \to R_\lambda$$

is called $\lambda^{-1}g$-Gaussian (mean value) if it is (λ^{-1}, ϕ)-adically continuous, and

(3.18) $$\langle \exp(\lambda^{-1} \sum_a C_a \phi^a) \rangle = \exp\left((2\lambda)^{-1} \sum_a C_a g^{ab} C_b\right).$$

3.4.2. Wick's Lemma. *If (C_a) are independent variables over R_λ, then we have:*

a) $\langle \phi^{a_1} \ldots \phi^{a_n} \rangle = 0$ *for* $n \equiv 1 \bmod 2$.

b) $\langle \phi^a \phi^b \rangle = \lambda g^{ab}$.

c) $\langle \phi^{a_1} \ldots \phi^{a_{2m}} \rangle = \lambda^m \sum g^{a_{i_1} a_{j_1}} \ldots g^{a_{i_m} a_{j_m}}$ *where the summation is taken over all unordered partitions of $\{1, \ldots, 2m\}$ into m unordered pairs $\{i_1, j_1\}, \ldots, \{i_m, j_m\}$ (pairings).*

Conversely, if a (λ^{-1}, ϕ)-adically continuous functional $\langle \cdot \rangle$ satisfies a), b), c), then it is $\lambda^{-1}g$-Gaussian.

Proof. We have

$$\langle \exp(\lambda^{-1} \sum_a C_a \phi^a) \rangle = \sum_{n=0}^{\infty} \frac{1}{\lambda^n n!} \sum_{a_1, \ldots, a_n \in A} C_{a_1} \ldots C_{a_n} \langle \phi^{a_1} \ldots \phi^{a_n} \rangle,$$

$$\exp\left((2\lambda)^{-1} \sum_a C_a g^{ab} C_b\right) = \sum_{m=0}^{\infty} \frac{1}{2^m \lambda^m m!} \sum_{a_i, b_i \in A} C_{a_1} C_{b_1} \ldots C_{a_m} C_{b_m} g^{a_1 b_1} \ldots g^{a_m b_m}.$$

Comparing the coefficients, we get the lemma.

Now put

$$\Phi_0(\phi) = -\frac{1}{2} \sum_{a,b \in A} g_{ab} \phi^a \phi^b, \quad \Phi_1(\phi) = \sum_{k=1}^{\infty} \frac{1}{k!} \sum_{a_1, \ldots, a_k \in A} C_{a_1, \ldots, a_k} \phi^{a_1} \ldots \phi^{a_k},$$

and denote by $w(\tau)$ the weight function (3.3).

Let Γ be the set of (isomorphism classes of) all finite graphs without tails, not necessarily connected, including the empty graph, and Γ_0 the subset of connected non–empty graphs. Let $\langle \cdot \rangle$ be the $\lambda^{-1}g$–Gaussian mean value. Denote by $\chi(\tau)$ the Euler characteristic of $\|\tau\|$.

3.4.3. Theorem. *We have*

$$(3.19) \qquad \sum_{\tau \in \Gamma} \frac{\lambda^{-\chi(\tau)}}{|\mathrm{Aut}\,\tau|} w(\tau) = \langle \exp\left(\lambda^{-1}\Phi_1(\phi)\right) \rangle,$$

$$(3.20) \qquad \sum_{\tau \in \Gamma_0} \frac{\lambda^{-\chi(\tau)}}{|\mathrm{Aut}\,\tau|} w(\tau) = \log \langle \exp\left(\lambda^{-1}\Phi_1(\phi)\right) \rangle.$$

Proof. The second equality follows from the first one in view of (3.2) and the additivity of the Euler characteristic wrt the disjoint union.

Let us now calculate the rhs of (3.19). By definition, it is

$$(3.21) \qquad \left\langle \sum_{n=0}^{\infty} \lambda^{-n} \frac{1}{n!} \left(\sum_{k=1}^{\infty} \frac{1}{k!} \sum_{a_1,\ldots,a_k \in A} C_{a_1,\ldots,a_k} \phi^{a_1} \ldots \phi^{a_k} \right)^n \right\rangle.$$

Choose some $(n; k_1, \ldots, k_n)$. A typical monomial in the decomposition of (3.21) will be

$$(3.22) \qquad \lambda^{-n} \frac{1}{n!} \prod_{i=1}^{n} \frac{1}{k_i!} C_{a_1^{(i)},\ldots,a_{k_i}^{(i)}} \langle \prod_{i=1}^{n} \phi^{a_1^{(i)}} \ldots \phi^{a_{k_i}^{(i)}} \rangle.$$

It vanishes if $k_1 + \cdots + k_n$ is odd. Otherwise, in view of Wick's Lemma (3.22) can be rewritten as

$$(3.23) \qquad \lambda^{-n+\frac{1}{2}\sum k_i} \frac{1}{n!} \prod_{i=1}^{n} \frac{1}{k_i!} C_{a_1^{(i)},\ldots,a_{k_i}^{(i)}} \left(\sum g^{a_{l_1}^{(i_1)} a_{m_1}^{(j_1)}} \ldots g^{a_{l_r}^{(i_r)} a_{m_r}^{(j_r)}} \right),$$

where $r = \frac{1}{2} \sum k_i$ and the inner sum is taken over all pairings of the set of ordered pairs $F = \bigcup_{i=1}^{n} \{(i,1),\ldots,(i,k_i)\}$.

Construct the family of graphs τ whose set of flags is $F_\tau := F$, $V_\tau = \{1,\ldots,n\}$, $\partial_\tau(i,l) = i$, and involutions bijectively correspond to various pairings in (3.23). If we color the flags of one such graph by the map $F_\tau \to A : (i,l) \mapsto a_l^{(i)}$, then the sum over all pairings will produce the same monomials as in (3.4). It remains to do the accurate bookkeeping in order to identify the coefficients.

The graphs constructed above bijectively correspond to all elements of Γ. In fact, a choice of $(n; k_1, \ldots, k_n)$ determines the number of vertices of any valence, and the choice of a pairing determines which pairs of flags become edges ($n = 0$ produces the empty graph). Moreover, a non–empty graph comes thus equipped with a total ordering of its vertices and all sets of flags belonging to one vertex. The sum over graphs does not take care of these orderings. The group $\mathrm{Aut}\,\tau$ effectively acts on the whole set of them consisting of $n! \prod_{i=1}^{n} k_i!$ elements. Summing over isomorphism classes, we may replace the numerical coefficient in (3.23) by $|\mathrm{Aut}\,\tau|^{-1}$.

Finally,
$$-n + \frac{1}{2}\sum_{i=1}^{n} k_i = -|V_\tau| + |E_\tau| = \chi(\tau).$$

3.4.4. Variants. Theorem 3.4.3 is only one of the several useful summation formulas extended over graphs of arbitrary topology. There are two more categories of graphs furnishing formulas important for the study of topology and intersection theory of $\overline{M}_{g,n}$.

a) *(Stable) modular graphs.* See [GeK2], [Ge3].

b) *Ribbon graphs.* See [P1], [P4], [Ko1], [Mul1], [Mul2], [MulP]. It would be interesting to study operads corresponding to ribbon graphs.

§4. Generating functions

In this section we calculate several generating functions related to moduli spaces and quantum cohomology, first representing them as sums over trees of the type treated in §3.

4.1. Virtual Poincaré polynomial. Let Y be an algebraic variety over $\mathbf{C}$, possibly non–smooth and non–compact. Following [FulMPh] we denote by $P_Y(q)$ the virtual Poincaré polynomial of Y which is uniquely defined by the following properties.

a) If Y is smooth and compact, then

(4.1) $$P_Y(q) = \sum_j \dim H^j(Y) q^j.$$

In particular

(4.2) $$\chi(Y) = P_Y(-1).$$

b) If $Y = \coprod_i Y_i$ is a finite union of pairwise disjoint locally closed strata, then

(4.3) $$P_Y(q) = \sum_i P_{Y_i}(q).$$

c) $P_{Y \times Z}(q) = P_Y(q) P_Z(q)$. It follows that if Y is a fibration over base B with fiber F locally trivial in the Zariski topology, then $P_Y(q) = P_B(q) P_F(q)$.

A definition of $P_Y(q)$ can be given using the weight filtration on the cohomology with compact support:

(4.4) $$P_Y(q) = \sum_{i,j} (-1)^{i+j} \dim (\mathrm{gr}_W^j H_c^i(Y, \mathbf{Q})) q^j.$$

4.2. Generating function for moduli spaces of genus zero. We put

(4.5) $$\varphi(q,t) := t + \sum_{n=2}^{\infty} P_{\overline{M}_{0,n+1}}(q) \frac{t^n}{n!} \in \mathbf{Q}[q][[t]],$$

(4.6) $$\chi(t) := \varphi(-1, t) = t + \sum_{n=2}^{\infty} \chi(\overline{M}_{0,n+1}) \frac{t^n}{n!} \in \mathbf{Q}[[t]].$$

4.2.1. Theorem. *a)* $\varphi(q,t)$ *is the unique root in* $t + t^2 \mathbf{Q}[q][[t]]$ *of any one of the following functional/differential equations in* t *with parameter* q :

(4.7) $$(1+\varphi)^{q^2} = q^4\varphi - q^2(q^2-1)t + 1,$$

(4.8) $$(1 + q^2 t - q^2\varphi)\varphi_t = 1 + \varphi.$$

b) χ *is the unique root in* $t + t^2 \mathbf{Q}[[t]]$ *of any one of the similar equations*

(4.9) $$(1+\chi)\log(1+\chi) = 2\chi - t,$$

(4.10) $$(1+t-\chi)\chi_t = 1 + \chi.$$

Equation (4.8) is equivalent to the following recursive formulas for the Poincaré polynomials. Put $p_n = p_n(q) = P_{\overline{M}_{0,n+1}}/n!$.

4.2.2. Corollary. *We have for* $n \geq 1$:

(4.11) $$(n+1)p_{n+1} = p_n + q^2 \sum_{\substack{i+j=n+1 \\ i \geq 2}} j p_i p_j,$$

(4.12) $$P_{\overline{M}_{0,n+2}}(q) = P_{\overline{M}_{0,n+1}}(q) + q^2 \sum_{\substack{i+j=n+1 \\ i \geq 2}} \binom{n}{i} P_{\overline{M}_{0,i+1}}(q) P_{\overline{M}_{0,j+1}}(q).$$

One can compare (4.11) with recursive formulas in [Ke], p. 550.

From (4.10) one sees that the function inverse to χ has a critical point at $t = e - 2$. From this one can derive the following asymptotical formula:

$$\chi(\overline{M}_{0,n+1}) \cong \frac{1}{\sqrt{n}} \left(\frac{n}{e^2 - 2e}\right)^{n-\frac{1}{2}}.$$

In order to prove Theorem 4.2.1, we will first apply the additivity formula (4.3) to the open boundary strata of $\overline{M}_{0,n}$ which are briefly described in III.2.8 and then use Theorem 3.3.1. However, the classes of trees involved in the labeling of stable curves, on the one hand, and the summation formula (3.9), on the other, are slightly different: we need tails in the first problem and do not allow them in the second. In order to unify the combinatorial pictures, and *only in this section*, we will eliminate tails by putting end–point vertices on them. This will lead to the following temporary modification of the conventions described in III.2.

A tree without tails is called *stable* if $|v| \neq 2$ for all vertices v. If $|v| = 1$ we call v an end vertex. Let V_τ^1 be the set of end vertices. An n-marking of τ is a bijection $\mu : V_\tau^1 \to \{1, \ldots, n\}$. We also put $V_\tau^0 = V \setminus V_\tau^1$ and refer to it as the set of interior vertices.

Now let $(C; x_1, \ldots, x_n)$ be a stable compact connected curve of arithmetical genus zero with $n \geq 3$ labeled non–singular points. The combinatorial structure of this curve is described by the following stable tree with n-marking (τ, μ): $V_\tau^0 = $ {irreducible components of C}, $V_\tau^1 = \{x_1, \ldots, x_n\}$; $\mu : x_i \mapsto i$; an edge connects two interior vertices if the respective components of C have non–empty intersection; an edge connects an interior vertex to an end vertex if the respective point belongs to the respective component.

§4. GENERATING FUNCTIONS

Denote now by $M(\tau,\mu) \subset \overline{M}_{0,n}$ the set of points parametrizing stable curves of the type (τ,μ). If τ has only one interior vertex, $M(\tau,\mu) := M_{0,n}$ is the big cell. The following statement summarizes the main properties of these sets; for a proof, see [Ke].

4.2.3. Proposition. *a) $M(\tau,\mu)$ is a locally closed subset of $\overline{M}_{0,n}$ depending only on (the isomorphism class of) (τ,μ).*

b) $\overline{M}_{0,n}$ is the union of pairwise disjoint strata $M(\tau,\mu)$ for all marked stable n-trees (τ,μ).

c) For any (τ,μ),

$$(4.13) \qquad M(\tau,\mu) \cong \prod_{v \in V_\tau^0} M_{0,|v|}.$$

Notice that there exists exactly one stable tree •———• which does not correspond to any stable curve.

We can now calculate Poincaré polynomials.

4.2.4. Proposition. *We have*

$$(4.14) \qquad P_{M(\tau,\mu)}(q) = \prod_{v \in V_\tau^0} P_{M_{0,|v|}}(q),$$

$$(4.15) \qquad P_{M_{0,k}}(q) = \binom{q^2 - 2}{k - 3}(k-3)!.$$

Proof. (4.14) follows from (4.13) and the multiplicativity of Poincaré polynomials.

To prove (4.15), one can use the following geometric facts. First, the morphism $\pi: \overline{M}_{0,n+1} \to \overline{M}_{0,n}$ forgetting the last marked point is (canonically isomorphic to) the universal curve. Second, the boundary of the source consists of structure sections and fibers at infinity of the target. Therefore, over the big cell $M_{0,n}$ this morphism is a Zariski locally trivial fibration with fiber $\mathbf{P}^1$, and $M_{0,n+1} = \pi^{-1}(M_{0,n}) \setminus \{\text{union of structure sections}\}$.

From the additivity of Poincaré polynomials it follows that

$$P_{M_{0,n+1}}(q) = P_{M_{0,n}}(q) P_{\mathbf{P}^1}(q) - n P_{M_{0,n}}(q) = (q^2 + 1 - n) P_{M_{0,n}}(q).$$

Since $P_{M_{0,3}}(q) = 1$, we get (4.15).

Summarizing, we have for $n \geq 3$:

$$(4.16) \qquad P_{\overline{M}_{0,n}}(q) t^n = \sum_{\substack{(\tau,\mu)/(iso) \\ |V_\tau^1| = n}} \prod_{v \in V_\tau^0} \binom{q^2 - 2}{|v| - 3}(|v|-3)! \prod_{v \in V_\tau^1} t,$$

where t is a new formal variable, and the sum is taken over n-marked stable trees.

We want to present (4.5) as a partition function. Comparing (4.16) to (3.3) and (3.5), we are more or less compelled to choose $A = \{*\}$ (one element set),

$g^{**} = 1$, $C_* = t$, $C_{**} = 0$ (this gives weight zero to non–stable trees), and finally, denoting by C_k the component with $k \geq 3$ subscripts, we get

(4.17) $$C_k = \binom{q^2 - 2}{k - 3}(k-3)!.$$

In particular, we can forget about $u: F_\tau \to \{*\}$.

If $|V_\tau^1| = n$, the set of all n–markings of τ consists of $n!$ elements and is effectively acted upon by the group Aut τ. We see finally that $\psi(q,t) = Z$ where

(4.18) $$\psi(q,t) := \frac{t^2}{2!} + \sum_{n \geq 3} \frac{t^n}{n!} P_{\overline{M}_{0,n}}(q),$$

(4.19) $$Z := \sum_{\tau/(iso)} \frac{1}{|\text{Aut } \tau|} \prod_{v \in V_\tau} C_{|v|}.$$

The summation in (4.19) is now taken over all trees, and the term $t^2/2$ in (4.18) comes from the two–vertex tree.

We will now use (3.8) in order to calculate

$$\frac{\partial Z}{\partial t} = \frac{\partial \psi(q,t)}{\partial t} =: \varphi(q,t).$$

From (3.5) and (4.17) one sees that

$$\Phi(\varphi) = -\frac{\varphi^2}{2} + t\varphi + \sum_{k \geq 3} C_k \frac{\varphi^k}{k!} = -\frac{\varphi^2}{2} + t\varphi + \sum_{k \geq 3} \binom{q^2 - 2}{k-3} \frac{\varphi^k}{k(k-1)(k-2)}.$$

This can easily be summed. We need only the derivative.

For generic q we have

(4.20) $$\frac{\partial}{\partial \varphi}\Phi(\varphi) = \frac{(1+\varphi)^{q^2} - 1 - q^4\varphi}{q^2(q^2 - 1)} + t,$$

and for $q = -1$,

(4.21) $$\frac{\partial}{\partial \varphi}\Phi(\varphi) = (1+\varphi)\log(1+\varphi) - 2\varphi + t.$$

We see now that (4.7), resp. (4.9), are equations for the critical point $d_\varphi \Phi = 0$. Differentiating them in t and eliminating $(1+\varphi)^{q^2}$, resp. $\log(1+\varphi)$, we get (4.8), resp. (4.10).

4.3. Generating function for configuration spaces. Let X be a smooth compact algebraic variety. The configuration space $X[n]$, $n \geq 2$, is defined in [FuIMPh] as the closure of its big cell $X^n \setminus (\bigcup_{i<j} \Delta_{ij})$ (Δ_{ij} is the diagonal $x_i = x_j$) in $X^n \times \prod_S \widetilde{X}^S$, where S runs over subsets $S \subset \{1,\ldots,n\}$, $|S| \geq 2$; X^S denotes the respective partial product of X's, and $\widetilde{X}^S$ is the blow up of the small diagonal Δ_S in X^S.

Every S determines a divisor at infinity $D(S) \subset X[n]$. Namely, let $\pi_S: X[n] \to X^S$ be the canonical projection. Then $\pi_S^{-1}(\Delta_S) = \bigcup_{T \supset S} D(T)$.

§4. GENERATING FUNCTIONS

The natural stratification of $X[n]$ described in [FulMPh] consists of (open subsets of) intersections $\overline{X(\mathcal{S})} = \bigcap_{i=1}^{r} D(S_i)$ corresponding to sets $\mathcal{S} = \{S_1, \ldots, S_r\}$ of subsets in $\{1, \ldots, n\}$ called *nests*.

We put
$$\psi_X(q,t) = 1 + \sum_{n \geq 1} P_{X[n]}(q) \frac{t^n}{n!} \in \mathbf{Q}[q][[t]],$$

$$\chi_X(t) = \psi_X(-1,t) = 1 + \sum_{n \geq 1} \chi(X[n]) \frac{t^n}{n!} \in \mathbf{Q}[[t]].$$

Put also
$$\kappa_m = \frac{q^{2m} - 1}{q^2 - 1} = P_{\mathbf{P}^{m-1}}(q).$$

4.3.1. Theorem. *Denote by $y^0 = y^0(q,t)$ the unique root in $t + t^2 \mathbf{Q}[q^2][[t]]$ of any one of the following equations:*

(4.22) $\qquad \kappa_m (1 + y^0)^{q^{2m}} = q^{2m}(q^{2m} + \kappa_m - 1)y^0 - q^{2m}(q^{2m} - 1)t + \kappa_m,$

(4.23) $\qquad \left[q^{2m}t + 1 - (q^{2m} - 1 + \kappa_m)y^0\right] y_t^0 = 1 + y^0.$

Then we have in $\mathbf{Q}[q][[t]]$:

(4.24) $\qquad\qquad\qquad \psi_X(q,t) = (1 + y^0)^{P_X(q)}.$

4.3.2. Theorem. *Denote by $\eta = \eta(t)$ the unique root in $t + t^2 \mathbf{Q}[[t]]$ of any one of the following equations:*

(4.25) $\qquad\qquad m(1 + \eta) \log(1 + \eta) = (m+1)\eta - t,$

(4.26) $\qquad\qquad (t + 1 - m\eta)\eta_t = 1 + \eta.$

Then we have in $\mathbf{Q}[[t]]$:

(4.27) $\qquad\qquad\qquad \chi_X(t) = (1 + \eta)^{\chi(X)}.$

We start with combinatorics of the strata.

4.4.1. Definition. *a) $\mathcal{S} = \{S_1, \ldots, S_r\}$ is a nest (or n–nest) if $|S_i| \geq 2$ for all i, and either $S_i \subset S_j$ or $S_j \subset S_i$ for all i,j such that $S_i \cap S_j \neq \emptyset$.*

In particular, $\mathcal{S} = \emptyset$ is a nest, and $\mathcal{S} = \{S\}$ is a nest, if $|S| \geq 2$.

b) A nest $\mathcal{S}$ is called whole (resp. broken) if $\{1, \ldots, n\} \in \mathcal{S}$ (resp. $\{1, \ldots, n\} \notin \mathcal{S}$).

Denote by $X(\mathcal{S}) \subset \overline{X(\mathcal{S})} = \bigcap_{S \in \mathcal{S}} D(S)$ the subset of points not belonging to smaller closed strata. The following facts are proved in [FulMPh].

4.4.2. Proposition. *a) For any $n \geq 2$ and n-nest $\mathcal{S}$, $X(\mathcal{S})$ is a locally closed subset of $X[n]$.*

b) $X[n]$ is the union of pairwise disjoint strata $X(\mathcal{S})$ for all n-nests $\mathcal{S}$.

Now we will show how to pass from nests to marked trees. As above, we consider a bijection $\mu: V_r^1 \to \{1, \ldots, n\}$ as a part of the appropriate marking for our problem. The remaining data is supplied by choosing *orientations of all edges*.

4.4.3. Definition. *A tree τ marked in this way is called admissible iff:*

a) Every vertex of τ except one has exactly one incoming edge.

b) The exceptional vertex has only outgoing edges, and their number is ≥ 2. This vertex is called the source.

c) All interior vertices with possible exception of the source have valency ≥ 3.

4.4.4. Proposition. *The following maps are (1,1):*

$$\{\text{broken } n\text{-nests}\} \to \{\text{ whole } n\text{-nests}\} \to \{\text{admissible marked } n\text{-trees}\}/(iso),$$

$$S \mapsto S \cup \{\{1,\ldots,n\}\} \mapsto \tau(S) = \tau(S \cup \{\{1,\ldots,n\}\}).$$

Here τ is defined by its sets of vertices and edges: if $S = \{S_1, \ldots, S_r\}$, then

$$V_\tau = \{\widetilde{S}_1, \ldots, \widetilde{S}_{n+r}\} := \{S_1, \ldots, S_r, \{1\}, \ldots, \{n\}\},$$

and an edge oriented from $\widetilde{S}_i$ to $\widetilde{S}_j$ connects these two vertices iff $\widetilde{S}_j \subset \widetilde{S}_i$ and no $\widetilde{S}_k$ lies strictly in between these two subsets.

This is proved by direct observation. The following facts are worth mentioning.

a) $\{1, \ldots, n\}$ is the source of $\tau(S)$ for any S.

b) $\{1\}, \ldots, \{n\}$ are all end vertices.

c) $i \in S_j$ iff one can pass from $S_j \in V_\tau$ to $\{i\} \in V_\tau$ in τ by always going in the positive direction.

The reader is advised to convince him– or herself that the source has valency ≥ 2 and all other interior vertices have valency ≥ 3.

Denote the source by s and the set of the remaining interior vertices by V_τ^0.

4.4.5. Proposition ([FulMPh]). *The virtual Poincaré polynomials of the strata $X(S)$ are given by the following formulas (we add a formal variable t):*

If S is a broken n-nest, $s \in V_{\tau(S)}$:

$$(4.28) \quad t^n P_{X(S)}(q) = \binom{P_X(q)}{|s|} |s|! \times \prod_{v \in V_{\tau(S)}^0} \kappa_m \binom{q^{2m}-2}{|v|-3}(|v|-3)! \times \prod_{v \in V_{\tau(S)}^1} t.$$

If S is a whole n-nest:

$$t^n P_{X(S)}(q) = P_X(q) \kappa_m \binom{q^{2m}-2}{|s|-2}(|s|-2)!$$

$$(4.29) \qquad \times \prod_{v \in V_{\tau(S)}^0} \kappa_m \binom{q^{2m}-2}{|v|-3}(|v|-3)! \times \prod_{v \in V_{\tau(S)}^1} t.$$

Comparing (4.28) and (4.29) one sees that one can express the joint contribution of two nests corresponding to an admissible marked tree τ as a product of local weights corresponding to all vertices of τ. The local weight of the source will be

$$\binom{P_X(q)}{|s|}|s|! + P_X(q)\kappa_m \binom{q^{2m}-2}{|s|-2}(|s|-2)!$$

and the remaining local weights coincide and depend only on the valency.

§4. GENERATING FUNCTIONS

In order to find the appropriate standard weights of marked trees (summands in (3.3)), we make the following choices.

Put $A = \{+,-\}$. Interpret a mark + (resp. −) on a flag as incoming (resp. outgoing) orientation of this flag. Thus, $f : F_\tau \to A$ is a choice of orientation of all flags.

Put $g^{+-} = g^{-+} = 1$, $g^{++} = g^{--} = 0$. This makes the standard weight of (τ, f) vanish unless all *edges* are unambiguously oriented by f.

Put $C_+ = t$ (see (4.22) and (4.23)) and $C_- = 0$. The last choice makes the standard weight vanish unless all end edges are oriented outwards.

Put $C_{+-} = C_{-+} = 0$. This excludes vertices of the type $\to \bullet \to$.

Put also $C_{a_1,\ldots,a_k} = 0$ if $\{+,+\} \subset \{a_1,\ldots,a_k\}$. This eliminates vertices with ≥ 2 incoming edges.

For tensors with $k \geq 2$ minuses among the indices we put

(4.30) $$C_{-\cdots-} = \binom{P_X(q)}{k} k! + \kappa_m P_X(q) \binom{q^{2m} - 2}{k - 2}(k-2)!$$

(because only the source has all outgoing edges), and

(4.31) $$C_{+-\cdots-} = \kappa_m \binom{q^{2m} - 2}{k - 2}(k-2)!$$

(cf. (4.28) and (4.29)).

The standard weight of a marked tree defined by this data again is independent on the part $\mu : V_\tau^1 \to \{1,\ldots,n\}$ of the initial marking, which accounts for the factor $\dfrac{n!}{|\mathrm{Aut}\,\tau|}$ below.

Summarizing, we put

(4.32) $$\psi_X(q,t) := \sum_{n \geq 2} \frac{t^n}{n!} P_{X[n]}(q),$$

(4.33) $$Z := \sum_{\tau/(iso)} \frac{1}{|\mathrm{Aut}\,\tau|} \sum_{f:F_\tau \to \{+,-\}} \prod_{\alpha \in E_\tau} g^{f(\partial\alpha)} \prod_{v \in V_\tau} C_{f(\sigma v)},$$

and get from the previous discussion

(4.34) $$Z = \psi_X(q,t), \quad \frac{\partial}{\partial t} Z := \phi_X(q,t).$$

The arguments in the potential will be denoted $\varphi_+ = x$, $\varphi_- = y$. We see that the potential is

$$\Phi(x,y) = -xy + tx + \kappa_m \sum_{k=2}^{\infty} \binom{q^{2m} - 2}{k - 2} \frac{xy^k}{k(k-1)}$$

(4.35) $$+ \sum_{k=2}^{\infty} \binom{P_X(q)}{k} y^k + \kappa_m P_X(q) \sum_{k=2}^{\infty} \binom{q^{2m} - 2}{k - 2} \frac{y^k}{k(k-1)}$$

(we have two arguments x, y but only one $t = t_+$ because $C_- = 0$).

We must solve the system

$$\frac{\partial \Phi}{\partial x}\Big|_{x^0,y^0} = \frac{\partial \Phi}{\partial y}\Big|_{x^0,y^0} = 0, \tag{4.36}$$

and (3.8) then tells us that

$$\frac{\partial}{\partial t} Z = \varphi_X(q,t) = x^0. \tag{4.37}$$

Again, $\Phi(x,y)$ can be easily summed. To write down the functional equation, we need only the x-derivative which for general q is

$$\frac{\partial \Phi}{\partial x} = -y + t + \kappa_m \frac{(1+y)^{q^{2m}} - 1 - q^{2m} y}{q^{2m}(q^{2m} - 1)}. \tag{4.38}$$

For $q = -1$:

$$\frac{\partial \Phi}{\partial x} = -y + t + m[(1+y)\log(1+y) - y]. \tag{4.39}$$

We now see that (4.22), resp. (4.25), are the equations defining y^0. Taking the derivative in t we get (4.23) and (4.26). And since $\Phi(x,y)$ is linear in x, the vanishing of the y-derivative provides an explicit expression of x^0 via y^0:

$$\varphi_X(q,t) = P_X(q) \frac{(1+y^0)^{P_X(q)} + (q^{2m} + \kappa_m - 1)y^0 - q^{2m} t - 1}{1 + (1 - q^{2m} - \kappa_m)y^0 + q^{2m} t}.$$

To see that this is equivalent to (4.24) one can derivate (4.24) in t and use (4.23).

CHAPTER V

Stable Maps, Stacks, and Chow Groups

§1. Prestable curves and prestable maps

This section is the direct continuation of Chapter III, §2. We keep the definitions and notation introduced there.

1.1. The dualizing sheaf and the sheaf of 1–forms. Let C be a prestable curve over a base field. We denote by Ω_C its sheaf of 1–forms, and by ω_C the dualizing sheaf. For more details, see e.g. [Kn1].

If C is smooth, Ω_C and ω_C can be identified. In the general case, they can be compared in two different ways.

i) Passage to the normalization. Let $f : \tilde{C} \to C$ be the normalization of C. Denote by D the divisor on $\tilde{C}$ which is the sum of inverse images of singular points. Then we have

$$\Gamma(U, \omega_C) = \{\nu \in \Gamma(f^{-1}(U), \Omega_{\tilde{C}}(\log D)) \,|\, \operatorname{res}_x \nu + \operatorname{res}_y \nu = 0$$

(1.1) $\hspace{4em}$ for all $x \neq y$ with $f(x) = f(y)\}$.

To explain the latter condition, notice that if the completed local ring of a double point of C is represented as $k[[u,v]]/(uv)$ so that u, v are formal parameters at the two branches, then from $d(uv) = 0$ we find $u^{-1}du = -v^{-1}dv$. On the other hand, regular local functions on the two branches can be glued iff their values at the double point coincide. This restricts residues as in (1.1).

ii) Comparison on C. Consider a surjection $\mathcal{E}_0 \to \Omega_C$ where $\mathcal{E}_0$ is a locally free sheaf. Then its kernel is locally free as well. Therefore we can construct the determinant sheaf

$$\det \Omega_C := \wedge^{\max} \mathcal{E}_0 \otimes \wedge^{\max} \mathcal{E}_1^{\vee}.$$

A different choice of the resolution produces a canonically isomorphic result.

This determinant sheaf is dualizing:

(1.2) $\hspace{10em} \omega_C = \det \Omega_C.$

As a corollary, we see that ω_C is invertible.

Moreover, let ν be a local section of Ω_C liftable to a local section $\bar{\nu}$ of $\mathcal{E}_0$. Let $\nu_1, \ldots, \nu_k$ be a local basis of $\mathcal{E}_1$ and $\{\nu^i\}$ the respective dual basis of $\mathcal{E}_1^{\vee}$. Then the map

$$\Omega_C \to \omega_C : \nu \mapsto \bar{\nu} \wedge \nu_1 \wedge \cdots \wedge \nu_k \wedge \nu^1 \cdots \wedge \nu^k$$

does not depend on arbitrary choices and is an isomorphism on the complement to the singular points.

By the definition of the dualizing sheaf, we have $H^1(C, \omega_C) \cong k$. This induces the perfect duality between $H^1(C, \mathcal{F})$ and $\operatorname{Hom}_{\mathcal{O}_C}(\mathcal{F}, \omega_C)$. Defining the degree of the invertible sheaf by the general formula $\deg L = \chi(L) - \chi(\mathcal{O}_X)$ we see that $\deg \omega_C = 2g - 2$.

1.2. Modified dualizing sheaf. Now let $(C, x_i \mid i \in S)$ be a prestable S-pointed curve over a field. We will call $\omega_C(\sum x_i)$ the *modified dualizing sheaf*.

The first important property is its behavior with respect to morphisms identifying pairs of points. More precisely, let $(C', y_j \mid j \in T)$ be another prestable curve, and $f: C' \to C$ a morphism identifying C with $C'/(\sim)$ where $\sim$ is an equivalence relation gluing certain pairs of T-labeled points. Then we have

$$(1.3) \qquad f^*(\omega_C(\sum_{i \in S} x_i)) = \omega_{C'}(\sum_{j \in T} y_j).$$

This follows from the description (1.1) applied to the common normalization of C and C'.

The second important property relates the modified dualizing sheaf to stability.

1.2.1. Lemma. *A prestable curve $(C, x_i \mid i \in S)$ is stable if the following equivalent conditions hold:*

a) $\mathrm{Aut}\,(C, (x_i))$ is a finite group.

b) Every irreducible component of the normalization of C of genus zero (resp. 1) contains ≥ 3 (resp. ≥ 1) special points.

c) $\omega_C(\sum_{i \in S} x_i)$ is ample, so that $\omega_C(\sum_{i \in S} x_i)^{\otimes 3}$ is very ample.

Proof. Assume first that C is irreducible. Then $\mathrm{Aut}\,(C, (x_i))$ is infinite iff its normalization has genus g, n special points, and satisfies the condition $2g - 2 + n \leq 0$. This means that $\deg \omega_C(\sum x_i) \leq 0$, that is, $\omega_C(\sum x_i)$ is not ample. If the degree is positive, it is the classical fact that the cube of the modified canonical sheaf is very ample. The passage to the general case can be made using (1.3) and the restriction to the irreducible components.

Here is another application of (1.3).

1.2.2. Proof of Proposition III.2.6. Consider the normalization morphism

$$f: \coprod_{v \in V_\tau} \tilde{C}_v \to C.$$

From (1.3) we get

$$f^*(\omega_C(\sum_{i \in S} x_i)) \cong \bigoplus_{v \in V_\tau} \omega_{\tilde{C}_v}(\sum_{j \in F_\tau(v)} y_j),$$

where $F_\tau(v)$ is the set of flags of τ adjoining the vertex v. Comparing the degrees of both sides, we get

$$2g - 2 + n = \sum_{v \in V_\tau}(2g_v - 2 + |v|) = \sum_{v \in V_\tau}(2g_v - 2) + 2|E_\tau| + n.$$

This proves (2.2) of loc. cit. Hence

$$g = \sum_{v \in V_\tau} g_v + 1 - |V_\tau| + |E_\tau|.$$

The last three terms amount to $-\chi(\|\tau\|) = -1 + h^1(\|\tau\|)$ which establishes (2.1). Finally, both sides of (2.3) count $|F_\tau|$.

1.3. Prestable maps. Let V be an algebraic scheme over the ground field. Whenever we speak about ample sheaves on V, we assume in addition that V is projective. As above, S is a finite set.

1.3.1. Definition. *A prestable S-pointed map to V is a family $(C, x_i \,|\, i \in S, f: C \to V)$ such that $(C, x_i \,|\, i \in S)$ is a prestable curve, and f a morphism.*

1.3.2. Lemma–Definition. *A prestable map as above is called stable, if the following equivalent conditions hold:*

a) The automorphism group of $(C, x_i \,|\, i \in S, f)$ (identical on V) is finite.

b) If C_v is an irreducible component of C contracted by f, then the pointed curve

$$(C_v, \{y_j\}) := (C_v, \{x_i \text{ lying on } \{C_v\} \cup C_v \cap \overline{(C \setminus C_v)}\})$$

is stable.

c) Let M be an ample invertible sheaf on V (here assumed projective). Then the sheaf $L := \omega_C(\sum_{i \in S} x_i) \otimes f^*(M)^{\otimes 3}$ is ample on C.

Proof of equivalence. The automorphism group in question is infinite iff some infinite group acts on an irreducible component C_v compatibly with other data. This component then must be contracted by f (otherwise it is "stabilized by f").

In this case the pointed curve

$$(C_v, \{x_i \text{ lying on } C_v\} \cup \{C_v \cap \overline{(C \setminus C_v)}\})$$

must be unstable. Then the restriction of L to this component cannot be ample, since on it $\deg \omega_{C_v}(\sum y_j) \leq 0$ (use (1.3)), and $f^*(M)$ is trivial.

Conversely, if for some C_v this restriction is not ample, then its degree is non-positive. But the condition $2g_v - 2 + n_v + 3\deg f_v^*(M) \leq 0$ (in obvious notation) implies $\deg f_v^*(M) = 0$ so that C_v is contracted.

1.3.3. Remark. (Pre)stable maps to a point are exactly tautological maps of (pre)stable curves.

1.4. Class of the prestable map. For a projective V we put

(1.4) $\qquad B(V) := \{\beta \in \text{Hom}\,(\text{Pic}\,(V), \mathbf{Z}) \,|\, \beta(L) \geq 0 \text{ for all ample } L\}.$

This is an algebraic geometric version of the group of classes of effective algebraic cycles in $H_2(V, \mathbf{Z})$.

It is known that $B(V)$ is a semigroup in which zero has no non–trivial decomposition, and any β has only finitely many decompositions.

Therefore we can define the ring of formal series in formal monomials $q^\beta, \beta \in B(V)$:

(1.5) $\qquad \Lambda = \Lambda_K(V) = K[[q^\beta]] := \{\sum_\beta a_\beta q^\beta \,|\, a_\beta \in K\},$

where K is a **Q**–algebra, for example, coefficient ring of a cohomology theory. We will call Λ or its localization with respect to some monomials q^β the *Novikov ring of V*. The map $\beta \mapsto q^\beta$ is a universal character of $B(V)$.

1.4.1. Definition. *The total class of the prestable map $f : C \to V$ is*

(1.6) $$\beta : \text{Pic}(V) \to \mathbf{Z}, \ \beta(L) := \deg f^*(L).$$

The class of f is the family (β_v) consisting of the classes of the restrictions of f to the irreducible components C_v of C.

1.4.2. Remark. The space (stack) of maps of curves of given genus to V generally consists of infinitely many components. The semigroup $B(V)$ is a convenient set of indices which breaks this space into subspaces of finite type. For this reason various numerical characteristics of this space should be considered as infinite sums over $B(V)$. Multiplying first the β–summand by q^β in order to ensure the formal convergence we get numerical characteristics with values in the Novikov ring of V. This explains the role of Λ.

1.5. Combinatorial type of the prestable map. Let $(C, x_i \,|\, i \in S, f : C \to V)$ be a prestable map. As in III.2.5, we can define the dual modular graph (τ, g) of $(C, x_i \,|\, i \in S)$. To take account of f we add to this picture the additional labeling of vertices by the class of f: this is the map $V_\tau \to B(V) : v \mapsto \beta_v$. This graph can be called *the combinatorial type of the map*. It belongs to the general class of $B(V)$–labeled modular graphs which can be considered as objects of several categories: cf. [BehM].

Any such graph has total class $\beta := \sum \beta_v$ and genus $g := \sum g_v + \dim H_1(\|\tau\|)$ (the latter definition is reasonable if τ is connected).

1.5.1. Lemma. *a) A prestable map is stable iff its graph is stable, that is, has no vertices labeled by $g_v = 0, \beta_v = 0$ with $|v| \leq 2$ or by $g_v = 1, \beta_v = 0, |v| = 0$.*

b) There is only a finite number of stable $B(V)$–labeled modular graphs of given genus and class (up to isomorphism).

Proof. The first assertion is a restatement of 1.3.2 b). The second one can be proved by induction on the number of vertices which become unstable when one forgets the $B(V)$–labeling. If there are no such vertices, the finiteness is proved in III.2.6.1. If there is one such v, it must be stabilized by the label $\beta_v \neq 0$, and typically has $g_v = 0, |v| = 2$ so that by simply erasing it we obtain a graph of the same genus and smaller class. We leave to the reader the remaining marginal cases.

1.6. Stabilization of prestable curves. Let $(C', (y_j))$ be a connected prestable curve. Consider all morphisms to stable curves $(C, (x_i))$

$$(C', (y_j)) \to (C, (x_i))$$

mapping (y_j) surjectively onto (x_i).

i) If the category of these morphisms is non–empty, it contains a minimal morphism (the initial object of the category). It is called the stabilization morphism.

In fact, if C' is smooth of genus one without labeled points, the minimal morphism is the empty map. Otherwise it contracts to a point every maximal chain of unstable components of genus zero. If such a chain does not coincide with C' and has a marked point (necessarily one), it maps to the intersection point of this chain with the remaining part of the curve.

ii) *The stabilization morphism coincides with the one defined by (a positive power of) the invertible sheaf* $\omega_{C'}(\sum y_j)$.

This essentially follows from (1.3). It is convenient to extend the definition of the stabilization morphism in order to include the marginal cases when it can be empty or contracting the whole curve. Then ii) can be taken as the definition.

This projective description is more important than the combinatorial one with which we have started because thanks to it the stabilization morphism can be constructed uniformly in flat families, using the relative dualizing sheaf.

1.7. Stabilization of prestable maps. Now let $(C', (y_j), g : C' \to V)$ be a connected prestable map. Again, consider all morphisms to stable maps identical on V
$$(C', (y_j), g) \to (C, (x_i), f)$$
mapping (y_j) surjectively onto (x_i).

i) *Among these morphisms there exists the minimal one. It is called the stabilization morphism.*

In fact, if C' is smooth of genus one without labeled points and C' is contracted by g, the stabilization morphism is the empty map. Otherwise it contracts to a point every maximal chain of unstable components of genus zero and class zero. If such a chain has a marked point (necessarily one), it maps to the intersection point of this chain with the remaining part of the curve.

ii) *The stabilization morphism coincides with the one defined by the invertible sheaf* $\omega_{C'}(\sum y_j) \otimes g^*(M)^{\otimes 3}$ *where M is an ample sheaf on V.*

This is a refinement of 1.3.2 c).

More generally, let $\phi : V \to W$ be a morphism. Consider the category of morphisms compatible with ϕ
$$(C', (y_j), g) \to (C, (x_i), f),$$
where the target is a stable map to W. Then the evident analogs of i) and ii) hold (in ii), $g^*(M)$ should be replaced by $(\phi g)^*(M)$ where M is an ample sheaf on W).

In particular, if W is a point, we get the notion of *"absolute stabilization"*.

1.8. A warning. In III.2.5 we remarked that any modular graph (τ, g) is the combinatorial type of a prestable curve (over any algebraically closed field). To see this, take a family (C_v) of smooth curves of genera (g_v) labeled by the vertices of τ. Choose on every C_v a family of pairwise distinct closed points labeled by flags of τ incident to v. Finally, on $\coprod C_v$ glue pairs of labeled points corresponding to the halves of the same edge of τ.

Contrary to this, not every $B(V)$–labeled modular graph is the combinatorial type of a prestable map. Moreover, when V varies in a flat system, $B(V)$ itself and the realizable graphs can change.

For example, let V be a blow up of $\mathbf{P}^2$ at three different points, $e_i, i = 1, 2, 3$, the classes of the respective exceptional curves, l the class of the lift of the generic line. Then $\beta := l - \sum e_i$ (as homology class) is generally not in $B(V)$ but lands in $B(V)$ if the three points are collinear, so that the one–vertex graph labeled by $(g = 0, \beta)$ sometimes is the type of a stable map and sometimes not.

Similarly, let V be a smooth cubic surface, and β_1, β_2 classes of two disjoint lines. Then one–vertex graphs labeled by $(0, \beta_1)$ and $(0, \beta_2)$ are realizable by stable maps, but the one–edge graph with these vertices is not.

1.9. Dimension and virtual dimension. Looking ahead and developing the remarks above, we see that the space of the prestable curves of a given combinatorial type (τ, g) heuristically can be assigned the dimension

$$(1.7) \qquad \dim(\tau, g) := \sum_{v \in V_\tau} (3g_v - 3 + |v|) = 3g - 3 + |S_\tau| - |E_\tau|.$$

The first representation assigns the dimension $3g_v - 3$ to the space of smooth curves of genus g_v. The second (which is a formal identity, cf. 1.2.2 above) can be interpreted as saying that acquisition of a double point (= an edge) is a codimension one condition.

The assignment (1.7) can be made precise in several contexts:

i) For stable curves, the relevant spaces are Deligne–Mumford stacks, and the dimension of them having transparent geometric meaning is exactly (1.7).

ii) For unstable types, the relevant spaces belong to the more general category of Artin stacks, and moreover they are highly non–separated. For example, if $g = 0, n = 0$, the dimension -3 is assigned by (1.7) to the stack containing the classifying space $BPGL(2)$ ("a point, class of $\mathbf{P}^1$, modulo the automorphism group of $\mathbf{P}^1$", hence -3). In addition, this stack contains infinitely many copies of various classifying spaces of subgroups of powers of $PGL(2)$ because if one has a family of prestable curves over a curve, one can blow up any number of smooth points of fibers and get a new family of this type. Therefore, the geometric meaning of the combinatorial dimension becomes much less transparent.

Turning to the (pre)stable maps, we are led to introduce the numerical character of the $B(V)$–labeled modular graph which will be called *the virtual dimension* of the relevant stack of maps:

$$(1.8) \qquad \begin{aligned} \dim(\tau, g, \beta) := & (\chi(\|\tau\|) - \sum g_v)(\dim V - 3) \\ & - \sum \beta_v(\omega_V) + |S_\tau| - |E_\tau|. \end{aligned}$$

In the stable case, it will be the dimension of an element of the homological Chow group of the relevant stack of maps. This element with remarkable properties will be called the *virtual fundamental class* of this stack.

Construction of this class involves the stacks of unstable maps as well.

§2. Flat families of curves and maps

2.1. Review of flatness. Many of the known universal relations between the Gromov–Witten invariants are proved by studying degenerations in the families of "good" curves and maps. In III.2.2 we formally introduced the class of families we will be concerned with: that of flat families.

The role of flat dependence of parameters is explained by the fact that it provides just the correct balance between continuity and discontinuity. Flat families are continuous enough so that some important invariants of the fibers still form continuous families. At the same time, they allow for enough discontinuity to include some degenerations sufficient to "fill holes" in the incomplete bases. For this reason bases of the universal flat families (in whatever category they may exist) tend to be proper.

However, flatness alone usually does not constrain the family strongly enough in order to make the limiting fibers unique. Technically speaking, the bases of universal families usually are not separated. As we remarked in 1.9 ii), bases of universal families of prestable curves are not separated at any point. It is here that the stability condition comes to the rescue. One of the first tasks in the theory of flat families of stable curves and stable maps is to show that the combination of flatness and stability makes the respective stacks simultaneously proper and separated.

Flatness allows a very simple algebraic definition but seemingly no complete geometric description. Perhaps for this reason its applications are confined to algebraic and analytic geometries, as opposed to the smooth one. We will start this section with a brief report on flat families. For more details and proofs, see [Ha2], pp. 253–276, [Mu1], Lectures 6 and 7, [Mu2], pp. 46–55.

2.1.1. Algebra: flat modules. A module M over a commutative ring A is called *flat* if any of the following equivalent properties holds:

i) Tensor product by M transforms any exact triple of modules into an exact triple.

In particular, free modules and projective modules are flat. Conversely, if A and M are noetherian and M is flat, then M is projective.

ii) M is a direct limit of free A-modules.

In particular, if A is a domain, then any localization of it, A_S, is flat over A. More generally, such a localization is flat, if S does not contain zero divisors.

This description shows also that if M is A-flat, then for any A-algebra B, $B \otimes_A M$ is B-flat.

iii) All linear relations in M follow from linear relations in A in the following sense. If $\sum_i a_i m_i = 0$ for some $a_i \in A$, $m_i \in M$, then we can write $m_i = \sum_j b_{ij} n_j$, $b_{ij} \in A$, $n_j \in M$, in such a way that for all j, $\sum_i b_{ij} a_i = 0$.

2.1.2. Geometry: flat families of schemes and sheaves. Let $f: X \to Y$ be a morphism of schemes, and $\mathcal{F}$ a quasicoherent sheaf on X.

In this situation $\mathcal{F}$ is called *flat over Y* (or $\mathcal{O}_Y$) if for all $x \in X$, $\mathcal{F}_x$ is flat over $\mathcal{O}_{f(x)}$. The morphism f itself is called flat, if $\mathcal{O}_X$ is flat over Y.

From the algebraic remarks made above one easily deduces that flat coherent sheaves over a noetherian Y ($f = \mathrm{id}$) are exactly locally free sheaves. In particular, a finite $f: X \to Y$ is flat iff $f_*(\mathcal{O}_X)$ is locally free.

2.1.3. Three criteria of flatness. i) Flatness over a base is stable with respect to any base extension, in particular, passing to fibers and geometric fibers.

ii) Let Y be a smooth curve. If X is reduced, then a morphism $f: X \to Y$ is flat iff it maps any generic point of X onto the generic point of Y. For general X, every associated point of X must be mapped onto the generic point of Y. (A point x is associated, if its maximal ideal consists only of zero divisors.)

iii) Let $f : X \to Y$ be a relatively projective morphism. This means that it is the composition of the closed embedding of X into a projectivized vector bundle over Y and the projection to Y. Then we have the induced sheaf $\mathcal{O}(1)$ on X.

In this situation, a coherent sheaf $\mathcal{F}$ on X is flat over Y iff for all large enough m, $f_*(\mathcal{F}(m))$ is locally free on Y. In particular, f itself is flat iff $f_*(\mathcal{O}(m))$ is locally free for large m.

This is equivalent to the requirement that the Hilbert polynomials of fibers do not depend of the base point. In particular, dimensions of the fibers are locally constant.

2.1.4. Continuity and discontinuity of cohomology in flat families.
Let $\mathcal{F}$ be a coherent sheaf on X, $f : X \to Y$. The single most important function on the base in this situation is the Euler characteristic:

$$y \mapsto \chi(\mathcal{F}_y) := \sum_i (-1)^i \dim H^i(X_y, \mathcal{F}_y).$$

If $\mathcal{F}$ is flat over Y, then $\chi(\mathcal{F}_y)$ is locally constant.

Dimensions of the separate cohomology groups can jump even in very smooth families. For example, let E be an elliptic curve. Consider a projection morphism $f : E \times E \to E$ and the sheaf $L := \mathcal{O}_{E \times E}(2(\Delta - D))$ where D is a constant section of f and Δ is the diagonal. Then $H^0(E, L_y) = 0$ outside of four points y in E where this group becomes one–dimensional.

Discontinuity of the cohomology of complexes, say, of finite–dimensional vector spaces with differential d depending on a parameter, occurs at those values of the parameter y where the rank of the matrix of the differential d_y drops. Then the dimension of the kernel jumps and that of the image drops, hence cohomology can only jump. It was Grothendieck's insight that behavior of the cohomology in flat families can be reduced to this transparent picture.

The discontinuity of the *dimension* of the cohomology spaces in principle is compatible with the following weaker property: for a fixed i and variable y, $H^i(X_y, \mathcal{F}_y)$ could be the y–fiber of the coherent sheaf $R^i f_*(\mathcal{F})$ on Y. In fact, we always have a natural morphism

$$(2.1)_i \qquad k(y) \otimes R^i f_*(\mathcal{F}) \to H^i(X_y, \mathcal{F}_y)$$

which however need not be an isomorphism even for flat families.

Here are two important cases when it is an isomorphism. Flatness is assumed below.

i) *For a fixed i, assume that the function $y \mapsto \dim H^i(X_y, \mathcal{F}_y)$ is locally constant, or equivalently, $R^i f_*(\mathcal{F})$ is locally free. Then $(2.1)_i$ is an isomorphism.*

ii) *In the situation of i), $(2.1)_{i-1}$ is an isomorphism as well.*

For flat families of (pre)stable curves, it suffices to check that $\dim H^i(X_y, \mathcal{F}_y)$ is locally constant for $i = 0$ or $i = 1$ because the Euler characteristic is always locally constant.

2.2. Flat families of curves and maps. Flat families of (pre)stable (resp. S–labeled (pre)stable) curves over a scheme T were formally defined in III.2.1 (resp. III.2.2) and called curves over T there. We will use both terms indiscriminately.

Because of the remark above, *genus* is a locally constant function on the base, as was stated in III.2.1.

A flat family of prestable maps to V over T (or simply a prestable map over T) consists of the data $(C/T, x_i \,|\, i \in S, f : C \to V)$ where $(C/T, x_i \,|\, i \in S)$ is a prestable curve and f a morphism. It is called stable iff its restrictions to all geometric fibers over T are stable.

Since $\deg f^*(L)$ in (1.6) is an Euler characteristic, *the total class* of the prestable map (cf. 1.4.1) to a projective V is a locally constant function $T \to B(V)$.

2.2.1. Lemma. *Assume that C/T has connected fibers. The total class of a prestable map*
$$(C/T, x_i \,|\, i \in S, f : C \to V)$$
vanishes iff f factors through T, and then this factorization is unique.

Proof. Let $\pi : C \to T$ be the structure morphism. From the definition of prestable curve and 2.1.4 i) it follows that $\pi_*(\mathcal{O}_C) = \mathcal{O}_T$. Vanishing of the total class of f implies that each geometric fiber C_t is contracted by f to a point. Therefore we can apply Lemma 8.11.1 of [GD], EGA II.

2.3. Gluing along pairs of sections. Let $(\pi : C \to T; s_1, s_2 : T \to C)$ be a prestable curve endowed with two structure sections. We want to show that they can be glued in a canonical way so that the constructions of the kind described in 1.8 can be done over a base.

2.3.1. Proposition. *In the category of T-morphisms $q : C \to D$ with $qs_1 = qs_2$ there is a universal (initial) one which we denote $p : C \to C'$. It is defined up to unique isomorphism and possesses the following properties:*

i) C'/T is a prestable curve, and p is a finite morphism.

ii) As a topological space, C' is the quotient of C under the equivalence relation $s_1(t) = s_2(t)$ for all $t \in T$. The structure sheaf of C' is determined by the condition
$$\Gamma(U, \mathcal{O}_{C'}) = \{f \in \Gamma(p^{-1}(U), \mathcal{O}_C) \,|\, s_1^*(f) = s_2^*(f)\}.$$

Assume furthermore that C' has connected geometric fibers. Then two cases are possible:

A. Geometric fibers of C are irreducible. In this case $g(C') = g(C) + 1$.

B. $C = C_1 \amalg C_2$ where geometric fibers of C_1 and C_2 are connected and $s_i : T \to C_i$. In this case $g(C') = g(C_1) + g(C_2)$.

A rather straightforward proof proceeds as follows. Assertion ii) describes C' as a ringed space over T. By a local analysis in the neighborhood of glued points, one checks that it is actually a scheme flat over T and its fibers are prestable curves. The genus can then be calculated using the combinatorial formula in 1.2.2. For more details, see [Kn1], pp. 181–183.

We can now similarly treat gluing prestable maps along a pair of sections.

2.3.2. Proposition. *In the situation above, consider a prestable map $f : C \to V$. If $fs_1 = fs_2$, then there exists a unique prestable map $f' : C' \to V$ such that $f = f'p$. The total class of it is (in the self-explanatory notation):*

Case A: $\beta(C') = \beta(C)$.

Case B: $\beta(C') = \beta(C_1) + \beta(C_2)$.

This follows from 2.3.1 and the additivity of Euler characteristic.

2.4. The relative dualizing sheaf and the sheaf of 1–forms. The results of 1.1–1.3 admit straightforward generalizations to the relative case. Following [Kn1], §3, we add some complements.

2.4.1. Cotangent complex of the prestable curve. Let C/T be a prestable curve. Then $\Omega_{C/T}$ is flat over T, and locally on C it allows free resolutions $\mathcal{E}_1 \to \mathcal{E}_0$ of length 2 defined up to a quasi–isomorphism. They represent the cotangent complex of C/T (cf. [LS] and [Il]).

The formula (1.2) still holds in the relative case so that $\omega_{C/T}$ is invertible.

As a refinement of (1.1), we have the following.

2.4.2. Proposition. *In the situation of 2.3.1, denote by $\mathcal{N}_i$ the sheaf on T which is induced by the conormal sheaf to $D_i = s_i(T)$ in C. Let $s : T \to C'$ be the common image of s_1, s_2. Then one can define the following exact sequence on C':*

$$0 \to s_*(\mathcal{N}_1 \otimes \mathcal{N}_2) \to \Omega_{C'/T} \to p_*(\Omega_{C/T}) \to 0.$$

This is [Kn1], Th. 3.5, p. 183.

§3. Groupoids and moduli groupoids

3.1. A preview. The main goal of this section is to provide the definitions of the moduli "spaces" $\overline{M}_{g,S}(V,\beta)$ and $\overline{C}_{g,S}(V,\beta)$ and their prestable counterparts as objects of the 2–category of groupoids.

In a certain sense what the algebraic geometry suggests in this case is a tautology: for example, $\overline{M}_{g,S}(V,\beta)$ is just *the category of all diagrams* of the type $(C/T, (x_i | i \in S); f : C \to V)$ which are stable maps to V with given (g, S, β). The more conservative approach would be to consider *the sets of isomorphism classes of such diagrams* over a variable T as defining the functor of points $T \mapsto \overline{M}_{g,S}(V,\beta)(T)$. If it were representable (as, say, for $g = 0, \beta = 0$) our moduli space $\overline{M}_{g,S}(V,\beta)$ would be the representing scheme. However, generally the representability is obstructed by the automorphisms of (pre)stable maps. Hence we step back and consider $\overline{M}_{g,S}(V,\beta)(T)$ not as a set but as *a category in which all morphisms are isomorphisms* over T, and the union of all $\overline{M}_{g,S}(V,\beta)(T)$ as a category, in which the additional morphisms are base changes.

We start with axiomatizing the basic properties of such categories (*groupoids*), and then proceed to the main examples. Morphisms between groupoids will be introduced in the next section.

The reader should keep in mind that the actual moduli groupoids $\overline{M}_{g,S}(V,\beta)$, $(C/T, x_i | i \in S; f : C \to V)$ and alike satisfy stronger restrictions which express the locality of the notion of a family. Their axiomatization requires introduction of a topology on the category of bases. Groupoids with these properties are called *stacks of groupoids*. We turn to stacks in §5.

3.2. Groupoids. Let $\mathcal{F}, \mathcal{S}$ be two categories and $b : \mathcal{F} \to \mathcal{S}$ a functor. If $F \in \mathrm{Ob}\,\mathcal{F}$, $b(F) = T$, we will sometimes call F *a family with the base T*, or a *T-family* (as we remarked, this is more justified in the context of stacks).

In order to form a groupoid, these data must satisfy two conditions which can be stated in two almost equivalent versions.

VERSION 1. First, for any base $T \in \mathcal{S}$, any morphism of families over T inducing identity on T must be an isomorphism. Let $\mathcal{F}_T$ be this subcategory of $\mathcal{F}$. It is called *the fiber of $\mathcal{F}$ over T*.

Second, for any arrow $\phi : T_1 \to T_2$ between the bases, there must be given the "base change" functor $\phi^* : \mathcal{F}_{T_2} \to \mathcal{F}_{T_1}$, and for two composable arrows ϕ, ψ there must be given functor isomorphisms $(\phi \circ \psi)^* \xrightarrow{\sim} \psi^* \circ \phi^*$ satisfying the cocycle condition expressing associativity.

Groupoids in real life often come in this form, especially when they are used to treat a moduli problem, where the natural notion of induced family (base change functor) can be defined. Sometimes, however, it is unreasonable to include in the definition the actual choice of the base change functors and isomorphisms between their compositions. If we want only to ensure their existence, then the following version is at hand.

VERSION 2. First, for any arrow $\phi : T_1 \to T_2$ between the bases and any family F_2 over the target T_2, there must exist a T_1-family F_1 and a morphism $F_1 \to F_2$ lifting ϕ.

Second, consider any commutative triangle of bases $\phi_{ij} : T_i \to T_j$, $i < j \in \{1, 2, 3\}$. Assume that we are given three families F_i over T_i and two arrows $F_1 \to F_3$, $F_2 \to F_3$ lifting respectively ϕ_{13}, ϕ_{23}. Then the remaining arrow lifts uniquely so that the triangle of F_i becomes commutative.

The passage from Version 2 to Version 1 runs as follows. If Version 2 holds, then any lift of an isomorphism in $\mathcal{S}$ is an isomorphism in $\mathcal{F}$, and if the target of an arrow in $\mathcal{S}$ is lifted to $\mathcal{F}$, then the source can be compatibly lifted to an object in $\mathcal{F}$ uniquely up to unique isomorphism. Making choices of such liftings, we get the base change functor of Version 1.

In a shorthand description of a concrete groupoid, we often restrict ourselves to defining a typical object of $\mathcal{F}$ (say, as a diagram of schemes) and specifying what the base of this diagram is. Base change functors are usually self-evident.

We now give basic examples.

3.2.1. Groupoids of S–labeled (pre)stable curves. Here $\mathcal{S}$ is the category of schemes (or schemes over a given base, eventually ground field of characteristic zero), objects of $\mathcal{F}$ are (pre)stable S-labeled curves over $T \in \mathcal{S}$, and a morphism $(C_1/T_1, x_{1i} \,|\, i \in S) \to (C_2/T_2, x_{2i} \,|\, i \in S)$ is a pair of compatible morphisms $\phi : T_1 \to T_2, \psi : C_1 \to C_2$ such that ψ induces an isomorphism of labeled curves

$C_1 \to \phi^*(C_2)$. Equivalently, the diagram

$$\begin{array}{ccc} C_1 & \xrightarrow{\psi} & C_2 \\ \downarrow & & \downarrow \\ T_1 & \xrightarrow{\phi} & T_2 \end{array}$$

is cartesian, and induces the bijection of the two families of S-labeled sections.

The groupoid of S-labeled stable (resp. connected prestable) curves of genus g is denoted $\overline{M}_{g,S}$, resp. $\mathcal{M}_{g,S}$.

3.2.2. Groupoids of S-labeled (pre)stable maps. We can similarly define the groupoids of (pre)stable maps to V's by allowing as morphisms

$$(C_1/T_1, x_{1i} \mid i \in S, f_1 : C_1 \to V) \to (C_2/T_2, x_{2i} \mid i \in S, f_2 : C_2 \to V)$$

commutative diagrams

$$\begin{array}{ccc} V & \xrightarrow{id} & V \\ f_1 \uparrow & & \uparrow f_2 \\ C_1 & \xrightarrow{\psi} & C_2 \\ \downarrow & & \downarrow \\ T_1 & \xrightarrow{\phi} & T_2 \end{array}$$

such that the lower square induces a morphism of the respective (pre)stable curves.

The groupoid of S-labeled stable (resp. prestable) maps to V of *connected* curves of genus g and class β is denoted $\overline{M}_{g,S}(V,\beta)$, resp. $\mathcal{M}_{g,S}(V,\beta)$. By definition, morphisms in these groupoids should restrict to identity on V.

3.2.3. Groupoids of universal curves. The groupoid $\overline{C}_{g,S}$, *the universal S-labeled stable curve of genus g*, has as its objects stable curves $(C/T, x_i \mid i \in S)$ as in 3.2.1 endowed with an additional section $\Delta : T \to C$ not constrained by any restrictions. Morphisms must be compatible with this additional data. Similarly one defines the prestable version $\mathcal{C}_{g,S}$ and the two versions for maps $\overline{C}_{g,S}(V,\beta)$, $\mathcal{C}_{g,S}(V,\beta)$.

3.2.4. Schemes as groupoids. Any scheme V produces a groupoid $\mathcal{V}$ whose objects over T are morphisms $T \to V$ and whose base change functor is the composition. Since this is the same as the functor represented by V we neither lose nor gain any information passing from V to $\mathcal{V}$, but only change the viewpoint: a scheme is considered as the category of families of its points, with identical morphisms.

Clearly, this argument is applicable to any category $\mathcal{S}$: any object of this category can be considered as a groupoid over $\mathcal{S}$ (with identical 2–morphisms, cf. §4).

3.2.5. Classifying groupoids. Again let $\mathcal{S}$ be the category of schemes over a given base. Consider a group object G in $\mathcal{S}$. *The classifying groupoid BG* is defined in the following way: a T-family in BG is a principal G-bundle over $T \in \mathcal{S}$. The base change functor is the standard one.

§4. Morphisms of groupoids and moduli groupoids

4.1. A preview. In this section we construct the basic structure morphisms between the moduli groupoids related to (pre)stable maps. The most important of them are listed below. First, we have the diagram

(4.1)
$$\begin{array}{ccc} \overline{C}_{g,S}(V,\beta) & \xrightarrow{f} & V \\ {\scriptstyle \pi}\downarrow & & \\ \overline{M}_{g,S}(V,\beta) & & \end{array}$$

where V is considered as a groupoid (denoted $\mathcal{V}$ in 3.2.4). Second, there are the structure sections of π:

(4.2) $\qquad x_i := x_{i;g,S}(V,\beta) : \overline{M}_{g,S}(V,\beta) \to \overline{C}_{g,S}(V,\beta),\ i \in S.$

We also have similar diagrams for prestable maps. Third, we have a series of non-trivial stabilization morphisms from prestable to stable moduli groupoids. They lead to the existence of the equivalence of groupoids

(4.3) $\qquad u := u_{g,S}(V,\beta) : \overline{M}_{g,S\cup\{*\}}(V,\beta) \xrightarrow{\sim} \overline{C}_{g,S}(V,\beta).$

Finally, we construct the boundary morphisms expressing the structure of degenerations. Before doing this, we must introduce the moduli groupoids of stable maps of a given combinatorial type.

To put all of this on a firm basis, we must first continue our discussion of abstract groupoids. Since they are categories rather than sets with structure, 1-morphisms between them are functors, and commutative diagrams must sometimes be replaced by weakly commutative ones where the equality of 1-morphisms is weakened to an isomorphism between them.

4.2. 1-morphisms of abstract groupoids. We will be considering only morphisms between groupoids over *the same category of bases* $\mathcal{S}$. By definition, such a morphism $\{b_1 : \mathcal{F}_1 \to \mathcal{S}\} \to \{b_2 : \mathcal{F}_2 \to \mathcal{S}\}$ is a functor $\Phi : \mathcal{F}_1 \to \mathcal{F}_2$ such that $b_2 \circ \Phi = b_1$. Composition of morphisms is evident.

Since we require here the strict equality $b_2 \circ \Phi = b_1$ and not just isomorphism of functors, in some contexts this definition may be too restrictive. However, moduli problems furnish plenty of morphisms between moduli groupoids which actually do not change the base. Such a morphism is described by a natural construction producing from a family of one type the family of another type and commuting with the base change.

These natural constructions are quite simple in the cases (4.1) and (4.2).

4.2.1. Morphisms f, π, x_j. A typical object of the groupoid $\overline{C}_{g,S}(V,\beta)$ is a diagram $(C/T, (x_i), \Delta; \phi : C \to V)$ (cf. 3.2.2 and 3.2.3 above). The natural construction π forgets Δ. The natural construction f produces from this diagram the T-point $\phi \circ \Delta : T \to V$ (cf. 3.2.4). The natural construction x_j produces from the stable map $(C/T, (x_i); \phi : C \to V)$ the diagram

$$(C/T, (x_i), \Delta := x_j; \phi : C \to V).$$

Similarly elementary is the action of the group of bijections of S on the diagrams (4.1), (4.2). It simply relabels the structure sections.

Replacing in this discussion stable maps by the prestable ones, we get the prestable versions of these morphisms.

4.2.2. Evaluation morphisms. The evaluation morphisms are defined in the context of diagrams (4.1) and (4.2): for any $j \in S$ we put

$$\mathrm{ev}_j = \mathrm{ev}_{j;g,S}(V,\beta) := f \circ x_j : \overline{M}_{g,S}(V,\beta) \to V.$$

4.2.3. Proposition. *There exists the canonical isomorphism of groupoids*

(4.4) $$V \times \overline{M}_{g,S} \xrightarrow{\sim} \overline{M}_{g,S}(V,0)$$

such that each evaluation morphism of the rhs groupoid becomes the projection to V.

There are similar isomorphisms for prestable maps and universal curves.

Proof. By definition, an object of $V \times \overline{M}_{g,S}$ consists of the data

$$(\phi : T \to V; \pi : C \to T, x_i \,|\, i \in S),$$

where C is a stable S-labeled curve. We produce from it the stable map $\phi \circ \pi : C \to V$. In order to construct the reverse map, we use Lemma 2.2.1. From this lemma it follows that any stable map of class zero is a map of stable curve. The remaining cases are treated similarly.

We now turn again to abstract groupoids.

4.3. 2–morphisms. By definition, a 2–morphism $F : \Phi_1 \to \Phi_2$ between two 1–morphisms of groupoids with the same source $\mathcal{F}_1$ and target $\mathcal{F}_2$ is *an isomorphism of these functors*. In more detail, it is a morphism (natural transformation) of functors such that for any $X \in \mathrm{Ob}\,\mathcal{F}_1$, the map $F(X) : \Phi_1(X) \to \Phi_2(X)$ is an isomorphism.

In practical terms, a context in which 2–morphisms are essential arises every time when a commutative diagram of 1–morphisms is considered: instead of *equality* of two compositions of 1–morphisms we usually can assure only their *isomorphism*, so that the definition of commutativity must be relaxed in this way. We will include the choice of such isomorphism(s) in the data defining the notion of the commutative square, and will sometimes speak about *weak* commutativity in order to be reminded about this.

Let us review the definition and the construction of cartesian squares of groupoids in this light.

4.3.1. Cartesian squares. Given two 1–morphisms of groupoids $\Phi_i : \mathcal{F}_i \to \mathcal{G}$, $i = 1,2$, we construct their fibered product in the following way.

The groupoid $\mathcal{F}_1 \times_{\mathcal{G}} \mathcal{F}_2$. By definition, its objects are triples

$$(X_1 \in \mathcal{F}_1, X_2 \in \mathcal{F}_2, a = a_{X_1, X_2} : \Phi_1(X_1) \to \Phi_2(X_2)),$$

where a is an isomorphism lifting the identity isomorphism of the common base, and its morphisms are pairs of morphisms in $\mathrm{Mor}\,\mathcal{F}_1 \times \mathrm{Mor}\,\mathcal{F}_2$ compatible with other data.

The weakly commutative cartesian square. This is the diagram

(4.5)
$$\begin{array}{ccc} \mathcal{F}_1 \times_\mathcal{G} \mathcal{F}_2 & \xrightarrow{\psi_2} & \mathcal{F}_2 \\ \psi_1 \downarrow & & \downarrow \phi_2 \\ \mathcal{F}_1 & \xrightarrow{\phi_1} & \mathcal{G} \end{array}$$

where ψ_i maps (X_1, X_2, a) to X_i and all a_{X_1, X_2} taken together constitute the 2-morphism $\phi_1 \circ \psi_1 \to \psi_2 \circ \phi_2$ establishing the weak commutativity.

The universal cartesian property. Consider any weakly commutative square:

$$\begin{array}{ccc} \mathcal{H} & \xrightarrow{\psi_2'} & \mathcal{F}_2 \\ \psi_1' \downarrow & & \downarrow \phi_2 \\ \mathcal{F}_1 & \xrightarrow{\phi_1} & \mathcal{G} \end{array}$$

For $Y \in \text{Ob}\,\mathcal{H}$, denote by $b = b_Y : \phi_1 \circ \psi_1'(Y) \to \psi_2' \circ \phi_2(Y)$ the weak commutativity isomorphism.

The universality of (4.5) expresses itself in the existence and uniqueness of the 1–morphism $\mathcal{H} \to \mathcal{F}_1 \times_\mathcal{G} \mathcal{F}_2$:

$$Y \mapsto (\psi_1'(Y), \psi_2'(Y), b)$$

whose compositions with ψ_i strictly coincide with ψ_i'.

This last property will be taken as the definition of cartesian squares.

4.3.2. 1–category of groupoids. Composition of 1–morphisms is compatible with 2–isomorphisms. Therefore sometimes it is convenient to consider the 1–category of groupoids whose arrows are isomorphism classes of 1–morphisms in the sense of 4.3.

Let us now review 1–morphisms connecting stable moduli groupoids to the unstable ones.

4.4. Stabilization. Considering a stable map as a prestable one over the same base, we have the natural 1–morphism of groupoids

$$m = m_{g,S}(V, \beta) : \overline{\mathcal{M}}_{g,S}(V, \beta) \to \mathcal{M}_{g,S}(V, \beta).$$

In 1.7 we described how to produce the stable map from an unstable map f over a point by blowing down all the components of the curve for which the restriction of f is unstable. We will show now that this construction can be globalized to define the stabilization morphism

(4.6) $$\text{st} = \text{st}_{g,S}(V, \beta) : \mathcal{M}_{g,S}(V, \beta) \to \overline{\mathcal{M}}_{g,S}(V, \beta).$$

In fact, st is left adjoint to m.

To be more precise, let $(\pi : C \to T, (x_i), f : C \to V)$ be a prestable map with connected geometric fibers. We will say that a map of T–schemes $g : C \to U$ *stabilizes* f if the following holds: for every geometric point t of T, the map of underlying topological spaces $g_t : C_t \to U_t$ contracts to a single point every f-unstable component of C_t. Notice that $(f, \pi) : C \to V \times T$ stabilizes f.

4.4.1. Proposition. *a) There exists the universal morphism $p : C \to \widetilde{C}$ stabilizing f (initial object of the appropriate category). We will call it the stabilization morphism. It is defined uniquely up to unique isomorphism identical on C. Moreover, $(\widetilde{C}/T, (x_i \circ p))$ is an S-labeled prestable curve. Choose one such object for each family.*

b) Let $\widetilde{f} : \widetilde{C} \to V$ be the canonical map of the initial object to (f, π) followed by the projection to V. Then $(\widetilde{C}/T, (x_i \circ p), \widetilde{f})$ is a stable map to V.

c) The maps

$$(4.7) \qquad (C/T, (x_i), f) \mapsto (\widetilde{C}/T, (x_i \circ p), \widetilde{f})$$

extend to the 1-morphism (4.6). Any two such extensions are connected by the unique 2-isomorphism.

Proof. Let $(\pi : C \to T, (x_i), f : C \to V)$ be a prestable map with connected geometric fibers. Following the lead of 1.7, we choose an ample invertible sheaf M on V and construct an invertible sheaf on C:

$$L = \omega_{C/T}(\sum_{j \in S} x_j) \otimes f^*(M)^{\otimes 3}.$$

Put

$$\widetilde{C} := \mathrm{Proj}_{\mathcal{O}_T}(\bigoplus_{k=0}^{\infty} \pi_* L^k).$$

For large enough k, $H^1(C_t, L^k) = 0$ on all geometric fibers, so that $\pi_*(L^k)$ is locally free and commutes with base change. Therefore, formation of $\widetilde{C}$ commutes with base change. To check that $\widetilde{C}$ is projective over T, notice that $\widetilde{C} = \mathrm{Proj}_{\mathcal{O}_T}(\bigoplus_{k=0}^{\infty} \pi_*(L^{dk}))$ for any $d \geq 1$. If d is sufficiently large, then the natural map $\pi_*(L^d)^{\otimes k} \to \pi_*(L^{dk})$ is fiberwise surjective and therefore surjective. It follows that $\bigoplus_{k=0}^{\infty} \pi_*(L^{dk})$ is generated by its degree one part so that $\widetilde{C}$ is projective over T. By the flatness criterion 2.1.3 iii), $\widetilde{C}$ is flat over T.

The map of graded algebras

$$\pi^*(\bigoplus_{k=0}^{\infty} \pi_*(L^k)) \to \bigoplus_{k=0}^{\infty} L^k$$

defines a morphism from an open subset of C to $\widetilde{C}$. It is everywhere defined because $\pi^* \pi_*(L^k) \to L^k$ is an epimorphism, which follows from the corresponding fact on geometric fibers. Denote it by $p : C \to \widetilde{C}$. Clearly, it is proper and dominant, therefore it is surjective. As was remarked in 1.7, p stabilizes f. More precisely, the inverse image of any geometric point $x \in \widetilde{C}$ with respect to p is either a point, or an f-unstable component of the respective fiber.

It follows that $H^1(p^{-1}(x), k(x) \otimes_{\mathcal{O}_{\widetilde{C}}} \mathcal{O}_C) = 0$. Therefore the formation of $p_*(\mathcal{O}_C)$ commutes with base change.

Now the same argument as in the proof of Lemma 2.2.1 shows that p is the universal morphism stabilizing f and that it is unique. Fiberwise analysis made above shows that $(\widetilde{C}/T, (x_i \circ p), \widetilde{f})$ is an S-labeled prestable map. Compatibility with base change establishes that (4.7) extends to a 1-morphism of groupoids.

For more details, cf. [BehM], pp. 24–26, and [Kn1].

4.5. Theorem. *Let $S \cup \{*\}$ be the set obtained by adding to S one element. In the case $\beta = 0$ assume that $2g + |S| \geq 3$. Then we have an equivalence of groupoids*

(4.8) $$u := u_{g,S}(V, \beta): \overline{M}_{g, S \cup \{*\}}(V, \beta) \xrightarrow{\sim} \overline{C}_{g,S}(V, \beta)$$

whose composition with $\pi = \pi_{g,S}(V, \beta)$ from (4.1) is isomorphic to forgetting x_* followed by stabilization.

Proof. We define (4.8) as the 1-morphism which produces from the prestable map
$$(C/T, (x_i), x_*; f: C \to V) \in \mathrm{Ob}\, \overline{M}_{g,S}(V, \beta)$$
an object of the groupoid of universal curves (cf. 3.2.3)
$$(C'/T, (x'_i), \Delta; f': C' \to V) \in \mathrm{Ob}\, \overline{C}_{g,S}(V, \beta)$$
in the following way. We forget x_* and stabilize the resulting S-labeled curve as in (4.7). In the notation of (4.7), we put $C' := \widetilde{C}, x'_i := x_i \circ p, f' := \widetilde{f}$. Finally, we put $\Delta := p \circ x_*$. Since π simply forgets Δ, the last statement holds. It remains to check that u is an equivalence of groupoids, i.e. that it induces bijection on morphisms and surjection on the isomorphism classes of objects.

We take for granted the case $V = a$ *point* treated by Knudsen in [Kn1] and show how to reduce the general situation to this one.

Let us start with morphisms. Since morphisms with fixed source and target in a groupoid form either an empty set, or a principal homogeneous space over the automorphism group of the source, it suffices to treat the following situation. Let $(C/T, (x_i), x_*, f)$ and $(D/T, (y_i), y_*, g)$ be two stable maps such that the stabilizations of $(C/T, (x_i), f)$ and $(D/T, (y_i), g)$ are explicitly identified. Denote this stabilization $(E, (z_i), h)$. Let $p: C \to E$ and $q: D \to E$ be the respective stabilization maps. We want to prove that if $p \circ x_* = q \circ x_* := \Delta$, then there exists a unique isomorphism of stable maps
$$r: (C/T, (x_i), x_*, f) \to (D/T, (y_i), y_*, g)$$
such that $p = q \circ r$.

By the descent theory, we may localize in the étale topology of the base. Hence we may assume that we can construct some additional sections $v_j: T \to E$, $j = 1, \ldots, N$, avoiding z_i and Δ such that $(E/T, (z_i, v_j))$ becomes a stable curve. More precisely, v_j are chosen in order to stabilize the h-unstable components of fibers. These sections then can be uniquely lifted to C and D making $(C/T, (x_i, x_*, v'_j))$ and $(D/T, (y_i, y_*, v''_j))$ stable $(S \cup \{*, 1, \ldots, N\})$-labeled curves which after forgetting x_*, y_* have the common stabilization $(E/T, (z_i, v_j))$. By this trick we got rid of V and reduced the problem for stable maps to that of stable curves. This was treated in [Kn1], which provides the morphism r we sought.

It remains to prove the surjectivity of u on objects. Let $(D/T, (z_i), \Delta, g)$ be an object of the groupoid of universal curve. We want to represent it as $(\widetilde{C}/T, (x_i \circ p), \widetilde{f})$ in the notation of (4.7). Since uniqueness is already established, it remains to prove the local existence. We choose additional sections as in the previous paragraph so that $(D/T, (z_i, v_j))$ becomes a stable curve. Again appealing to Knudsen's theorem in the $V = a$ *point* case, we can find a stable curve $(D'/T, (z'_i, z_*, v'_j))$ whose stabilization after forgetting z_* is $(D/T, (z_i, v_j))$ and such that $z_* \circ q = \Delta$ where $q : D' \to D$ is the stabilization morphism. We then have the stable map

$$(D'/T, (z'_i, z_*, v'_j), g \circ q).$$

Now forget (v'_j) and stabilize. This will give the map we seek.

4.6. Absolute stabilization. Another important morphism of moduli groupoids is the *absolute stabilization map*

(4.9) $$\text{st} : \overline{M}_{g,S}(V, \beta) \to \overline{M}_{g,S}.$$

By definition, it is the composition of the following three morphisms: first we treat a stable map as a prestable one ($m_{g,S}(V, \beta)$ defined in 4.4), then forget the map to V landing in $\mathcal{M}_{g,S}$, and finally apply $\text{st}_{g,S}$ for the case $V =$ a point.

4.7. Groupoids $\overline{M}(V, \tau)$ and boundary morphisms. We start with a description of the groupoid $\overline{M}(V, \tau)$ where τ is a stable V-marked modular graph. In 1.5 above we used the notation (τ, g) but henceforth we will often omit g. Put

(4.10) $$N(V, \tau) := \prod_{v \in V_\tau} \overline{M}_{g_v, F_\tau(v)}(V, \beta_v),$$

where (g_v, β_v) are labels at the vertex v and $F_\tau(v)$ is the set of flags incident to v. This product is endowed with the morphism

(4.11) $$\text{ev} := (\text{ev}_{h; g_v, F_v(\tau)}(V, \beta_v) \circ \text{pr}_v | v \in V_\tau, h \in F_\tau(v)) : N(V, \tau) \to V^{F_\tau}.$$

Consider also the partial diagonal map

$$\Delta_\tau := (\Delta_e \circ \text{pr}_e \,|\, e \in E_\tau) : V^{E_\tau} \to V^{F_\tau^0} \cong (V \times V)^{E_\tau},$$

where F_τ^0 is the set of all flags which are parts of edges, E_τ is the set of all edges, and Δ_e maps V to the partial diagonal corresponding to the edge e. Then $\overline{M}(V, \tau)$ can be defined as the fibered product

(4.12)
$$\begin{array}{ccc} \overline{M}(V, \tau) & \xrightarrow{D_\tau} & N(V, \tau) \\ \downarrow & & \downarrow \text{evopr} \\ V^{E_\tau} & \xrightarrow{\Delta_\tau} & V^{F_\tau^0} \end{array}$$

in which the right vertical arrow is the vector of the obvious projections followed by the evaluation.

In plain words, $\overline{M}(V, \tau)$ classifies the following data: families of stable maps to V, $(C_v; x_{v,h} | h \in F_\tau(v); f_v), v \in V_\tau$, such that genus of C_v is g_v, class of $f_v(C_v)$ is β_v, and $f_v(x_{v,h}) = f_w(x_{w,k})$ whenever flags h and k belong to the same edge. The universal curves lifted from the factors of $N(V, \tau)$ can be glued together along pairs of sections corresponding to edges (cf. Proposition 2.3.2 above). Hence $\overline{M}(V, \tau)$ carries an S_τ-labeled curve

(4.13) $$\pi(V, \tau) : C(V, \tau) \to \overline{M}(V, \tau)$$

together with a stable map

(4.14) $$C(V, \tau) \to V.$$

This map is induced by the *boundary morphism*

(4.15) $$b(V,\tau): \overline{M}(V,\tau) \to \overline{M}_{g,S_\tau}(V,\beta),$$

where $g = \sum g_v + \dim H_1(\|\tau\|)$ is *the genus* of the marked modular graph, $\beta = \sum \beta_v$ is its *class*, and S_τ is the set of tails of τ. The reader is invited to reformulate this in the language of natural constructions used so far.

If τ has only one vertex, we get the familiar moduli groupoids, and (4.15) is identical.

As a slightly more sophisticated example (virtual codimension one), consider the one-edge graphs τ of genus g and class β. If such a graph has one vertex, it must be marked by $(g-1,\beta)$. If it has two vertices, they must be marked by members of any 2-partition $(g,\beta) = (g_1,\beta_1) + (g_2,\beta_2)$ subject only to the stability restriction taking into account tails at both vertices.

4.7.1. Exercise. After the identification of $C_{g,S}(V,\beta)$ with $\overline{M}_{g,S\cup\{*\}}(V,\beta)$ (Theorem 4.5), the map $x_{i;g,S}$ becomes the boundary morphism $b(V,\tau_i)$ associated with the one-edge, two-vertex graph, whose vertices are marked by $(0,0)$, (g,β), and support the tails $(*,i)$ and $S \setminus \{i\}$, respectively. In particular,

$$N(V,\tau) \cong \overline{M}_{0,3} \times V \times \overline{M}_{g,S}(V,\beta),$$

and so $\overline{M}(V,\tau)$ can be canonically identified with $\overline{M}_{g,S}(V,\beta)$, as expected.

Maps (4.13) and (4.14) generalize, respectively, π and f from (4.1). There are also section maps

(4.16) $$x_i(V,\tau): \overline{M}(V,\tau) \to C(V,\tau),\ i \in S_\tau,$$

generalizing (4.2), and the evaluation maps

(4.17) $$\mathrm{ev}_i(V,\tau): \overline{M}(V,\tau) \to V,\ i \in S_\tau.$$

Finally, for any τ, the groupoids $\overline{M}(V,\tau)$ are targets of the generalized boundary morphisms which can be constructed from the boundary morphisms of the factors. In addition, there are morphisms induced by stably forgetting any subset of points labeled by tails, and finally, one can compose the generalized boundary morphisms with stably forgetful morphisms. The resulting category of the stacks $\overline{M}(V,\tau)$ is fibered over a category of marked modular graphs which is a fairly complex combinatorial object. We refer the reader to [BehM] for more details.

§5. Stacks

In the proof of Theorem 4.5, we have already used the fact that families of stable maps over a base T can be dealt with by localizing with respect to the étale coverings of T. We will now describe the respective formalism in the abstract setting.

5.1. Definition. *A stack (of groupoids) is a quadruple*

$$(b: \mathcal{F} \to \mathcal{S},\ \text{Grothendieck topology}\ \mathcal{T}\ \text{on}\ \mathcal{S})$$

satisfying the following conditions.

a) $b: \mathcal{F} \to \mathcal{S}$ *is a groupoid. Each contravariant representable functor* $\mathcal{S}^{op} \to$ *Sets is a sheaf on* $\mathcal{T}$.

b) *Any isomorphism between families over a given base is uniquely defined by its restrictions to the elements of any covering of the base. Such local data can be glued iff they are compatible on pairwise intersections. Formally speaking, given* X_1, X_2 *over* T, *the functor* $T' \mapsto Iso_{T'}(X_1 \to X_2)$ *is a sheaf.*

c) *Any family over a given base is uniquely defined by its local restrictions. Such local data can be glued iff they satisfy the cocycle compatibility condition.*

1-morphism of stacks (over the same base category) is a 1-morphism of the relevant groupoids. The same applies to 2-morphisms.

Here is a brief reminder and explication. We describe the topology $\mathcal{T}$ by the set of its *coverings* $\text{Cov}\,\mathcal{T}$. One covering (of an object T in $\mathcal{S}$) is a family of maps $\{T_i \to T\}$. This set must be stable with respect to the arbitrary base changes $T' \to T$, and contain all families consisting of one identity map. Moreover, any composition of a covering $\{T_i \to T\}$ and a family of coverings of all T_i must be again a covering of T. The change of base axiom presupposes that fibered products exist in $\mathcal{S}$.

A *presheaf* on $\mathcal{T}$ with values in a category $\mathcal{C}$ with products is a contravariant functor $G : \mathcal{S}^{op} \to \mathcal{C}$. It is *a sheaf* if for any covering $\{T_i \to T\}$ the following diagram is exact:

$$G(T) \to \prod_i G(T_i) \rightrightarrows \prod_{i,j} G(T_i \times_T T_j).$$

Condition b) above rephrases this definition for the presheaf of isomorphisms.

Condition c) refers to a typical descent situation. Namely, suppose that a family F over a base T is given, and consider a covering $\{\phi_i : T_i \to T\}$. Applying to F the base change functors ϕ_i^* (cf. 3.2, Version 1) we get the localized families F_i over T_i, and similarly F_{ij} over $T_{ij} := T_i \times_T T_j$, F_{ijk} over T_{ijk}, etc. They come together with the descent data, that is, isomorphisms

$$f_{ij} : \text{pr}_{ji,i}^* F_i \widetilde{\to} \text{pr}_{ji,j}^* F_j$$

which in turn satisfy the cocycle condition

$$f_{ki} = f_{kj} \circ f_{ji} \text{ on } T_{kji}.$$

Conversely, suppose that F_i over T_i and f_{ij} are given satisfying the cocycle condition above. Then the statement c) of Definition 5.1 requires the existence of an object F in $\mathcal{F}$ and isomorphisms $f_i : \phi_i^*(F) \widetilde{\to} F_i$ such that

$$f_j = f_{ji} \circ f_i \text{ on } T_{ji}.$$

Their uniqueness up to unique isomorphism (effectivity of the descent) follows from b).

The second part of condition a) in Definition 5.1 ensures that any object of $\mathcal{S}$ considered as a groupoid (i.e. the representable functor, cf. 3.2.4) is in fact a stack.

5.2. Proposition. *Endow the category of schemes* $\mathcal{S}$ *by étale or fppf topology. Then the groupoids* $\overline{M}_{g,S}(V,\beta), \overline{C}_{g,S}(V,\beta), \mathcal{M}_{g,S}(V,\beta), \mathcal{M}_{g.S}(V,\beta),$ *and* V *become stacks.*

This follows from the Grothendieck descent theory.

§5. STACKS

5.3. Definition. *Let $f : \mathcal{F} \to \mathcal{G}$ be a morphism of groupoids over a category S. It is called representable, if for any morphism of groupoids $h : Y \to \mathcal{G}$ where $Y \in \operatorname{Ob} S$, the fibered product $Z := Y \times_{\mathcal{F}} \mathcal{G}$ is (equivalent to) an object of S.*

The explicit construction of the fibered product of groupoids in 4.3.1 shows that it preserves the sheaf properties characterizing stacks. So we can speak of representable morphisms of stacks.

To see the meaning of the definition, notice that to give a morphism $h : Y \to \mathcal{G}$ is the same as to choose a family in $\mathcal{G}$ with the base Y, namely the h–image of id_Y. Similarly, the morphism $Z \to \mathcal{F}$ comes from a family in $\mathcal{F}$ over Z. Thus the representability essentially says that after the base change $Z \to Y$, the natural construction f establishes a bijection between the respective families.

5.3.1. Example. Identical morphism of groupoids is representable.

5.3.2. Example. The structure morphism $b : \mathcal{F} \to S$ considered as the morphism of groupoids is representable iff $\mathcal{F}$ is (equivalent to) an object of the base category.

5.3.3. Example. The morphism π in (4.1) is representable. In fact, if $Y \to \overline{M}_{g,S}(V,\beta)$ corresponds to the stable map $(p : C \to Y, (x_i), f : C \to V)$, the morphism
$$Y \times_{\overline{M}_{g,S}(V,\beta)} \overline{C}_{g,S}(V,\beta) \to \overline{C}_{g,S}(V,\beta)$$
corresponds to the stable map over the base C obtained by the base change $p : C \to Y$, endowed with an extra section, the relative diagonal (cf. 3.2.3),
$$(\operatorname{pr}_1 : C \times_Y C \to C; \ (x_{i,C}), \ \Delta, \ f \circ \operatorname{pr}_2).$$

5.3.4. Example. Let $\mathcal{G}$ be a groupoid over S. Assume that S has finite fibered products. The following two properties of $\mathcal{G}$ are equivalent:

(i) Any morphism $X \to \mathcal{G}$, $X \in \operatorname{Ob} S$, is representable.

(ii) The diagonal morphism $\Delta : \mathcal{G} \to \mathcal{G} \times \mathcal{G}$ is representable.

In fact, assume (i). We want to prove that for any two morphisms $f, g : X \to \mathcal{G}$, X in S, the upper left corner in the Cartesian square

$$\begin{array}{ccc} Z & \longrightarrow & X \\ \downarrow & & \downarrow {\scriptstyle (f,g)} \\ \mathcal{G} & \xrightarrow{\Delta} & \mathcal{G} \times \mathcal{G} \end{array}$$

is in S. But this diagram is the exterior square in the join of two Cartesian squares

$$\begin{array}{ccc} Z & \longrightarrow & X \\ \downarrow & & \downarrow {\scriptstyle \Delta_X} \\ X \times_{\mathcal{G}} X & \longrightarrow & X \times X \\ \downarrow & & \downarrow {\scriptstyle f \times g} \\ \mathcal{G} & \xrightarrow{\Delta} & \mathcal{G} \times \mathcal{G} \end{array}$$

in which $X \times_{\mathcal{G}} X$ is in $\mathcal{S}$ because of (i), and so Z is in $\mathcal{S}$ because of representability of fibered products in $\mathcal{S}$.

Now assume (ii). We have to check that for any two morphisms $f: X \to \mathcal{G}$, $g: Y \to \mathcal{G}$, the product $X \times_{\mathcal{G}} Y$ is in $\mathcal{S}$ whenever X, Y are. But because of (ii) we know that Z in the Cartesian square

$$\begin{array}{ccc} Z & \longrightarrow & X \times Y \\ \downarrow & & \downarrow f \times g \\ \mathcal{G} & \stackrel{\Delta}{\longrightarrow} & \mathcal{G} \times \mathcal{G} \end{array}$$

belongs to $\mathcal{S}$. We omit the check that Z is $X \times_{\mathcal{G}} Y$.

5.4. Uses of representability. When the base category $\mathcal{S}$ is the category of schemes (or algebraic spaces), it is important to distinguish and establish various properties P of morphisms: flatness, smoothness, properness, etc. Considering stacks over schemes as generalized schemes, we need machinery for transporting these properties to the morphisms of stacks. Representable morphisms are well suited for this.

More precisely, assume that P is stable under base change and local with respect to the topology we use to define stacks. Then we say that a *representable* morphism of groupoids $f: \mathcal{F} \to \mathcal{G}$ satisfies P, if for any morphism $Y \to \mathcal{G}$, Y a scheme, the morphism $Y \times_{\mathcal{G}} \mathcal{F} \to Y$ satisfies P.

In some cases, one can define a further extension of P to non-necessarily representable morphisms, but this is done in a more *ad hoc* way. This applies to proper morphisms.

5.5. Definition. *An algebraic stack $\mathcal{F}$ is a stack over the category of schemes over a ground field $\mathcal{S}$, with étale or fppf topology, satisfying the following conditions:*

(i) The diagonal morphism $\Delta: \mathcal{F} \to \mathcal{F} \times \mathcal{F}$ is representable, quasicompact, and separated.

(ii) There is a scheme U and a surjective morphism $U \to \mathcal{F}$ (atlas) which is either étale (then $\mathcal{F}$ is called a Deligne–Mumford or DM–stack), or more generally smooth (then $\mathcal{F}$ is called an Artin stack).

5.5.1. Comments. In view of 5.3.4, the representability of the diagonal implies the representability of the atlas, hence it is legitimate to impose on it the conditions of étaleness or smoothness.

The coverings in the fppf topology are finite families of flat finitely presented morphisms whose images are set theoretic coverings.

Quasicompactness is a finiteness condition. A scheme is quasicompact if any open covering of it contains a finite subcovering. A morphism of schemes $X \to T$ is quasicompact if for any open quasicompact subscheme $Y \to T$, $X \times_T Y$ is quasicompact.

Finally, a morphism of schemes $X \to T$ is separated, if the relative diagonal $X \to X \times_T X$ is a closed immersion.

One can prove that the diagonal of any DM–stack is unramified, that is, its relative diagonal is an open immersion (cf. [Vis], p. 665, (7.15)).

5.5.2. Atlases and equivalence groupoids. An intuitively and technically useful picture of an algebraic stack is that of a "quotient object of a scheme with respect to an equivalence groupoid". In this subsection we explain the basics of this approach.

First, let U be a set. Classically, an equivalence relation $\sim$ on U is given by its graph $R \subset U \times U$, $R := \{(a,b) \mid a \sim b\}$ which satisfies the three conditions:

Reflexivity:
$$a \sim a \iff \Delta_U \subset R;$$

Symmetry:
$$(a \sim b \Leftrightarrow b \sim a) \iff s_{12}(R) = R;$$

Transitivity:
$$((a \sim b) \& (b \sim c) \Rightarrow a \sim c) \iff \operatorname{pr}_{13}[(R \times U) \cap (U \times R)] \subset R.$$

All of this can be rephrased as follows: there exists a category with the set of objects U and set of morphisms R, such that $R \to U \times U$ is the map $f \mapsto$ (source of f, target of f), and in addition, every morphism is an isomorphism, and all automorphism groups are trivial.

Consider now a diagram $R \to U \times U$ satisfying this description with the last condition deleted so that the automorphism groups can now be arbitrary. We will call such a diagram *an equivalence groupoid* (on the set U) in order to distinguish it from the notion of groupoid given in 3.2. Of course, an equivalence groupoid $R \to U \times U$ comes together with the identity map $U \to R : a \mapsto \operatorname{id}_a$ and the associative multiplication map $R \times_U R \to R$ satisfying the usual categorical axioms which reduce to reflexivity, symmetry, and transitivity for the usual equivalence relations. Notice that the image of R is in fact an equivalence relation, and the respective quotient is the set of isomorphism classes of objects.

Thus the basic difference between equivalence groupoids and equivalence relations on sets can be demonstrated on one–point sets $U = \{*\}$: in this case R is simply a group. In the framework of homotopy theory, the respective quotient object $\{*\}/R$ is represented by the classifying space BR. Stacks provide a categorical context for constructing such quotients (cf. 3.2.5).

In fact, the notion of equivalence groupoid was formulated in such a way that it readily generalizes to the case when $R \to U \times U$ is a diagram in an arbitrary category with products, e.g. schemes.

If $\mathcal{F} \to \mathcal{S}$ is an algebraic stack and $U \to \mathcal{F}$ a surjective morphism of a scheme to $\mathcal{F}$, we get an equivalence groupoid on U by putting $R := U \times_{\mathcal{F}} U$ which "is" a scheme because the diagonal of $\mathcal{F}$ is representable. The stack $\mathcal{F}$ itself should be considered as the quotient object of U with respect to this equivalence groupoid. Such a morphism is sometimes called *a presentation* of $\mathcal{F}$.

In particular, an étale atlas (for DM–stacks) or smooth atlas (for Artin stacks) is a presentation.

If $\mathcal{F}$ is a DM–stack admitting an étale atlas whose equivalence groupoid is in fact an equivalence relation (that is, $R \to U \times U$ is an imbedding), then $\mathcal{F}$ is called an *algebraic space*. Algebraic spaces are studied in [Knut].

5.5.3. Algebraic spaces and Artin's version of representability. In [LaM–B] the base category $\mathcal{S}$ is the category of *affine* schemes over a fixed ground scheme G.

The category of algebraic spaces is very close to that of schemes, and in particular most properties of morphisms of schemes can be more or less directly extended to algebraic spaces as is shown in [Knut]. Using this, [LaM–B] introduce a weaker notion of representable morphisms than that of Definition 5.3: in our notation, they require $Z := Y \times_{\mathcal{F}} \mathcal{G}$ to be only an algebraic space. This still allows them to use the prescription of 5.4.

In the following we adopt Definition 5.5 and even mostly its DM–version, because the intersection theory in [Vis] is treated for algebraic stacks in this sense, and because our stable moduli stacks are DM–stacks of this type. However, for some crucial constructions involving the unstable moduli spaces the framework of [Vis] turns out to be insufficient. In the recent preprint [Kr] many basic techniques of the intersection theory are extended to Artin stacks.

5.6. (Quasi)coherent sheaves on algebraic stacks. Let $\mathcal{F}$ be an algebraic stack. A quasicoherent sheaf $\mathcal{E}$ on $\mathcal{F}$ is by definition a quasicoherent sheaf on the étale (for the DM–case) or smooth (for the Artin case) topology of $\mathcal{F}$ (cf. [LaM–B], sec. 6). In more down to earth terms, $\mathcal{E}$ is given by a family of quasicoherent sheaves $\mathcal{E}_U$ on each atlas $U \to \mathcal{F}$ connected by a family of isomorphisms $\alpha_\phi : \mathcal{E}_U \to \phi^* \mathcal{E}_V$ defined for every morphism of atlases $\phi : U \to V$ over $\mathcal{F}$ and satisfying the standard cocycle condition $\alpha_{\psi \circ \phi} = \alpha_\phi \circ \phi^* \alpha_\psi$.

A quasicoherent sheaf $\mathcal{E}$ is called coherent, resp. locally free, resp. invertible, iff all $\mathcal{E}_U$ have these properties. A morphism of sheaves is given by a family of morphisms on atlases compatible with the sheaf data.

As 1–morphisms of stacks, (quasi)coherent sheaves and their morphisms are often given by a natural construction.

5.7. Theorem. *Let τ be a stable $B(V)$–marked modular graph. Then the groupoid $\overline{M}(V, \tau)$ (cf. 4.7 above) is a stack. If the ground field is of characteristic zero, then $\overline{M}(V, \tau)$ is an algebraic DM–stack.*

The same is true in the characteristic $p > 0$ case, if there exists a very ample sheaf L on V such that $(\beta_v, L) < p$ for all $v \in V_\tau$.

Sketch of proof. Since products and fibered products of stacks are stacks, and $\overline{M}_{g,S}(V, \beta)$ are stacks, $\overline{M}(V, \tau)$ are stacks as well.

The case $V =$ a point, $g \geq 2$, $n = 0$ was treated in [DeM]. The general case $V =$ a point, (g, n) stable, was treated in [Kn1]. Taking these results for granted, we will deduce the general statement following [BehM], pp. 28–31.

We must check that the diagonal is representable, finite, and unramified. We may and will assume that all maps to V which we classify are separable (this follows from $(\beta_v, L) < p$).

Unraveling the definitions, one can convince oneself that it suffices to check the following statement.

Let $(C/T, (x_i), f)$ and $(D/T, (y_i), h)$ be two T–families of S–labeled stable maps. Then every point of T has an étale neighborhood T' such that the induced families $(C_{T'}, (x_{i,T'}))$, resp. $(D_{T'}, (y_{i,T'}))$, admit additional sections (x'_j), resp. (y'_j), labeled by a common set, with the following properties. The curves $(C_{T'},(x_{i,T'},x'_j))$, $(D_{T'},(y_{i,T'},y'_j))$ are stable. There is a closed immersion of sheaves on T'–schemes

(5.1) $\quad Iso\,((C,(x_i),f),(D,(y_i),h))_{T'} \to Iso\,((C_{T'},(x_{i,T'},x'_j)),(D_{T'},(y_{i,T'},y'_j)))$.

In order to prove this, we may and will assume that the initial stable maps are of the same genus g and class β. If V is embedded in $\mathbf{P}^r$, the source of (5.1) constructed for maps to V admits a natural closed embedding into the similar object constructed for maps to $\mathbf{P}^r$. Hence it suffices to consider the case $V = \mathbf{P}^r, d =$ (the degree of) β. Put $N = d(r+1)$. The additional sections of C and D will be indexed by $1,\ldots,N$. To define them, choose linearly independent hyperplanes $H_0,\ldots,H_r$ in $\mathbf{P}^r$ in general position. The generality is defined with respect to a chosen geometric point t of T for which we construct an étale neighborhood with the desired properties: all H_i must miss images of the special points of C_t and D_t and must be transversal to f_t and h_t. Additional multisections of C, D are defined as intersections of C, D with these hyperplanes. Over a local étale covering they become sections with the necessary properties.

5.8. Theorem. *The stack $\overline{M}(V,\tau)$ is proper.*

Sketch of proof. First, we discuss the definition of proper morphisms of algebraic stacks. If a morphism $f : \mathcal{F} \to \mathcal{G}$ is representable, then properness can be defined as in 5.4. More generally, we will call f proper if it is separated of finite type and locally over $\mathcal{G}$ we can find a representable proper morphism $h : \mathcal{H} \to \mathcal{G}$ such that $h = f \circ g$ for an appropriate surjective g.

To check properness, it is convenient to utilize the generalization of Grothendieck's valuative criterion:

(i) Separatedness of f. Let R be a discrete valuation ring, $\operatorname{Spec} R \to \mathcal{G}$ a morphism, and $g_i : \operatorname{Spec} R \to \mathcal{F}$, $i = 1, 2$, two lifts of this morphism to $\mathcal{F}$. Then any isomorphism between the restrictions of g_1 and g_2 to the generic point $\operatorname{Spec} K$ of $\operatorname{Spec} R$ can be extended to an isomorphism between g_1, g_2.

(ii) Properness of f. A morphism of finite type f is proper, if it is separated and if for any commutative diagram (in the same notation as above)

$$\begin{array}{ccc} \operatorname{Spec} K & \xrightarrow{g} & \mathcal{F} \\ \downarrow & & \downarrow f \\ \operatorname{Spec} R & \longrightarrow & \mathcal{G} \end{array}$$

there exists a finite integral extension R' of R (integral closure of R in a finite integral extension K' of K) such that the induced morphism $\operatorname{Spec} R' \to \mathcal{G}$ lifts to $\mathcal{F}$. (By separatedness, this extension is unique.)

Translating this into the context of the moduli stack of stable maps we get the following picture. For brevity, we restrict ourselves to the case $\overline{M}_{g,S}(V,\beta)$. We start with the extension to the closed point. Let $(C_K,(x_i,K),f_K)$ be a stable map over the generic point of $\operatorname{Spec} R$. The curve (C_K) can be extended to a prestable curve (C/R') over a finite extension of R, together with sections (x_i) and map f, by starting with a model over R, using the theorem on the prestable reduction and resolution of singularities, and blowing up the points where the extension of f was not defined. The uniqueness part can then be treated using the stabilization map as in 4.4.1 and 4.5.

5.9. Further properties of moduli stacks. We list here some further properties of moduli stacks of curves and maps and their morphisms established in [Beh].

Let τ be a modular graph. The stack $\mathcal{M}(\tau)$ of prestable curves of the type τ is a smooth Artin stack whose dimension is given by the formula (1.7) ([Beh], Prop. 2).

The tautological map $\overline{M}(\tau) \to \mathcal{M}(\tau)$ is an open imbedding ([Beh], Lemma 1). The stabilization morphism $\mathcal{M}(\tau) \to \overline{M}(\tau^s)$ which can be defined as in 4.4 is flat ([Beh], Prop. 3).

Now let τ be a $B(V)$-marked modular graph and $\mathcal{C}(\tau)$ the universal curve over the stack of prestable curves with forgotten $B(V)$-marking. One can construct an algebraic stack of maps $\mathrm{Mor}\,(\mathcal{C}(\tau), V)$ of the class defined by the marking. Its structure morphism to $\mathcal{M}(\tau)$ is representable. In the conditions of Theorem 5.7, $\overline{M}(V, \tau)$ is an open substack of $\mathrm{Mor}\,(\mathcal{C}(\tau), V)$ ([Beh], Lemma 4).

§6. Homological Chow groups of schemes

6.1. Homological Chow groups of algebraic schemes. We start with a brief report on homological Chow groups for schemes, based upon Fulton's book [Ful], and then review Vistoli's extension to the context of DM–stacks.

We use here Fulton's terminology. A *scheme* means a scheme of finite type over the fixed ground field. A *variety* is a reduced and irreducible scheme, *imbedding* without qualification means closed imbedding, the same for *subschemes* and *subvarieties*.

6.1.1. Cycles. Let X be a scheme. The group of k-dimensional cycles $Z_k(X)$ is the group freely generated by the symbols $[V]$ for all closed subvarieties of dimension k in X.

For any closed subscheme $Y \subset X$ one can define its fundamental cycle $[Y] = \sum m_i [Y_i]$ where Y_i runs over reduced irreducible components of the support of Y, and m_i is the length of the local ring of the generic point of Y_i.

6.1.2. Rationally equivalent to zero k-cycles. This is the subgroup $R_k(X)$ generated by those k-cycles which are divisors of rational functions on $(k+1)$-dimensional subvarieties $W \subset X$. More precisely,

$$K(W) \ni \varphi \mapsto \mathrm{div}\,\varphi = \sum_{V^k \subset W} \mathrm{ord}_V(\varphi)\,[V] \in Z_k(X),$$

where for $\varphi = a/b$, $a, b \in \mathcal{O}_{W,v} := A$, we put

$$\mathrm{ord}_V(\varphi) := l_A(A/aA) - l_A(A/bA),$$

l_A being the length function.

6.1.3. Homological Chow groups. They are defined by

$$A_k(X) = Z_k(X)/R_k(X), \quad A_*(X) = \bigoplus_{k \geq 0} A_k(X).$$

6.1.4. Proper pushforward. Let $f : X \to Y$ be a proper (in particular, separated) morphism. Then for any closed subvariety $V \subset X$, $f(V) := W$ is closed in Y. On the level of cycles, define $f_* : Z_k(X) \to Z_k(Y)$ by

$$f_*([V]) := [K(V) : K(W)][W]$$

interpreting the degree $[K(V) : K(W)]$ as zero, if $\dim W < \dim V$. Using separatedness of f, one can prove that $f_*(R_k(X)) \subset R_k(Y)$. The essential fact is that $f_*(\operatorname{div} \varphi) = \operatorname{div} N(\varphi)$ where N is the norm map.

It follows that we can define the functorial proper pushforward map of the homological Chow groups of grading degree zero

$$f_* : A_*(X) \to A_*(Y), \ (fg)_* = f_* g_*.$$

In particular, since the spectrum of the base field K is a point whose A_*-group is $A_0 = \mathbf{Z}$, for a proper scheme $f : X \to \operatorname{Spec} K$ we have the degree map:

$$\deg := f_* : A_*(X) \to \mathbf{Z}.$$

6.1.5. Flat pullback. Now let $f : X \to Y$ be a flat morphism *of constant relative dimension* n. The composition of the schematic pullback (fiber product) and the fundamental cycle operation (see 6.1.1) produces the flat pullback on the level of cycles

$$f^* : Z_k(Y) \to Z_{k+n}(X): \ f^*[V] := [f^{-1}(V)].$$

It preserves the divisors of functions so that we get the functorial flat pullback morphisms

$$f^* : A_*(Y) \to A_{*+n}(X), \ (fg)^* = g^* f^*.$$

An important special case. If X is a vector bundle over Y, then f^* is an isomorphism (shifting the dimension).

6.2. Gysin maps. Let $i : X \to Y$ be a regular closed imbedding of codimension d. This means that for any $x \in X$, there exists an open neighborhood of x in Y in which the ideal of X is generated by a *regular* sequence of length d.

6.2.1. Theorem. *For any Cartesian square*

(6.1)
$$\begin{array}{ccc} X' & \xrightarrow{i'} & Y' \\ g \downarrow & & \downarrow f \\ X & \xrightarrow{i} & Y \end{array}$$

one can define the Gysin map

$$i^! = i^!_f : A_*(Y') \to A_{*-d}(X')$$

satisfying a series of compatibility and functoriality properties which will be spelled out below.

Sketch of the construction. Assume first that i is the zero section imbedding of X into the vector bundle $p : Y \to X$. As we remarked above, p^* is then an isomorphism, and we define $i^! := (p^*)^{-1}$. Geometrically, every cycle class in Y has a "cylindric representative" of the form $p^{-1}(Z)$, and $i^!$ is the operation of intersecting this representative with the zero section.

We reduce the general case to this one by a partial linearization. More precisely, suppose that we want to define $i^!([V])$ where V is a subvariety of Y'. Build upon (6.1) one more Cartesian square:

(6.2)
$$\begin{array}{ccc} W & \longrightarrow & V \\ {\scriptstyle h}\downarrow & & \downarrow{\scriptstyle k} \\ X' & \xrightarrow{i'} & Y' \\ {\scriptstyle g}\downarrow & & \downarrow{\scriptstyle f} \\ X & \xrightarrow{i} & Y \end{array}$$

It defines the exterior Cartesian square

(6.3)
$$\begin{array}{ccc} W & \longrightarrow & V \\ {\scriptstyle gh}\downarrow & & \downarrow{\scriptstyle fk} \\ X & \xrightarrow{i} & Y \end{array}$$

which we then partly linearize in the following sense.

Replace $i: X \to Y$ by the imbedding of X as the zero section of its *normal bundle*:

$$X \to N_X Y := \operatorname{Spec}_{\mathcal{O}_X} \bigoplus_{n \geq 0} S^n(I_X/I_X^2) = \operatorname{Spec}_{\mathcal{O}_X} \bigoplus_{n \geq 0} I_X^n/I_X^{n+1}.$$

The last expression defines *the normal cone* $C_X Y$, which in the case of regular imbedding coincides with the normal bundle.

Replace $W \to V$ by the normal cone

$$W \to C_W V = \operatorname{Spec}_{\mathcal{O}_W} \bigoplus_{n \geq 0} I_W^n/I_W^{n+1}.$$

We cannot directly replace $fk: V \to Y$ by a morphism $C_W V \to N_X Y$, instead we should first construct $N = (gh)^*(N_X Y)$ and then use the closed imbedding $C_W V \to N$ induced by the obvious map $(fk)^* I_X \to I_W$.

Let $s: W \to N$ be the composite map

$$W \to C_W V \to N$$

coinciding with the zero section map, and $s^!: A_*(N) \to A_*(W)$ be the relevant Gysin morphism. We finally put

(6.4)
$$i^!([V]) := s^!([C_W V]).$$

6.3. Properties of Gysin maps.
Gysin maps fit into a series of commutative diagrams very useful in various calculations.

(i) Compatibility with proper pushforwards and flat pullbacks. Consider the double Cartesian square

(6.5)
$$\begin{array}{ccc} X'' & \xrightarrow{i''} & Y'' \\ q\downarrow & & \downarrow p \\ X' & \xrightarrow{i'} & Y' \\ g\downarrow & & \downarrow f \\ X & \xrightarrow{i} & Y \end{array}$$

in which i is a regular imbedding of codimension d.

If p is proper, $y \in A_*(Y'')$, we have:

(6.6) $$i^! p_*(y) = q_* i^!(y).$$

If p is flat of constant relative dimension, $y \in A_*(Y')$, then

(6.7) $$i^! p^*(y) = q^* i^!(y).$$

For certain compositions of regular imbeddings and flat morphisms one can define pullback which is independent on the choice of the decomposition and functorial. The following two situations are typical. Consider a diagram of two Cartesian squares

$$\begin{array}{ccccc} X' & \xrightarrow{i'} & Y' & \xrightarrow{p'} & Z' \\ \downarrow & & \downarrow & & \downarrow \\ X & \xrightarrow{i} & Y & \xrightarrow{p} & Z \end{array}$$

in which i is regular of codimension d.

Assume first that p is flat of relative dimension n, and pi is flat of relative dimension $n - d$. Then i' is regular of codimension d, p' and $p'i'$ are flat, and

$$(p'i')^*(y) = i'^* p^*(y) = i^! p'^*(y).$$

If p is smooth of relative dimension n and pi is regular of codimension $d - n$, then

$$(pi)^!(y) = i^!(p'^*(y)).$$

(ii) Compatibility between two Gysin morphisms. If i' as well is regular of the same codimension d, $y \in A_*(Y'')$, then

(6.8) $$i^!(y) = i'^!(y) \in A_{*-d}(X'').$$

When i' is regular, but not of the same codimension as i, this simple version is replaced by the more sophisticated *excess intersection formula*; cf. 8.8 below.

(iii) Commutativity. This time we start with the diagram consisting of three Cartesian squares

(6.9)
$$\begin{array}{ccccc} X'' & \longrightarrow & Y'' & \longrightarrow & S \\ \downarrow & & \downarrow & & \downarrow j \\ X' & \longrightarrow & Y' & \xrightarrow{g} & T \\ \downarrow & & \downarrow f & & \\ X & \xrightarrow{i} & Y & & \end{array}$$

in which i and j are regular imbeddings. Our claim is that the following diagram is commutative:

(6.10)
$$\begin{array}{ccc} A_*(X'') & \xleftarrow{i^!} & A_*(Y'') \\ j^! \uparrow & & \uparrow j^! \\ A_*(X') & \xleftarrow{i^!} & A_*(Y') \end{array}$$

(iv) Functoriality. Consider the diagram consisting of two Cartesian squares in which i and j, and hence ji, are regular imbeddings:

(6.11)
$$\begin{array}{ccccc} X' & \longrightarrow & Y' & \longrightarrow & Z' \\ \downarrow & & \downarrow & & \downarrow \\ X & \xrightarrow{i} & Y & \xrightarrow{j} & Z \end{array}$$

Then

(6.12) $$(ji)^! = i^! j^!.$$

(v) Gysin maps for locally complete intersection morphisms. Assume $f : X \to Z$ is a *local complete intersection (l.c.i.)* morphism, i.e. a morphism representable as pi, where $i : X \to Y$ is a closed regular imbedding of constant codimension, and $p : Y \to Z$ is smooth of constant relative dimension. Consider any Cartesian square:

$$\begin{array}{ccc} X' & \xrightarrow{f'} & Y' \\ h' \downarrow & & \downarrow h \\ X & \xrightarrow{f} & Y \end{array}$$

Extend it to the Cartesian diagram

$$\begin{array}{ccccc} X' & \xrightarrow{i'} & Y' & \xrightarrow{p'} & Z' \\ h' \downarrow & & \downarrow g & & \downarrow h \\ X & \xrightarrow{i} & Y & \xrightarrow{p} & Z \end{array}$$

as in (i). For $y \in A_*(Y')$ we can define
$$f^!(y) = i^!(p'^*(y)).$$
This definition is independent on the choice of the decomposition $f = pi$. If f is also flat, then $f^!$ coincides with the flat pullback. Gysin maps for l.c.i. morphisms satisfy the appropriate generalizations of 6.3 (i) (compatibility with proper pushforwards and flat pullbacks) and 6.3 (iii) (commutativity).

6.4. Complements on the homological Chow groups. In this subsection we collect a list of additional properties of Chow groups from Fulton's book [Ful] which in some form extend to the category of DM–stacks.

6.4.1. Exterior product and intersection multiplication for smooth schemes. The map
$$Z_*(X) \otimes Z_*(Y) \to Z_*(X \times Y) : [V] \otimes [W] \to [V \times W]$$
induces the map
$$A_*(X) \otimes A_*(Y) \to A_*(X \times Y)$$
transforming $f_* \times g_*, f^* \times g^*$, and $i^! \times j^!$ into $(f \times g)_*, (f \times g)^*$, and $(i \times j)^!$, respectively (cf. [Ful], p. 111, Ex. 6.5.2).

If X is smooth, $\Delta : X \to X \times X$ is a regular imbedding, and the exterior product followed by $\Delta^!$ defines on $A_*(X)$ the structure of commutative associative algebra with intersection product.

6.4.2. Cartesian squares produce commutative diagrams of correspondences. Given a Cartesian square

$$\begin{array}{ccc} X' & \xrightarrow{g'} & X \\ f' \downarrow & & \downarrow f \\ Y' & \xrightarrow{g} & Y \end{array}$$

in which f is proper and g is flat, hence f' is proper and g' is flat, we have the following identity already on the level of cycles, therefore of Chow groups as well:

(6.13) $$f'_* g'^* = g^* f_* : Z_*(X) \to Z_*(Y').$$

6.4.3. An exact sequence. Given a Cartesian square

$$\begin{array}{ccc} Y' & \xrightarrow{j} & X' \\ q \downarrow & & \downarrow p \\ Y & \xrightarrow{i} & X \end{array}$$

in which i is a closed imbedding and p is a proper morphism inducing an isomorphism
$$X' \setminus Y' \xrightarrow{\sim} X \setminus Y,$$
we have the exact sequence
$$A_*(Y') \to A_*(Y) \oplus A_*(X') \to A_*(X) \to 0$$

where the first two arrows are respectively

$$y \mapsto (q_*(y), -j_*(y)), \ (y,z) \mapsto i_*(y) + p_*(z).$$

§7. Homological Chow groups of DM–stacks

7.1. Homological Chow groups. Following Vistoli ([Vis]), we present here a report on the intersection theory for DM–stacks along the same lines as in Fulton's book. The reader is invited to read this section referring to the previous one for parallels.

7.1.1. Cycles. Let $\mathcal{F}$ be a DM–stack. The group $Z_k(\mathcal{F})$, by definition, is freely generated by the symbols $[\mathcal{G}]$ for all closed irreducible and reduced (integral) k–dimensional substacks $\mathcal{G}$ of $\mathcal{F}$.

We will now explain the terms used in this definition.

An (open, closed, dense, ...) substack of $\mathcal{F}$ is a representable morphism of stacks $\mathcal{G} \to \mathcal{F}$ which is represented by imbeddings of schemes of the respective type. Two such morphisms define the same substack if they are connected by an isomorphism identical on $\mathcal{F}$.

Let $U \to \mathcal{F}$ be an étale atlas of $\mathcal{F}$, and $p_i : R := U \times_{\mathcal{F}} U \to U, i = 1, 2$, its equivalence groupoid (see 5.5.2). Then we have a bijective correspondence

$$\{\text{substacks } \mathcal{F}' \text{ of } \mathcal{F}\} \longleftrightarrow \{\text{subschemes } U' \subset U \text{ such that } p_1^{-1}(U') = p_2^{-1}(U')\}$$

given by

$$\mathcal{F}' \mapsto U' := \mathcal{F}' \times_{\mathcal{F}} U, \ U' \mapsto U'/R',$$

where

$$R' = p_1^{-1}(U') = p_2^{-1}(U').$$

A substack is *irreducible* if it is not a union of two closed proper substacks.

A substack is *reduced* if one (and then every) atlas is reduced.

The *dimension* of a stack is the dimension of its atlas.

As in 6.1.1, we can associate to every closed substack $\mathcal{G}$ of $\mathcal{F}$ a cycle $[\mathcal{G}]$. To do this, consider an atlas $U \to \mathcal{F}$ and its equivalence groupoid $p_i : R \to U$. Since U is a scheme, we can define $[\mathcal{G} \times_{\mathcal{F}} U] \in Z_*(U)$. The two inverse images of this cycle in $Z_*(R)$ coincide with $[\mathcal{G} \times_{\mathcal{F}} R]$. Therefore $[\mathcal{G} \times_{\mathcal{F}} U]$ descends to a cycle $[\mathcal{G}] \in Z_*(\mathcal{F})$.

7.1.2. Rationally equivalent to zero k–cycles. This is the subgroup $R_k(\mathcal{F})$ generated by those k cycles which are divisors of functions on integral $(k + 1)$-substacks of $\mathcal{F}$.

Here *a rational function* on an integral stack $\mathcal{G}$ is an equivalence class of morphisms $\mathcal{G}' \to \mathbf{A}^1$ where $\mathcal{G}'$ is an open dense substack of $\mathcal{G}$, and two morphisms are equivalent if they coincide on a common open dense substack of their domains. Rational functions form the field $K(\mathcal{G})$.

It would be awkward to define the divisor of a rational function directly, as in 6.1.2. Since we have to localize wrt étale topology, at this point we might as well proceed consistently.

The étale topology of $\mathcal{F}$ is given by the following data (cf. 5.1). It is the category whose objects (open sets) are étale morphisms $x : X \to \mathcal{F}$ from schemes to $\mathcal{F}$, and morphisms $(X, x) \to (Y, y)$ are pairs $(f : X \to Y, 2$–(iso)morphism

$\varphi : x \to y \circ f$). A *covering family* is a family of morphisms $f_i : (X_i, x_i) \to (Y, y)$ such that $\coprod_i X_i \to Y$ is surjective.

We now have a morphism of presheaves on the étale topology of $\mathcal{F}$ given on an open set $x : X \to \mathcal{F}$ by

$$\text{div} : W_k(X) = \bigoplus_{\text{integral } Z^{k+1} \subset X} K(Z) \to Z_k(X).$$

The restriction map corresponding to $(f, \varphi) : (X, x) \to (Y, y)$ is given by the obvious lift of functions and cycles.

The descent theory shows that the presheaves W_k and Z_k are actually sheaves. Their groups of global sections are isomorphic to $\bigoplus_{Z^{k+1} \subset \mathcal{F}} K(Z)$ and $Z_k(\mathcal{F})$, respectively. Let $R_k(\mathcal{F})$ be the image of div. Elements of this image are called rationally equivalent to zero.

7.1.3. Homological Chow groups. As in 6.1.3, they are now defined as

$$A_k(\mathcal{F}) = Z_k(\mathcal{F})/R_k(\mathcal{F}), \quad A_*(\mathcal{F}) = \bigoplus_{k \geq 0} A_k(\mathcal{F}).$$

7.1.4. Proper pushforward. Let $f : \mathcal{F} \to \mathcal{G}$ be a proper morphism of stacks. We define the proper pushforward of cycles by the formula similar to that of 6.1.4: for an integral substack $\mathcal{F}'$ of $\mathcal{F}$,

(7.1) $$f_*[\mathcal{F}'] := \deg(\mathcal{F}'/\mathcal{G}')[\mathcal{G}'],$$

where $\mathcal{G}'$ is the image of $\mathcal{F}'$. However, some of the terms in this prescription need redefinition and comments. In particular, the coefficient $\deg(\mathcal{F}'/\mathcal{G}')$ involves orders of the automorphism groups of the generic points of $\mathcal{F}', \mathcal{G}'$ and generally is a rational number so that the proper pushforward is only defined for rational Chow groups $A_{*,\mathbf{Q}}$.

Image of a closed substack. Let $V \to \mathcal{G}$ be an atlas of $\mathcal{G}$. If f is representable, then $V' := \mathcal{F}' \times_\mathcal{G} V$ is a scheme whose image in V has the same lift to $V \times_\mathcal{G} V$ with respect to the two projections. Therefore it defines the closed substack of $\mathcal{G}$ which is called the image of $\mathcal{F}'$. If f is proper but not necessarily representable, there exists a surjective morphism $X \to \mathcal{F}$ such that $X \to \mathcal{G}$ is proper, and we argue in the same way using X.

The image of an integral substack is integral.

Degree. Let $\mathcal{F} \to \mathcal{G}$ be a separated dominant morphism so that we have an imbedding $f^* : K(\mathcal{G}) \to K(\mathcal{F})$. If $\mathcal{F}$ is integral, put

(7.2) $\delta(\mathcal{F}) :=$ the degree of the automorphism group of the generic point of $\mathcal{F}$

(a point here is a geometric point). Finally, define

(7.3) $$\deg(\mathcal{F}/\mathcal{G}) := \frac{\delta(\mathcal{G})}{\delta(\mathcal{F})}[K(\mathcal{F}) : K(\mathcal{G})],$$

interpreting the last factor as zero if it is not finite.

Formula (7.1) for the proper pushforward of the fundamental cycle of an integral substack can be written as

$$f_*(\delta(\mathcal{F}')[\mathcal{F}']) = [K(\mathcal{F}') : K(\mathcal{G}')]\delta(\mathcal{G}')[\mathcal{G}'].$$

It takes the same form as for the schemes in 6.1.4, if one replaces the fundamental cycles $[\mathcal{F}]$ by the *normalized fundamental cycles* $\delta(\mathcal{F})[\mathcal{F}]$.

The proof of the functoriality of this definition is based upon the multiplicative property
$$\deg(\mathcal{F}/\mathcal{H}) = \deg(\mathcal{F}/\mathcal{G})\deg(\mathcal{G}/\mathcal{H}).$$

For general Artin stacks, automorphism groups of generic points can be infinite so that the adopted definition of the pushforward must be modified. If the automorphism groups are algebraic, a natural idea is to replace $\delta(\mathcal{F})$ by $\langle G \rangle$ where G is the respective automorphism group and $\langle G \rangle$ its class in the K_0-group of Grothendieck's motives.

Passage to $A_{*,\mathbf{Q}}$. If f is proper, f_* commutes with div and therefore descends to $A_{*,\mathbf{Q}}$.

Example. Let τ be an S–marked stable modular graph. Recall that its automorphism group $\operatorname{Aut}\tau$ consists of permutations of flags preserving the S–labeling, the boundary map, edges, and labeling $v \mapsto g_v$. Consider the boundary morphism (4.15) of the moduli spaces of stable curves: $b(\tau) : \overline{M}(\tau) \to \overline{M}_{g,S}$. Denote by $\widetilde{M}(\tau)$ the closed substack of $\overline{M}_{g,S}$ which is the image of $\overline{M}(\tau)$. Then we have

(7.4) $$b(\tau)_*[\overline{M}(\tau)] = |\operatorname{Aut}\tau|\,[\widetilde{M}(\tau)].$$

To prove this, we must check that
$$\deg \overline{M}(\tau)/\widetilde{M}(\tau) = |\operatorname{Aut}\tau|.$$

For example, if $g = 2, S = \emptyset$, and τ consists of two vertices of genus zero connected by three edges, we get the morphism mapping one point to one point, which stack–theoretically has degree 12.

Generally, this is how it happens. First of all, $\delta(\overline{M}_{g,S})$ is 1 unless $g = 1, |S| = 1$ or $g = 0, S = \emptyset$ in which case $\delta(\overline{M}_{g,S}) = 2$ because the generic curve of one of these types has the automorphism group $\mathbf{Z}_2$. We will assume that τ is not a one–vertex graph of genus two without tails, because then the statement is evident. Denote by m the number of vertices of τ of genus one with precisely one flag. Then
$$\delta(\overline{M}(\tau)) = 2^m$$
because $\overline{M}(\tau) = \prod_{v \in V_\tau} \overline{M}_{g_v,n_v}$. On the other hand, the natural extension of $\operatorname{Aut}\tau$ by $\mathbf{Z}_2^m$ acts upon the universal curve $C(\tau)$ compatibly with its projection to the moduli space, and the resulting quotient stack is the universal curve over $\widetilde{M}(\tau)$. It follows that
$$\delta(\widetilde{M}(\tau))\,[K(\overline{M}(\tau)):K(\widetilde{M}(\tau))] = 2^m|\operatorname{Aut}(\tau)|.$$
Combined with (7.3), this proves (7.4).

7.1.5. Flat pullback.
Flat pullback can be defined on the level of integral cycles and Chow groups in the same way as above, but simpler.

Namely, for a flat morphism of constant fiber dimension $f : \mathcal{F} \to \mathcal{G}$ and a closed integral substack $\mathcal{G}'$ of $\mathcal{G}$ we put $f^*[\mathcal{G}'] = [\mathcal{G}' \times_\mathcal{G} \mathcal{F}]$. In the right hand side one should use the general definition of the cycle associated to a closed substack from 7.1.1, because $\mathcal{G}' \times_\mathcal{G} \mathcal{F}$ is not necessarily integral.

One checks then that f^* is compatible with rational equivalence and induces the functorial pullback homomorphisms $f^* : A_*(\mathcal{F}) \to A_*(\mathcal{G})$.

Moreover, the commutativity (6.13) extends to the (rational) Chow groups of DM–stacks.

7.2. Gysin maps. The initial setting for the construction of Gysin maps for DM–stacks differs from that of 6.2 in the following respect: the lower line $i: \mathcal{F} \to \mathcal{G}$ of the Cartesian square is assumed to be only *a local regular imbedding (l.r.i.)*. This means that i is representable of finite type such that there exist atlases $U \to \mathcal{F}$, $V \to \mathcal{G}$, and a regular imbedding $U \to V$ inducing i. An important example of l.r.i. is the diagonal of a smooth stack.

7.2.1. Theorem. *If i is an l.r.i. of codimension d, then for any Cartesian square of DM-stacks*

(7.5)
$$\begin{array}{ccc} \mathcal{F}' & \xrightarrow{i'} & \mathcal{G}' \\ g \downarrow & & \downarrow f \\ \mathcal{F} & \xrightarrow{i} & \mathcal{G} \end{array}$$

one can define the Gysin map

$$i^! = i^!_f : A_*(\mathcal{G}') \to A_{*-d}(\mathcal{F}')$$

satisfying the compatibility and functoriality properties spelled out for schemes in 6.3.

For proofs, cf. [Vis].

7.3. The case of Artin stacks. In a recent preprint [Kr] A. Kresch developed intersection theory for Artin stacks. We will briefly describe here his definition of $A_*(\mathcal{F})$. It consists of several steps.

STEP 1. In the construction of 7.1.2 one should replace étale topology by smooth topology. The divisor map div locally in the smooth topology sends a rational function to its Weil divisor. The group of algebraic cycles modulo rational equivalence can be defined as in 7.1.3. It is now called *the naive Chow group* and denoted $A_*^\circ(\mathcal{F})$.

STEP 2. Calculating the naive Chow group for the classifying stack BG where G is a linear algebraic group, we get the rather disappointing answer. Imitating the topological description of BG as the quotient of a contractible space by the free action of G, D. Edidin and W. Graham, following the idea of B. Totaro, consider a linear representation of G containing a G–invariant Zariski open subset U of big codimension N such that G acts on U freely. Then $A_i(U/G)$ for $i < N$ does not depend on U and is the natural candidate for $A_i(BG)$ (see [EdG2]).

Generalizing this to arbitrary Artin stacks $\mathcal{F}$, A. Kresch considers the set $\mathcal{B} = \mathcal{B}_\mathcal{F}$ of isomorphism classes of vector bundles E over $\mathcal{F}$ partially ordered by the relation $E \leq F$ iff there is a surjection of vector bundles $F \to E$. Flat pullback of naive Chow groups produces the direct system of the naive Chow groups, and Kresch defines *the Edidin-Graham-Totaro group of $\mathcal{F}$* as

(7.6)
$$\widehat{A}_*(\mathcal{F}) = \varinjlim_\mathcal{B} A^\circ_{*+\mathrm{rk}\,E}(E)$$

for connected $\mathcal{F}$, and extends this by additivity to general $\mathcal{F}$.

STEP 3. $\widehat{A}_*(\mathcal{F})$ turns out to be the correct Chow group not only for classifying spaces, but more generally for the stacks of the type X/G where X is a scheme or an algebraic space with the action of a linear algebraic group G (global quotients). However, $\widehat{A}_*(\mathcal{F})$ does not behave well with respect to pushforwards. For example, it may happen that a stack $\mathcal{F}$ does not admit non–trivial vector bundles, but contains a closed substack $i : \mathcal{G} \to \mathcal{F}$ which is a global quotient, and therefore has a large Chow group. Since the classes $\alpha \in \widehat{A}_*(\mathcal{G})$ are represented by cycles in the vector bundles over $\mathcal{G}$ which do not extend to $\mathcal{F}$, there seems to be no way to define $i_*(\alpha)$ in $\widehat{A}_*(\mathcal{F})$. As a remedy, Kresch suggests to take one more limit, this time over the following directed set $\mathcal{A} = \mathcal{A}_\mathcal{F}$.

Elements of $\mathcal{A}$ are 2-isomorphism classes of projective morphisms of stacks $f : \mathcal{G} \to \mathcal{F}$. We put $(\mathcal{G}, f) \le (\mathcal{G}', f')$ iff there is an $\mathcal{F}$-morphism $(\mathcal{G}, f) \to (\mathcal{G}', f')$ which is an isomorphism of $\mathcal{G}$ onto a union of connected components of $\mathcal{G}'$.

For any morphism $f : \mathcal{G} \to \mathcal{F}$ with connected $\mathcal{F}$ put as in (7.6)

$$(7.7) \qquad \widehat{A}^f_*(\mathcal{F}) = \varinjlim{}_{\mathcal{B}_\mathcal{F}} A^\circ_{*+\mathrm{rk}\,E}(f^*E).$$

Extend this by additivity to not necessarily connected $\mathcal{F}$. Denote by $i_f : \widehat{A}^f_*(\mathcal{F}) \to \widehat{A}_*(\mathcal{F})$ the natural map. For a projective morphism $f : \mathcal{G} \to \mathcal{F}$ one can define the natural pushforward $f_* : \widehat{A}^f_*(\mathcal{G}) \to \widehat{A}_*(\mathcal{F})$.

For a relative stack $f : \mathcal{G} \to \mathcal{F}$, consider pairs of projective morphisms $p_1, p_2 : \mathcal{H} \to \mathcal{G}$ such that $f \circ p_1$ and $f \circ p_2$ are 2–isomorphic. Denote by $\widehat{B}_*(\mathcal{G})$ the subgroup of $\widehat{A}_*(\mathcal{G})$ generated by all elements of the type

$$\{p_{2*}\beta_2 - p_{1*}\beta_1 \mid \beta_i \in \widehat{A}^{p_i}_*, i_{p_1}(\beta_1) = i_{p_2}(\beta_2)\}.$$

Finally, put

$$(7.8) \qquad A_*(\mathcal{F}) = \varinjlim{}_{\mathcal{A}_\mathcal{F}}(\widehat{A}_*(\mathcal{G})/\widehat{B}_*(\mathcal{G})).$$

As is shown in [Kr], with this definition many results of [Vis] extend from DM-stacks to Artin stacks, sometimes satisfying an additional restriction, like admitting stratification by locally closed substacks which are global quotients.

§8. Operational Chow groups of schemes and DM–stacks

In this section letters $X, Y, \mathcal{F}, \mathcal{G}$, etc., generally denote DM–stacks which may or may not be schemes, and A_* denotes the homological Chow theory with rational coefficients.

8.1. Definition. *Let $f : \mathcal{F} \to \mathcal{G}$ be a representable morphism of DM-stacks, $p \in \mathbf{Z}$. Then an element α of the operational Chow group $\bar{A}^p(\mathcal{F} \to \mathcal{G})$ is a family of maps*

$$\alpha_g \cap : A_*(Y) \to A_{*-p}(Y \times_\mathcal{G} \mathcal{F})$$

indexed by arbitrary morphisms $g : Y \to \mathcal{G}$ and satisfying the following compatibilities (cf. 6.2):

§8. OPERATIONAL CHOW GROUPS OF SCHEMES AND DM-STACKS

(i) *Compatibility with proper pushforwards and flat pullbacks.* Consider the double Cartesian square:

$$\begin{array}{ccc} X' & \longrightarrow & Y' \\ p\downarrow & & \downarrow q \\ X & \longrightarrow & Y \\ \downarrow & & \downarrow g \\ \mathcal{F} & \xrightarrow{f} & \mathcal{G} \end{array}$$

If q is proper, $y' \in A_*(Y')$, we have:

(8.1) $$\alpha_g \cap q_*(y') = p_*(\alpha_{gq} \cap y').$$

If q is flat of constant relative dimension, $y \in A_*(Y)$, then

(8.2) $$\alpha_{gq} \cap q^*(y) = p^*(\alpha_g \cap y).$$

(ii) *Compatibility with Gysin maps.* Start with the diagram consisting of three Cartesian squares

(8.3) $$\begin{array}{ccccc} X'' & \longrightarrow & Y'' & \longrightarrow & S \\ \downarrow & & \downarrow q & & \downarrow j \\ X' & \longrightarrow & Y' & \longrightarrow & T \\ \downarrow & & \downarrow g & & \\ \mathcal{F} & \xrightarrow{f} & \mathcal{G} & & \end{array}$$

in which j is a regular local imbedding of stacks. Then the following diagram must be commutative:

(8.4) $$\begin{array}{ccc} A_*(X'') & \xleftarrow{\alpha_{gq}\cap} & A_*(Y'') \\ j^!\uparrow & & \uparrow j^! \\ A_*(X') & \xleftarrow{\alpha_g\cap} & A_*(Y') \end{array}$$

Clearly, $\bar{A}^p(\mathcal{F} \to \mathcal{G})$ is a **Q**–linear space. Put $\bar{A}^*(\mathcal{F} \to \mathcal{G}) = \bigoplus_p \bar{A}^p(\mathcal{F} \to \mathcal{G})$.

8.1.1. Example: Orientation class of the local regular imbedding. Let $i : \mathcal{F} \to \mathcal{G}$ be a local regular imbedding. For any $g : Y \to \mathcal{G}$, define $\alpha_g \cap (\dots) := i_g^!(\dots)$. Comparing 8.1 with the properties of the Gysin maps from 6.3, one sees that in this way we defined an element $[i] \in \bar{A}^*(\mathcal{F} \to \mathcal{G})$ which is called *the orientation class of this morphism.*

8.1.2. Example: Orientation class of a flat morphism. Let $f : \mathcal{F} \to \mathcal{G}$ be flat with fibers of constant dimension. For any $g : Y \to \mathcal{G}$, define $f_g : \mathcal{F} \times_{\mathcal{G}} Y \to Y$ and then $\alpha_g \cap (\dots) := f_g^*(\dots)$. Again one sees that in this way we defined an element $[f] \in \bar{A}^*(\mathcal{F} \to \mathcal{G})$ which is called *the orientation class* of this f.

8.2. Product. For two representable morphisms $f : \mathcal{F} \to \mathcal{G}$ and $g : \mathcal{G} \to \mathcal{H}$, we can define the bilinear product

$$\bar{A}^*(g) \otimes \bar{A}^*(f) \to \bar{A}^*(gf) : \quad \alpha \otimes \beta \mapsto \alpha.\beta,$$

where

(8.5) $$(\alpha.\beta) \cap z := \alpha \cap (\beta \cap z).$$

It is compatible with grading and associative. In particular, $A^*(\mathcal{F}) := \bar{A}^*(\mathrm{id}_\mathcal{F})$ is an associative algebra with identity.

8.3. Proper pushforward. With the same notation, assume that f is proper and define
$$f_* : \bar{A}^*(gf) \to \bar{A}^*(g)$$
by

(8.6) $$f_*(\alpha) \cap z := f_*(\alpha \cap z).$$

This is a linear map of degree zero. If f, k are proper and composable, then $(kf)_* = k_* f_*$.

Moreover, let $h : \mathcal{H} \to \mathcal{K}$ be one more representable morphism, $\alpha \in \bar{A}^*(gf)$, and $\beta \in \bar{A}^*(h)$. Then

(8.7) $$f_*(\alpha).\beta = f_*(\alpha.\beta) \in \bar{A}^*(hg)$$

(compatibility of pushforward and product).

8.4. Pullback. Let $f : \mathcal{F} \to \mathcal{G}$ be a representable morphism and $g : \mathcal{G}' \to \mathcal{G}$ a morphism. Consider the fiber square:

$$\begin{array}{ccc} \mathcal{F}' & \xrightarrow{f'} & \mathcal{G}' \\ \downarrow & & \downarrow g \\ \mathcal{F} & \xrightarrow{f} & \mathcal{G} \end{array}$$

Then one can define a linear map of degree zero

(8.8) $$g^* : A^*(f) \to A^*(f') : g^*(\alpha) \cap y := \alpha \cap y.$$

We have the usual functoriality $(fk)^* = k^* f^*$.

Moreover, if we have a double Cartesian square

(8.9) $$\begin{array}{ccccc} \mathcal{F}' & \xrightarrow{f'} & \mathcal{G}' & \xrightarrow{h'} & \mathcal{H}' \\ \downarrow & & \downarrow g & & \downarrow g' \\ \mathcal{F} & \xrightarrow{f} & \mathcal{G} & \xrightarrow{h} & \mathcal{H} \end{array}$$

then for $\alpha \in \bar{A}^*(f)$, $\beta \in \bar{A}^*(h)$

(8.10) $$g^*(\alpha.\beta) = g'^*(\alpha).g^*(\beta)$$

(compatibility of pullback and product).

Finally, with the same notation, if f is proper, pushforward and pullback commute: for $\alpha \in \bar{A}^*(hf)$ we have

(8.11) $$g^* f_*(\alpha) = f'_*(g^*(\alpha)) \in \bar{A}^*(h').$$

8.5. A projection formula.
Consider now the diagram

(8.12)
$$\begin{array}{ccc} \mathcal{F}' & \xrightarrow{f'} & \mathcal{G}' \\ \downarrow{g'} & & \downarrow{g} \\ \mathcal{F} & \xrightarrow{f} & \mathcal{G} & \xrightarrow{h} & \mathcal{H} \end{array}$$

with a proper g and the Cartesian square. Then for $\alpha \in \bar{A}^*(f)$, $\beta \in \bar{A}^*(hg)$ we have

(8.13) $\qquad \alpha . g_*(\beta) = g'_*(g^*(\alpha).\beta) \in \bar{A}^*(hf).$

Notice that since we also have exterior multiplication between $\bar{A}^*$ and A_* and various pullback and pushforward operations on both kinds of groups, we can expect existence of other versions of the projection formula as well; cf. subsection 8.9 below.

8.6. Chern classes.
Chern classes of a vector bundle (locally free coherent sheaf) E on a DM–stack $\mathcal{F}$ furnish important elements $c_i(E) \in \bar{A}^*(\mathcal{F})$ of the operational Chow ring. As in [Ful], they can be constructed in the following way.

(i) Invertible sheaves. Let L be an invertible sheaf on $\mathcal{F}$. We define the expression $c_1(L) \cap [\mathcal{F}] \in A_{n-1}(\mathcal{F})$ as the Chow class of the Cartier divisor on $\mathcal{F}$ defined by L.

More generally, for a closed substack $f: \mathcal{G} \to \mathcal{F}$ we put

(8.14) $\qquad c_1(L) \cap [\mathcal{G}] := f_*(c_1(f^*L) \cap [\mathcal{G}]) \in A_*(\mathcal{F}).$

This defines $c_1(L)$ as an operation of degree -1 on $A_*(\mathcal{F})$.

Finally, for any morphism of stacks $g: \mathcal{G} \to \mathcal{F}$ define the action of $c_1(L)_g$ on $A_*(\mathcal{G})$ by

(8.15) $\qquad c_1(L)_g \cap y := c_1(g^*L) \cap y.$

One checks that it is a well–defined element of $\bar{A}^1(\mathcal{F})$.

(ii) Sheaves of arbitrary rank. Now let E be a vector bundle of rank $e+1$ on $\mathcal{F}$. The projectivization morphism $p: \mathcal{P} := P(E) \to \mathcal{F}$ (where $P(E)$ is the space of lines in E) is flat and proper. Moreover, $\mathcal{P}$ carries the standard invertible sheaf $\mathcal{O}_\mathcal{P}(1)$.

Define the operation $s_i(E) \cap$ first on $A_*(\mathcal{F})$ by

(8.16) $\qquad s_i(E) \cap y := p_*(c_1(\mathcal{O}_\mathcal{P}(1))^{e+i} \cap p^*y)$

and then extend it to arbitrary $A_*(\mathcal{G})$, $g: \mathcal{G} \to \mathcal{F}$ by

(8.17) $\qquad s_i(E)_g \cap y := s_i(g^*E) \cap y.$

In this way we get well–defined elements $s_i(E) \in \bar{A}^i(\mathcal{F})$, $i \geq 0$, called *the Segre classes of E*.

Finally, Chern classes are universal polynomials in $s_i(E)$ (which pairwise commute, cf. below) defined by

(8.18) $\qquad c_t(E) := \sum_i c_i(E) t^i = (s_t(E))^{-1} := (\sum_i s_i(E) t^i)^{-1}.$

One checks that they satisfy the usual formal properties.

8.7. Central operations. In principle, the ring $\bar{A}^*(\mathcal{F})$ may well be non-commutative, and it is important to know when the usual identities of numerical geometry of smooth manifolds hold without paying attention to the order of terms.

More generally, call an element $\alpha \in \bar{A}^*(\mathcal{F} \to \mathcal{G})$ *central*, if in the situation of 8.1(i), for every $\beta \in \bar{A}^*(Y' \to Y)$ and every $y \in A_*(Y)$ we have
$$\alpha \cap (\beta \cap y) = \beta \cap (\alpha \cap y).$$

The central elements form the linear space $C^*(\mathcal{F} \to \mathcal{G})$. They are compatible with products, proper pushforwards, and pullbacks.

Orientation classes of flat and r.l.i. morphisms are central. Segre and Chern classes are all central as well.

In the category of schemes this is clearly explained in [Ful], the end of p. 53, as the consequence of the theory developed in Chs. 2 and 3. For stacks, this is somewhat hidden in [Vis]. Basically, [Vis] remakes Ch. 6 of [Ful], which is a generalization of Chs. 2 and 3.

8.8. Excess intersection formula. Using the appropriate Chern class, one can now generalize formula (6.8).

Consider the Cartesian diagram

$$\begin{array}{ccc} X' & \xrightarrow{i'} & Y' \\ {\scriptstyle g'}\downarrow & & \downarrow{\scriptstyle g} \\ X & \xrightarrow{i} & Y \end{array}$$

in which i, resp. i', is an l.r.i. morphism of codimension d, resp. d'. Then we have

(8.19) $$g^*[i] = c_r(E).[i'] \in \bar{A}^d(i').$$

Here $r = d - d'$ and E is the so-called excess normal bundle. For the case when i is a regular imbedding and N, N' are normal bundles to i, i' we have $E = g'^*N/N'$. We omit the general definition.

8.9. Pushforward of the operational Chow rings and a projection formula. In this subsection, we work with operational Chow groups for identical morphisms $A^*(M) := \bar{A}^*(\mathrm{id}_M)$. For convenience, we reproduce the definition of $A^*(M)$ specializing 8.1.

An element $\alpha \in A^*(M)$ is a family of linear maps indexed by all morphisms $h : L \to M$:

(8.20) $$\alpha_h \cap\, :\, A_*(L) \to A_*(L),$$

and satisfying the following conditions.

(i) Commutation with proper pushforwards: if $p : P \to L$ is proper and $h : L \to M$ arbitrary, then

(8.21) $$\alpha_h \cap p_*(y) = p_*(\alpha_{hp} \cap y)$$

for all $y \in A_*(P)$.

(ii) Commutation with flat pullbacks: if $p: P \to L$ is flat and representable, and $h: L \to M$ arbitrary, then

$$\alpha_{hp} \cap p^*(y) = p^*(\alpha_h \cap y) \tag{8.22}$$

for all $y \in A_*(L)$.

(iii) Commutation with Gysin pullbacks: if $p: P \to L$ is a regular local embedding of stacks and $h: L \to M$ arbitrary, then

$$\alpha_{hp} \cap p^!(y) = p^!(\alpha_h \cap y) \tag{8.23}$$

for all $y \in A_*(L)$.

Linear combination and composition of operations defines on $A^*(M)$ the structure of an associative algebra. In principle, it might be non–commutative. Homological Chow groups become moduli over this ring.

We sometimes write simply $\alpha \cap y$ instead of $\alpha_g \cap y$.

As a particular case of (8.8), for any morphism of stacks $f: N \to M$, we can define the functorial pullback $f^*: A^*(M) \to A^*(N)$:

$$f^*(\alpha)_g \cap y := \alpha_{fg} \cap y \tag{8.24}$$

for any $g: P \to N$ and $y \in A_*(P)$. It is a ring homomorphism.

Now we will define two pushforward operations which are not special cases of the previously constructed ones.

8.9.1. Proper flat pushforward. Let $f: N \to M$ be a proper flat morphism of stacks. For any $\alpha \in A^*(N)$, $h: L \to M$, and $y \in A_*(L)$ put

$$f_\bullet(\alpha)_h \cap y := f_{L*}(\alpha_{h_N} \cap f_L^*(y)), \tag{8.25}$$

where $f_L: L \times_M N \to L$, $h_N: L \times_M N \to N$ are the canonical morphisms, and f_{L*}, f_L^* are respectively the proper pushforward and the flat pullback of the homological Chow groups A_*.

8.9.2. Gysin pushforward. Similarly, let $f: N \to M$ be a regular local imbedding of stacks. For any $\alpha \in A^*(N), h: L \to M$, and $y \in A_*(L)$ put

$$f_\bullet(\alpha)_h \cap y := f_{L*}(\alpha_{h_N} \cap f^!(y)), \tag{8.26}$$

where $f_{L*}, f^!$ are respectively the proper pushforward and the Gysin pullback of the homological Chow groups A_*.

We introduced the notation $f_\bullet$ in order to avoid confusion with f_* which has a different meaning in the full operational theory and in general has different source and target than f_*.

8.9.3. Proposition. *The pushforwards $f_\bullet(\alpha)$ are well defined as elements of $A^*(M)$. Moreover, $(fg)_\bullet = f_\bullet g_\bullet$ whenever f, g are simultaneously as in 8.9.1 or 8.9.2.*

Proof. Let us check that $f_\bullet(\alpha)$ defined by (8.25) satisfies the requirements (8.21)–(8.23).

Start with (8.21).

Keeping the notation of 8.9 (i) and 8.9.1 we must check that

(8.27) $$f_\bullet(\alpha)_h \cap p_*(y) = p_*(f_\bullet(\alpha)_{hp} \cap y).$$

According to (8.25), the left hand side of (8.27) is $f_{L*}(\alpha_{h_N} \cap f_L^* p_*(y))$, whereas the right hand side is $p_*(f_{P*}(\alpha_{(hp)_N} \cap f_P^*(y)))$. Here f_P (resp. $(hp)_N$) are projections of $P \times_M N$ to P, resp. N. Because of the associativity of the base change, we can identify $P \times_M N$ with $P \times_L (L \times_M N)$. Denoting by $q : P \times_M N \to L \times_M N$ the respective projection, we have $(hp)_N = h_N q$.

This is summarized in the following diagram, all of three squares of which are Cartesian:

$$\begin{array}{ccccc} P \times_M N & \xrightarrow{q} & L \times_M N & \xrightarrow{h_N} & N \\ \downarrow{f_P} & & \downarrow{f_L} & & \downarrow{f} \\ P & \xrightarrow{p} & L & \xrightarrow{h} & M \end{array}$$

Now we can rewrite the left hand side of (8.27) using [Vis], Lemma (3.9)(ii):

$$f_{L*}(\alpha_{h_N} \cap f_L^* p_*(y)) = f_{L*}(\alpha_{h_N} \cap q_* f_P^*(y)).$$

On the other hand, we can rewrite the right hand side of (8.27) using first the commutativity of the cartesian square involving f_L and p and then (8.21) for α:

$$p_*(f_{P*}(\alpha_{(hp)_N} \cap f_P^*(y))) = f_{L*} q_*(\alpha_{h_N q} \cap f_P^*(y)) = f_{L*}(\alpha_{h_N} \cap q_* f_P^*(y)).$$

Now we will check (8.22) for $f_\bullet(\alpha)$.

Keeping the notation of 8.9 (ii), 8.9.1, and the reasoning above, we must check that

(8.28) $$f_\bullet(\alpha)_{hp} \cap p^*(y) = p^*(f_\bullet(\alpha)_h \cap y).$$

According to (8.25), the left hand side of (8.28) is

$$f_{P*}(\alpha_{(hp)_N} \cap f_P^* p^*(y)) = f_{P*}(\alpha_{h_N q} \cap q^* f_L^*(y))$$

because of the commutativity of the Cartesian square involving f_L and p. Similarly, the right hand side is

$$p^*(f_{L*}(\alpha_{h_N} \cap f_L^*(y))) = f_{P*} q^*(\alpha_{h_N} \cap f_L^*(y))$$

(use [Vis], Lemma (3.9)(ii). The two expressions coincide in view of (8.22) for α.

Similar reasoning establishes (8.23). We omit the check of multiplicativity.

8.9.4. Projection formula. *Let $f : N \to M$ be a morphism of stacks which is either proper, flat and representable, or a regular local imbedding. Then we have for any $\alpha \in A^*(N)$, $\beta \in A^*(M)$:*

(8.29) $$f_\bullet(\alpha f^*(\beta)) = f_\bullet(\alpha)\beta.$$

Proof. In fact, let $h : L \to M$, $y \in A_*(L)$ be as above, and let f be, say, flat. Then we have in view of (8.25):

$$f_\bullet(\alpha f^*(\beta))_h \cap y = f_{L*}(\alpha f^*(\beta)_{h_N} \cap f_L^*(y)) = f_{L*}(\alpha_{h_N} \cap f^*(\beta)_{h_N} \cap f_L^*(y)).$$

On the other hand,

$$(f_\bullet(\alpha)\beta)_h \cap y = f_\bullet(\alpha)_h \cap (\beta_h \cap y) = f_{L*}(\alpha_{h_N} \cap f_L^*(\beta_h \cap y)).$$

Combining (8.24) and (8.22), we see that both sides coincide.

The Gysin case is checked similarly.

8.9.5. Commutativity with pullbacks. *Keeping the same assumptions and notation as in 8.9.4, we have*

(8.30) $$h^* f_\bullet = f_{L\bullet} h_N^*.$$

Proof. In fact, we must check that for any $p : P \to L$, $y \in A_*(P)$, $\alpha \in A^*(N)$ we have

(8.31) $$h^* f_\bullet(\alpha)_p \cap y = f_{L\bullet} h_N^*(\alpha)_p \cap y.$$

We will treat the proper flat pushforward; the Gysin case is similar. Applying first (8.24) and then (8.25) we rewrite the left hand side of (8.31) as

$$f_\bullet(\alpha)_{hp} \cap y = f_{P*}(\alpha_{h_N q} \cap f_P^*(y)).$$

In view of (8.25), the right hand side is $f_{P*}(h_N^*(\alpha)_q \cap f_P^*(y))$. These expressions coincide because of (8.24).

8.9.6. Splitting morphisms. We will call $f : N \to M$ a *splitting morphism* if $f_* f^* = \mathrm{id}$ as an endomorphism of $A_*(M)$. The most important examples of splitting morphisms are blow ups of regularly embedded substacks (for schemes, see [Ful], Prop. 6.7 (b), p. 115).

8.9.7. Lemma. *Let $f : N \to M$ be a proper flat morphism and $h : L \to M$ such a morphism that $f_L : L \times_M N \to L$ is splitting. Then*

(8.32) $$f_\bullet f^*(\alpha)_h \cap = \alpha_h \cap .$$

Proof. Applying consecutively (8.25), (8.24), and (8.22), we find:

$$f_\bullet f^*(\alpha)_h \cap y = f_{L*}(f^*(\alpha)_{h_N} \cap f_L^*(y)) = f_{L*}(\alpha_{fh_N} \cap f_L^*(y))$$
$$= f_{L*}(\alpha_{hf_L} \cap f_L^*(y)) = f_{L*} f_L^*(\alpha_h \cap y) = \alpha_h \cap y.$$

CHAPTER VI

Algebraic Geometric Introduction to the Gravitational Quantum Cohomology

§1. Virtual fundamental classes

1.1. Sketch of the construction. Let τ be a stable V-marked modular graph and $\overline{M}(V,\tau)$ the respective stack of stable maps defined (as groupoid) in V.4.7.

In this subsection we review the construction of a Chow class

(1.1) $$J(V,\tau) \in A_{D(\tau)}(\overline{M}(V,\tau)),$$

where *the virtual dimension* is defined by

(1.2) $$\begin{aligned}D(\tau) &:= (1-g)(\dim V - 3) + c_1(\omega_V)(\beta) + |S_\tau| - |E_\tau| \\ &= \operatorname{rk} R\pi_*(f^*T_V) + \dim \mathcal{M}(\tau).\end{aligned}$$

Here $\pi : C(V,\tau) \to \overline{M}(V,\tau)$ is the universal curve, and $f : C(V,\tau) \to V$ is the universal map.

This element is called *the virtual fundamental class*. Another notation for it is $[\overline{M}(V,\tau)]^{virt}$. When τ is a one–vertex graph with label (g,β) and tails S, we also write $J_{g,S}(V,\beta)$ and put

(1.3) $$I_{g,S}(V,\beta) := (\operatorname{ev}_S, \operatorname{st})_* J_{g,S}(V,\beta) \in A_{D(\tau)}(V^S \times \overline{M}_{g,S}).$$

Here ev refers to the evaluation morphism at all labeled sections corresponding to the tails S of τ, and st is the absolute stabilization morphism defined in V.4.6.

The Gromov–Witten invariants $I^V_{g,n,\beta}$ discussed axiomatically in III.5 are produced from $I_{g,n}(V,\beta)$ by treating this class as a correspondence.

One can also define $I(V,\tau)$ for general stable $B(V)$-labeled graphs. The exact definition is however a bit awkward. It involves the combinatorics of several graph categories: cf. [BehM], Property V (Isogenies) on pp. 46–47 and pp. 34–47, and 1.2 below.

The second line of (1.2) can be loosely interpreted in the following way. In the virtual world, when one wants to deform a stable map, one can freely deform the underlying curve keeping its combinatorial type, whereas the deformation of the map of a fixed curve is governed exactly by the first order obstruction theory.

In order to produce $J(V,\tau)$ of the required dimension and with controlled functorial properties, [BehF] and [Beh] suggest the following three–step construction.

(i) Stack theoretic version of $R\pi_*(f^*\mathcal{T}_V)$. First, it is proved that in the derived category one can construct an isomorphism

$$(1.4) \qquad R\pi_*(f^*\mathcal{T}_V) \xrightarrow{\sim} [E^0 \to E^1],$$

where E^i are locally free coherent sheaves on $\overline{M}(V,\tau)$.

Now replace the complex $E^0 \to E^1$ by the quotient stack denoted interchangeably as $h^1/h^0(E^*)$ or $[E^1/E^0]$. The fiber of this stack over an étale open U is the category of pairs (P,φ) where P is a principal homogeneous E^0-bundle and $\varphi : P \to E^1|U$ is an E^0-equivariant morphism of sheaves. It is proved in [BehF] that any quasi-isomorphism of two-term complexes $E^* \to F^*$ induces an isomorphism of stacks $[E^1/E^0] \to [F^1/F^0]$.

These stacks are only Artin, generally not DM-stacks. In special cases they reduce to the classifying spaces of (relative) vector spaces. Since vector spaces are contractible, this step leads to the relatively mild extension of the category of DM-stacks.

(ii) Intrinsic relative normal cone. Consider first a model situation. Let M be a scheme over a field K admitting a closed imbedding $j : M \to U$ into a smooth U. Then we can construct the normal bundle and the normal cone imbedded in it, $C_M U \to N_M U$ (see proof of V.6.2.1). Both are schemes over M depending on j. In order to make them independent, one can "subtract" the tangent sheaf of U restricted to M. More precisely, there exists the group scheme T_U over M whose sheaf of local sections is $j^*\mathcal{T}_U$ and which naturally acts on $C_M U \to N_M U$. In the category of Artin stacks, we can then form the natural quotient of this diagram with respect to T_U and get the *intrinsic normal cone and normal bundle* $\mathcal{C}_M \to \mathcal{N}_M$. Notice that our normal bundles generally are relative abelian group schemes rather than vector bundles.

One can check that this construction is independent on j, can be made local and relative, and eventually extended to stacks so that finally it becomes applicable to *a relative DM-stack over an Artin stack*.

The final product we need then is $\mathcal{C}_{\overline{M}(V,\tau)/\mathcal{M}(\tau)}$.

To get some feeling about this construction, consider again a scheme M over a field. If M is itself smooth, we can start with $j = \mathrm{id}_M$ and get the relative classifying stack of the tangent bundle $B\mathcal{T}_M = \mathcal{C}_M = \mathcal{N}_M$. According to the formal counting rules, its dimension is $\dim M - \mathrm{rk}\,\mathcal{T}_M = 0$. The compatibility with localization shows that this simple picture holds at the points where the scheme (or the morphism in the relative case) is smooth, so that generally the intrinsic normal cone is essentially an invariant of the singular locus.

In the relative case, the dimension of the intrinsic normal cone equals the dimension of the base.

(iii) Obstruction theory. A relativization of the Kodaira–Spencer obstruction theory for maps allows one to construct a closed imbedding

$$(1.5) \qquad \mathcal{C}_{\overline{M}(V,\tau)/\mathcal{M}(\tau)} \to h^1/h^0(R\pi_*f^*\mathcal{T}_V).$$

Moreover, there is the zero section morphism

$$(1.6) \qquad i : \overline{M}(V,\tau) \to h^1/h^0(R\pi_*f^*\mathcal{T}_V).$$

We would now like to define $J(V,\tau) := i^![\mathcal{C}_{\overline{M}(V,\tau)/\mathcal{M}(\tau)}]$. This involves an extension of the Gysin formalism to certain morphisms of DM–stacks to Artin stacks; see [Kr]. At this point such an extension can be avoided as in [Beh] by constructing first a Cartesian square

$$\begin{array}{ccc} \mathcal{C}' & \longrightarrow & \mathcal{C}_{\overline{M}(V,\tau)/\mathcal{M}(\tau)} \\ \downarrow & & \downarrow \\ E_1 & \longrightarrow & [E_1/E_0] \end{array}$$

involving a choice of global resolution as in (i) (and passage to vector bundles). It remains then to put

(1.7) $$J(V,\tau) := 0^![\mathcal{C}'] \in A_{D(\tau)}(\overline{M}(V,\tau)),$$

where this time $0 : \overline{M}(V,\tau) \to E_1$ is the zero section morphism of DM–stacks.

We will now give a useful list of universal properties of $J(V,\tau)$ which were formulated in [BehM] and established in [Beh] and which imply the axioms for Gromov–Witten invariants from III.5. In many cases, using this list allows one to avoid referring to the construction very sketchily described above. It would be important to find an extension of this list which would determine the virtual fundamental classes uniquely, so that the construction could be considered as an existence proof. However, no such extension is known, even on the level of homology/cohomology. The same remark applies to the Gromov–Witten correspondences.

1.2. Properties of the virtual fundamental classes. We list the basic relations in approximately the same order as the respective axioms of Gromov–Witten invariants in III.5.2 and III.5.3. The latter are of course derived from the former.

i) Effectivity. $J_{g,n}(V,\beta) = 0$ for $\beta \notin B = B(V)$. More generally, $J(V,\tau) = 0$ unless all vertices of τ are marked by effective classes.

This is obvious since otherwise $\overline{M}(V,\tau)$ is empty: see V.(4.10) and V.(4.12). We have in fact included this effectivity condition in the definition of the $B(V)$–marking.

ii) Covariance. Let $\Phi : \tau \to \sigma$ be an isomorphism of $B(V)$–marked modular graphs inducing the isomorphism $\overline{M}(\Phi) : \overline{M}(V,\tau) \to \overline{M}(V,\sigma)$. Then we have $\overline{M}(\Phi)_*(J(V,\tau)) = J(V,\sigma)$.

This is a trivial consequence of the functoriality of the construction with respect to the relabeling. In fact, the subtler behavior of $J(V,\tau)$ with respect to the forgetful and boundary morphisms can also be stated as functoriality over appropriate graph categories; cf. [BehM]. Below for brevity we treat only the simplest morphisms generating these categories.

iii) Degeneration. Here we describe the behavior of virtual fundamental classes with respect to the boundary morphisms

$$b(V,\tau): \overline{M}(V,\tau) \to \overline{M}_{g,S}(V,\beta).$$

They were defined in V.4.7. Here τ is a connected stable $B(V)$–marked modular graph with tails S.

Fix a stable modular graph σ of genus g with the set of tails S (without $B(V)$-marking). This implies that $2g - 2 + |S| > 0$.

Denote by $\mathbf{S}$ the set of all $B(V)$-markings of σ of the class β. We will be considering any element of $\mathbf{S}$ as a $B(V)$-marked graph τ whose underlying modular graph is explicitly identified with σ.

We have the commutative diagram

(1.8)
$$\begin{array}{ccc} \coprod_{\tau \in \mathbf{S}} \overline{M}(V,\tau) & \xrightarrow{\coprod b(V,\tau)} & \overline{M}_{g,S}(V,\beta) \\ \downarrow & & \downarrow \\ \overline{M}(\sigma) & \xrightarrow{b(\sigma)} & \overline{M}_{g,S} \end{array}$$

where the vertical arrows forget the map to V and stabilize the curve as in V.4.7. This diagram generally is not Cartesian. It induces the proper morphism

(1.9)
$$\coprod_{\tau \in \mathbf{S}} \overline{M}(V,\tau) \xrightarrow{\coprod h_\tau} \overline{M}(\sigma) \times_{\overline{M}_{g,S}} \overline{M}_{g,S}(V,\beta).$$

The stack $\overline{M}_\sigma \times_{\overline{M}_{g,S}} \overline{M}_{g,S}(V,\beta)$ carries two classes that can be produced from the virtual fundamental classes. First, $b(\sigma)$ is a l.r.i. morphism of smooth stacks of pure dimension (cf. [Beh], p. 611, where the more general boundary morphisms are treated). Hence we can form the Gysin pullback $b(\sigma)^! J_{g,S}(V,\beta)$. Second, using the diagrams V.(4.12) and (1.9) above, we can construct

$$h_{\tau*} \Delta_\tau^! \left(\bigotimes_{v \in V_\tau} J_{g_v, F_\tau(v)}(V, \beta_v) \right).$$

1.2.1. Proposition. *We have*

(1.10)
$$b(\sigma)^! J_{g,S}(V,\beta) = \sum_{\tau \in \mathbf{S}} h_{\tau*} \Delta_\tau^! \left(\bigotimes_{v \in V_\tau} J_{g_v, F_\tau(v)}(V, \beta_v) \right),$$

where the τ-summand comes from the Chow group of the τ-component of the upper left stack in (1.8).

This is a particular case of the Isogenies Axiom for $J(V,\tau)$ in [BehM]: see p. 38, Definition 5.8, Case I.

Formulas III.(5.2) and III.(5.3) can be deduced from (the suitable particular cases of) (1.10).

We pass now to the next property implying in particular the Identity and Divisor Axioms III.(5.6) and III.(5.8) for GW–invariants. (The Dimension Axiom directly follows from (1.2).)

iv) Forgetting a point. Let $*$ be a tail of the stable $B(V)$–marked modular graph σ, and S the set of remaining tails. Forgetting $*$ and stabilizing the obtained graph, we get another marked modular graph τ with tails S. In this situation we have the forgetful morphism

(1.11)
$$\phi : \overline{M}(V,\sigma) \to \overline{M}(V,\tau)$$

which is in fact isomorphic to the universal curve morphism: cf. Theorem V.4.5 for one–vertex σ from which the general case can be deduced formally. Hence ϕ is flat and proper.

The relevant property of the virtual fundamental class is simply

(1.12) $$J(V, \sigma) = \phi^*(J(V, \tau)),$$

where ϕ^* means the flat pullback.

v) Mapping to a point. Let τ be a connected stable labeled modular graph of class zero, that is, $\beta_v = 0$ for all $v \in V_\tau$. Then τ without labeling is absolutely stable so that any stable map of the type τ is a map of stable curve. An easy generalization of Proposition V.4.2.3 shows that there is a canonical isomorphism

(1.13) $$\overline{M}(V, \tau) \xrightarrow{\sim} V \times \overline{M}(\tau)$$

induced by evaluation and forgetting the map. Let $\pi : C \to \overline{M}(\tau)$ be the universal curve. The identification (1.13) induces the locally free sheaf $\mathcal{T}^{(1)}$ of rank $D := g \dim V$ on $\overline{M}(V, \tau)$, g being the genus of C.

With this notation, we have

(1.14) $$J(V, \tau) = c_D(\mathcal{T}^{(1)}) \cap [\overline{M}(V, \tau)].$$

We recall that Chern classes act on the homological Chow groups. Of course, in this particular case we work with smooth DM–stacks where the two types of Chow groups can be identified.

The Mapping to a Point Axiom III.(5.9) for GW–invariants follows from this.

1.3. Virtual fundamental classes vs Gromov–Witten correspondences. In III.5 we have shown that the main structures of Quantum Cohomology can be derived from the properties of GW–correspondences (1.3).

Stress upon $I_{g,S}(V, \beta)$ instead of $J(V, \tau)$ has some advantages.

(i) GW–correspondences furnish an additional structure on the explicitly known orbifolds $V^n \times \overline{M}_{g,n}$, whereas virtual fundamental classes live in the Chow groups of utterly uncontrollable stacks $\overline{M}(V, \tau)$.

(ii) The basic numerical invariants of quantum cohomology, *the correlators,* with or without gravitational descendants, are defined directly in terms of virtual fundamental classes J, but after some work can be rewritten via Gromov–Witten correspondences: cf. the next sections.

(iii) The Gromov–Witten correspondences form a much richer structure than just the supply of correlators. Namely, they define the (co)representation of the motivic modular (co)operad formed by the motives of $\overline{M}_{g,n}$ on the motive of any smooth projective manifold.

The virtual fundamental classes seemingly form a looser collection which is not a special case of some known structure.

(iv) The major shortcoming of I in comparison with J is that we do not know how to define and study these classes without recourse to J, and any time we need to establish a new property of GW–correspondences not covered by the "axioms", we have to invoke the actual construction of the virtual fundamental classes.

(v) The last drawback of GW–correspondences which might seem minor but is in fact the source of interesting developments, is the existence of the *unstable range* $(g, n) \in \{(0,0), (0,1), (0,2), (1,0)\}$ where I vanish whereas J may well be non–trivial.

This is reflected in the existence of the unstable, mainly genus zero two–points correlators, which play a very special role in the theory (see §7 below, [Giv2], and [DZh2]). It is tempting to conjecture that one could complete the geometric picture by introducing the unstable range Gromov–Witten correspondences belonging to $A_*(\mathcal{M}_{g,n} \times V^n)$ where A_* is the Chow group of Artin stacks, for example, defined in [Kr]. For more details, cf. 7.6.

§2. Gravitational descendants and Virasoro constraints

2.1. Notation. As above, let V be a smooth projective manifold over $\mathbf{C}$. *Correlators with gravitational descendants* form a family of polylinear functions on the rational cohomology of V. One such function

$$H^*(V)^{\otimes n} \to \mathbf{Q}$$

is defined by fixing $n \geq 0, g \geq 0, d_1, \ldots, d_n \geq 0$ and an algebraic homology class $\beta \in B(V)$ (or else in $H_2(V, \mathbf{Z})/(tors)$). The physical notation for such a function is

$$H^*(V)^{\otimes n} \ni \gamma_1 \otimes \cdots \otimes \gamma_n \mapsto \langle \tau_{d_1} \gamma_1 \ldots \tau_{d_n} \gamma_n \rangle_{g, \beta}$$

and the mathematical definition is

$$\langle \tau_{d_1} \gamma_1 \ldots \tau_{d_n} \gamma_n \rangle_{g, \beta}$$

(2.1) $$:= \int_{J_{g,n}(V,\beta)} c_1(L_1)^{d_1} \cup \mathrm{ev}_1^*(\gamma_1) \cup \cdots \cup c_1(L_n)^{d_n} \cup \mathrm{ev}_n^*(\gamma_n).$$

Here

$$L_i = L_{i; g,n}(V, \beta) := x_i^*(\omega_{\overline{C}/\overline{M}}),$$

in the notation of V.4.1 and VI.1.1.

We recall once again that since $\overline{M}_{g,n}(V, \beta)$ generally is not smooth, we need simultaneously two types of Chow groups: A_* spanned by the classes of closed substacks and A^* defined as a version of the bivariant operational theory in V.8. In (2.1), the virtual class lies in A_*, whereas the monomial in Chern classes belongs to A^*. If we replace $\gamma_1 \otimes \cdots \otimes \gamma_n$ by an element of $A^*(V^n)$, its pullback with respect to ev will also belong to A^*, and the meaning of (2.1) will be the degree of an A^*–class capped with an A_*–class. Notice that the dimension of (the irreducible components of) $\overline{M}_{g,n}(V, \beta)$ in general varies wildly and uncontrollably. Hence we have no idea what the dimension of $c_1(L_i)$ or the codimension of $J_{g,n}(V, \beta)$ might be. This determines the kind of Chow groups we have to put these classes in.

If we want to use arbitrary cohomology classes as arguments in (2.1), we must first extend to the context of stacks much of the contents of [Ful], Ch. 19, dealing with a pair of homology and cohomology theories H^*, H_* endowed with class maps from A^*, A_*, respectively, and satisfying a host of compatibilities with various products, Chern classes, and functorial morphisms. I do not know a reference for this. In the framework of the basic motivic formalism of smooth DM–stacks this problem can be altogether avoided when $2g - 2 + n \geq 1$: we can then replace (2.1) by the formula involving intersection theory and cohomology only for smooth Deligne–Mumford stacks (cf. 2.2 below). However, in order to treat the marginal cases and to keep a connection with the physical notation, I will also use (2.1) explicitly assuming in these cases that the relevant extension is possible. In the recent preprint [GrPa], it is remarked that the stack of stable maps admits a closed embedding into a smooth Deligne–Mumford stack. This might be helpful for developing the formalism of the Borel–Moore homology used in [Ful] for schemes.

The theory of the correlators (2.1) with $d_i = 0$ is quantum cohomology. Sometimes this term refers more restrictively only to the $g = 0$ correlators. In physics, (operators related to) the cohomology classes γ_i are called *primary fields*, whereas $\tau_d \gamma$ are referred to as *gravitational descendants*.

Hence the name *gravitational quantum cohomology* for the full theory.

2.2. Motivic correlators in the stable range. The stable range condition $2g - 2 + n \geq 1$ assures that $\overline{M}_{g,n}$ is non–empty. With this assumption, the theory of correlators can be enriched and stripped of some of the technical complications in the following way. Put

$$I_{g,n}(V, \beta; d_1, \ldots, d_n)$$

(2.2) $\quad := (\mathrm{ev}, \mathrm{st})_*(c_1(L_1)^{d_1} \ldots c_1(L_n)^{d_n} \cap J_{g,n}(V, \beta)) \in A_d(V^n \times \overline{M}_{g,n}),$

where

(2.3) $\quad d = (1 - g)(\dim V - 3) + (c_1(V), \beta) + n - d_1 - \cdots - d_n.$

In the smooth case, we can canonically identify $A_* = A^*$, and these groups are endowed with completely satisfactory class maps to various pairs of the (co)homology theories $H_* = H^*$ which are identified via the Poincaré duality (eventually with the Tate twist). In particular, choosing such a pair, we can consider the cohomology class of $I_{g,n}(V, \beta; d_1, \ldots, d_n)$ as a map, for brevity denoted by the same symbol,

(2.4) $\quad I_{g,n}(V, \beta; d_1, \ldots, d_n) : H^*(V^n) \to H^*(\overline{M}_{g,n}).$

On the algebraic classes in $H^*(V^n)$, one recovers (2.1) by following this map with the integration over $[\overline{M}_{g,n}]$. This is proved by an easy calculation using a projection formula: see Lemma 3.1.1 below. The same is true generally, provided we have the extension of H_*, H^* with necessary properties. So in the stable range we can forget about this problem and just take

(2.5) $\quad \langle \tau_{d_1} \gamma_1 \ldots \tau_{d_n} \gamma_n \rangle_{g,\beta} := \int_{[\overline{M}_{g,n}]} I_{g,n}(V, \beta; d_1, \ldots, d_n)(\gamma_1 \otimes \cdots \otimes \gamma_n)$

as an alternative definition of the correlators.

In general, the set of maps (2.4) is a richer structure than (2.5). We may call (2.4) *cohomological correlators*.

Still more generally, we can consider $I_{g,n}(V, \beta; d_1, \ldots, d_n)$ as a family of correspondences defining morphisms of the classical (Grothendieck) Chow motives extended to the 1-category of the smooth Deligne–Mumford stacks (see [BehM], pp. 53–56):

(2.6) $\quad h(V^n)((1 - g) \dim V) \to h(\overline{M}_{g,n})(d_1 + \cdots + d_n - (c_1(V), \beta))$

(this time the Tate twists are registered; cf. [Sch]: $h(X)(-d) := h(X) \otimes \mathbf{L}^{\otimes d}$). The morphisms (2.6) are our *motivic correlators*. The physical correlators (2.5) also have their motivic counterparts: the integration over the fundamental class of the moduli space must be replaced by the motivic projection

$$h(\overline{M}_{g,n}) \to \mathbf{L}^{3g-3+n}.$$

This refined theory of the stable range correlators shows their compatibility with Hodge structure, étale Galois representations, and so on. (Unfortunately, the situation in finite characteristic is still incompletely understood because of the lack of foundations.)

2.3. Relations between correlators. There is a vaguely defined notion of the "universal identities" between the physical correlators which can be heuristically deduced from the path integral formulation of various topological quantum field theories.

In practical terms, the physicists' correlators are coefficients of various generating functions satisfying (partly conjecturally, partly demonstrably) some remarkable differential equations, and/or serving as asymptotical series of interesting functions. Quite often these properties are equivalent to the families of identities between the correlators, and in some cases these identities have direct algebraic geometric meaning. This is the case of the Associativity (WDVV) Equations which is the central object of this book.

An example of the more mysterious structure constitutes the so-called *Virasoro constraints* to which we turn below.

In the final count we expect to upgrade all the universal identities between the physical correlators to their motivic counterparts, as Nakajima ([Na1], [Na2]) and Grojnowski ([Groj]) did in another context (see also [Bara], [Le]).

2.4. The generating functions. The gravitational potential is a generating series for the correlators (2.1) or (2.5) considered as a formal function on the *large phase (super)space* $\mathcal{H}(V) = \bigoplus_{d=0}^{\infty} H^*(V)[d]$. The d-th copy of $H^*(V)$ accommodates $\tau_d \gamma$'s.

To be more precise, we choose a basis $\{\Delta_a \mid a \in A\}$ of $H^*(V, \mathbf{C})$ such that $\Delta_a \in H^{p_a, q_a}(V)$. Denote by $\{x_d^a\}$ the dual coordinates to $\{\tau_d \Delta_a\}$ (a is a superscript, not a power) and by $\Gamma = \sum_{a,d} x_d^a \tau_d \Delta_a$ the generic even element of the extended phase superspace. As usual, x_d^a has the same $\mathbf{Z}_2$-degree as Δ_a, and the odd coordinates anticommute. Notice that the formal symbol τ_d acquires an independent meaning as the linear operator on $\mathcal{H}(V)$ identifying each $H^*(V)[e]$ to $H^*(V)[e+d]$ so that one can write $\tau_d = \tau_1^d$. We can as well put $\mathcal{H}(V) = H^*(V)[t]$ and identify τ_d with multiplication by t^d.

One more formal (even) variable is the genus expansion parameter λ.

Finally, as in III.5.2.1 we need the universal character $B(V) \to \Lambda : \beta \mapsto q^\beta$ with values in the Novikov ring Λ which is the completed semigroup ring of $B(V)$ eventually localized with respect to the multiplicative system q^β. It is topologically spanned by the monomials $q^\beta = q_1^{\beta_1} \ldots q_r^{b_m}$ where $\beta = (b_1, \ldots, b_m)$ in a basis of the numerical class group of 1-cycles, and $(q_1, \ldots, q_m)$ are independent formal variables.

We now put formally

(2.7)
$$F(x) = \sum_g \lambda^{2g-2} F_g(x) = \sum_{g,\beta} \lambda^{2g-2} q^\beta \langle e^\Gamma \rangle_{g,\beta} = \sum_{g,\beta} \lambda^{2g-2} q^\beta \sum_n \frac{\langle \Gamma^{\otimes n} \rangle_{g,\beta}}{n!}$$
$$= \sum_{n,(a_1,d_1),\ldots,(a_n,d_n)} \epsilon(a_1,\ldots,a_n) \frac{x_{d_1}^{a_1} \cdots x_{d_n}^{a_n}}{n!} \sum_{g,\beta} \lambda^{2g-2} q^\beta \langle \tau_{d_1} \Delta_{a_1} \ldots \tau_{d_n} \Delta_{a_n} \rangle_{g,\beta},$$

where ϵ is the standard sign in superalgebra. One can avoid this sign replacing the monomial in x_d^a in (2.7) by its reverse version $x_{d_n}^{a_n} \ldots x_{d_1}^{a_1}$.

Define the $(\mathbf{Z},\mathbf{Z})$-grading of all the monomials in (2.7) by the following rules:

(2.8)
$$\deg(\lambda^2) = (\dim V - 3, \dim V - 3),$$
$$\deg q^\beta = ((K_V, \beta), (K_V, \beta)),$$
$$\deg x_d^a = (d + p_a - 1, d + q_a - 1).$$

From (2.1), (2.3) it follows that $\langle \tau_{d_1} \Delta_{a_1} \ldots \tau_{d_n} \Delta_{a_n} \rangle_{g,\beta}$ can be non–zero only if

(2.9) $$\sum_i (p_i + d_i - 1) = \sum_i (q_i + d_i - 1) = (1-g)(\dim V - 3) + (c_1(V), \beta).$$

This can be expressed by saying that $F(x)$ is homogeneous of bidegree $(0,0)$.

Witten and others have conjectured that the *partition function* $Z(x) := e^{F(x)}$ is annihilated by certain differential operators L_i, $i \geq -1$, satisfying the Virasoro commutation relations (notice that for $i \geq 1$, L_i are of order two). For $V = $ *a point* this follows from Kontsevich's proof of another of Witten's conjecture ([Ko1]): see e.g. explanations in [Lo1]. In the general case, the problem is still open. Moreover, the general form of L_i's was only recently suggested by T. Eguchi, K. Hori, Ch.-Sh. Xiong ([EHX2]), and by Sh. Katz (lectures at the Mittag–Leffler Institute [K]). According to Katz, there are actually two natural algebras spanned by L_i and $\overline{L}_i$, respectively, which we will now describe.

2.5. Witten's and Eguchi–Hori–Xiong–Katz's operators. Put $\partial_{d,a} = \frac{\partial}{\partial x_d^a}$. It is convenient to shift the coordinate with $d = 1$ corresponding to the identity (fundamental class) in $H^*(V)$, so we will assume that $1 = \Delta_0$ and put $s_1^0 = x_1^0 - 1$, $s_d^a = x_d^a$ for all the other values of (d,a). Finally, we can raise and lower the a–indices using the Poincaré form $g_{ab} = \int_V \Delta_a \cup \Delta_b$.

With this notation, put

(2.10) $$L_{-1} = \overline{L}_{-1} = \frac{1}{2\lambda^2} \sum_{a,b} x_0^a g_{ab} x_0^b + \sum_{a,d} s_d^a \partial_{d-1,a}.$$

For other generators, we need the nilpotent matrix K of cup multiplication by $-K_V : -K_V \cup \Delta_a = \sum K_a^b \Delta_b$.

(2.11)
$$L_0 = C_0 + \frac{1}{2\lambda^2} \sum_{a,b,c} K_a^b g_{bc} x_0^a x_0^c + \sum_{a,d} \left(p_a + d + \frac{1 - \dim V}{2} \right) s_d^a \partial_{d,a} + \sum_{a,d} K_a^b s_d^a \partial_{d-1,b},$$

where

(2.12) $$C_0 = \deg \frac{(3-\delta)c_\delta(V) - 2c_1(V)c_{\delta-1}(V)}{48}, \quad \delta = \dim V.$$

Replacing in (2.11) p_a by q_a, we will get $\overline{L}_0$.

Finally, for $n \geq 1$ Katz puts

(2.13)
$$L_n = \frac{1}{2\lambda^2} \sum_{a,b,c} (K^{n+1})_a^b g_{bc} x_0^a x_0^c$$
$$+ \sum_{r=0}^{n+1} \sum_{d=0}^{\infty} \sum_{a,b} \sigma_{n+1-r,n+1}\left(p_a + d + \frac{1-\dim V}{2}\right)(K^r)_a^b s_d^a \partial_{d+n-r,b}$$
$$- \frac{\lambda^2}{2} \sum_{r=0}^{n+1} \sum_{d=0}^{n-r-1} \sum_{a,b,c} (-1)^d \sigma_{n+1-r,n+1}\left(p_a - d - \frac{1+\dim V}{2}\right)(K^r)_c^b g^{ac} \partial_{d,a} \partial_{n-1-r-d,b}$$

where

(2.14) $$\sigma_{k,n+1}(t) := \sigma_k(t, t+1, \ldots, t+n),$$

σ_k denoting the k–th elementary symmetric function. To get $\overline{L}_n$, one should replace p_a by q_a here.

2.6. Proposition. $[L_n, L_m] = (n-m)L_{n+m}$ for all $n, m \geq -1$, and similarly for $\overline{L}_n$.

For a proof, see [Ge6], Theorem 2.11. It extends the calculations of [EHX2] and [EJX] and is based on a version of the Sugawara construction, embedding the Virasoro algebra into the Heisenberg algebra so that the Virasoro generators become normally ordered quadratic expressions in the Heisenberg generators. As E. Getzler remarks, the formal part of the calculation shows that $[L_n, L_m] - (n-m)L_{n+m}$ is a constant, whereas the fact that this constant vanishes follows from a non–trivial identity which can be deduced from the Hirzebruch–Riemann–Roch theorem. More precisely, let $\mathcal{V}$ be the operator on $H^*(V, \mathbf{C})$ multiplying $H^{p_a,q_a}(V)$ by $\frac{\dim V}{2} - p_a$. Then the supertrace of $\mathcal{V}^2$ (the difference of traces on even and odd subspaces of $H^*(V)$) equals
$$\operatorname{str} \mathcal{V}^2 = \deg \frac{\delta c_\delta(V) + 2c_1(V)c_{\delta-1}(V)}{12}.$$

This identity was proved in [LibW].

In [DZh3] this construction of the Virasoro algebra is considerably generalized. Dubrovin and Zhang produce the Virasoro algebra starting with any quadruple $(H, g, \mathcal{V}, K)$ where H is a finite–dimensional vector (super)space endowed with a non–degenerate quadratic form g, $\mathcal{V}: H \to H$ is an antisymmetric linear operator, and $K = K_1 \oplus K_2 \oplus \ldots : H \to H$ is an operator such that $[K_m, \mathcal{V}] = mK_m$ for all $m \geq 1$. If M is a Frobenius manifold with trivial local system of flat vector fields, one can take for H global sections of this system and define $\mathcal{V}$ as in II.(1.2). The construction of K requires additional conditions on M. If M is of qc–type (see III.5.4), one should take for K the cup multiplication by the H^2–component of E.

Moreover, Dubrovin and Zhang, again extending [EHX2], show how to define the generators L_n with $n < -1$ in the case when the spectrum of $\mathcal{V}$ does not contain half–integers. In this way one gets the full Virasoro algebra with the central charge str Id_H.

The main motivation for introducing the Virasoro operators (2.10)–(2.13) is the following

2.7. Conjecture. $L_n(e^F) = \overline{L}_n(e^F) = 0$ *for all* $n \geq -1$.

Only partial results in this direction are known at the time of writing. Here is the list of the most important particular cases.

(i) For the case $V =$ a point, 2.7 was conjectured by Witten and proved by Kontsevich in [Ko1]. In III.6 we discussed the natural generalization of the function (2.7) which, after a coordinate change, becomes an averaged character of the rank one invertible CohFT's considered as a function on the subspace of the appropriate moduli space. It would be important to compute this function on the total moduli space and extend the Virasoro constraints appropriately. This problem has a natural generalization to arbitrary V. Some of the characteristic numbers appearing in this context were recently calculated in [GePa], [FabP], [Pa3].

One obstacle which hinders the extension of Kontsevich's proof to the quantum cohomology case is that he uses a non–algebraic cell–decomposition of the moduli spaces of stable curves whose counterpart for stable maps is not known. For this and other reasons, an algebraic geometric proof of the Witten–Kontsevich theorem would be very desirable.

(ii) The next case involves general V but only the maps of the class $\beta = 0$, that is, maps to a point in V. Let us show first that the Virasoro constraints admit a meaningful restriction to this case.

One can directly check that $L_n, \overline{L}_n$ are *homogeneous operators of bidegree* $(-n, -n)$ in the sense of (2.8). To do this, one must use (2.8) and the homogeneity of the Poincaré form and canonical class:

$$g_{ab} \neq 0 \Rightarrow p_a + p_b = q_a + q_b = \dim V, \ (K^r)_a^b \neq 0 \Rightarrow p_b - p_a = q_b - q_a = r.$$

Now, since L_n and $\overline{L}_n$ do not depend on q^β, it makes sense to put $F = \sum_{k \in \mathbb{Z}} F^{(k)}$ where $F^{(k)}$ involves only terms with $(-K_V, \beta) = k$, and similarly for $Z = e^F$, $Z = \sum_{k \in \mathbb{Z}} Z^{(k)}$.

Then Conjecture 2.7 is equivalent to the separate vanishing of all $L_n(Z^{(k)})$ and $\overline{L}_n(Z^{(k)})$.

Consider the case when, say, $-K_V$ is ample. The $k = 0$ terms of $F^{(0)}$ then consist of $\beta = 0$ summands of (2.7). So the equations we get from the conjecture are:

(2.15)
$$(?) \quad L_n(e^{F^{(0)}}) = 0, \ L_n(e^{F^{(0)}} F^{(1)}) = 0, \ L_n\left(e^{F^{(0)}} \left(\frac{(F^{(1)})^2}{2} + F^{(2)}\right)\right) = 0, \ \ldots$$

and similarly for $\overline{L}_n$. The first series of these equalities is the reduction we promised.

In 6.3 below we will list all non–vanishing correlators with $\beta = 0$. If $\dim V \geq 4$, they appear only in genus zero. Since the genus zero case can be treated separately, it remains to consider only curves, surfaces, and threefolds. For a detailed discussion and reduction of this case to some identities between characteristic numbers, partly proved, partly still conjectural, see [GePa]. It is interesting that the relevant numbers constitute a part of those that were mentioned in (i) and III.6.

(iii) The Virasoro constraints for L_1 (E. Witten) and L_0 (E. Witten and K. Hori) can now be proved mathematically using the properties of the virtual fundamental classes established in §3 and §4: see §5 below.

If the case $n = 2$ could be treated similarly, the rest would also follow since (the half of) the Virasoro algebra is generated by $L_{-1}, L_0,$ and L_2. Hence the conjectural formulas $L_2(e^F) = \overline{L}_2(e^F) = 0$ hide a host of universal relations between the physical correlators, which are not at all understood geometrically.

The main source of the universal relations is the analysis of the degenerations of the stable maps, that is, the geometry of the moduli stacks $\overline{M}(V, \tau)$ where τ is a $B(V)$-marked modular graph. These stacks are interconnected by many morphisms described in Chapter V, of which the most important are forgetful morphisms (of marked points), boundary morphisms, and absolute stabilization morphisms. The graphs of these morphisms determine correspondences which together with the tautological classes $c_1(L_i)$ satisfy various motivic identities.

(iv) Finally, for any fixed g, the Virasoro constraints can be meaningfully reduced to the genera $\leq g$ and proved for genus 0 ([EHX2], [LiuT], [Ge6]) and, with a semisimplicity restriction, for genera ≤ 1 ([DZh3]). We will restrict ourselves here to the brief explanation of this reduction, following [Ge6]. Put

$$e^{-F}L_n(e^F) = \sum_g z_{n,g}\lambda^{2g-2}.$$

Then the straightforward calculation shows that $z_{n,g}$ can be expressed as a formal series in the derivatives of F_h, $h \leq g$, and of s_d^a ([Ge6], formula (17)). Hence it makes sense to say that the Virasoro constraints are valid up to genus g.

§3. Correlators and forgetful maps

3.1. Virtual classes with descendants. We start by extending the formalism of correlators to arbitrary combinatorial types of maps. Let σ be a stable $B(V)$-marked modular graph. Since the tails of σ correspond to the labeled points of the curve whose combinatorial type is σ, the type of the gravitational descendants (former $d_1, \ldots, d_n$) is encoded by the additional marking $\mathbf{d} : S_\sigma \to \mathbf{Z}_{\geq 0} : j \mapsto d_j$. By analogy with (2.2), put

$$(3.1) \qquad J(V, \sigma, \mathbf{d}) := \left(\prod_{j \in S_\sigma} c_1(L_j(V, \sigma))^{d_j} \right) \cap J(V, \sigma) \in A_*(\overline{M}(V, \sigma)).$$

Here

$$L_j(V, \sigma) := x_j^*(\omega_\pi),$$

where ω_π is the relative dualizing sheaf of the universal curve $\pi : C(V, \sigma) \to \overline{M}(V, \sigma)$ and $x_j : \overline{M}(V, \sigma) \to C(V, \sigma)$, $j \in S_\sigma$, are the structure sections. We also write $\psi_j(V, \sigma)$ or simply ψ_j for $c_1(L_j(V, \sigma))$. These Chern classes lie in $A^1(\overline{M}(V, \sigma))$.

When σ has one vertex of genus g and class β, and S tails, we write $J_{g,S}(V,\beta,\mathbf{d})$ for (3.1) and similarly for other objects of interest.

If σ is in the stable range, we can also define the general counterpart of (2.2):

$$(3.2) \qquad I(V,\sigma^s,\mathbf{d}) \in A_*(V^{S_\sigma} \times \overline{M}(\sigma^s)).$$

It is the sum of the pushforwards of $J(V,\sigma,\mathbf{d})$ taken over all σ with the same absolute stabilization σ^s. For an exact definition, see [BehM], p. 58.

Taking (3.2) for granted, choose $\gamma \in A^*(V^{S_\sigma})$ or $H^*(V^{S_\sigma})$ where H^* is one of the standard cohomology theories extendable to the smooth DM–stacks so that it remains functorial with respect to the correspondences. Then we put

$$I(V,\sigma^s,\mathbf{d})(\gamma) := p_{2*}(p_1^*(\gamma) \cap I(V,\sigma^s,\mathbf{d}))$$

$$(3.3) \qquad = \sum_\sigma (\mathrm{ev},\mathrm{st})_*(\mathrm{ev}^*(\gamma) \cap J(V,\sigma,\mathbf{d})) \in A_*(\overline{M}(\sigma^s)),$$

where p_i are the two projections of $V^{S_\sigma} \times \overline{M}(\sigma^s)$. The zero–dimensional part of (3.3) for which we use several notations extends to all graphs in the stable range the definition of the physical correlators (2.5):

$$(3.4) \qquad \langle \tau_{\mathbf{d}} \gamma \rangle_{\sigma^s} = \langle I(V,\sigma^s,\mathbf{d})(\gamma) \rangle = \int_{[\overline{M}_{g,n}]} I(V,\sigma^s,\mathbf{d})(\gamma) = \deg I(V,\sigma^s,\mathbf{d})(\gamma),$$

where all these expressions are set equal to zero when the dimensions do not match.

The physical correlators can also be expressed through $J(V,\sigma)$ generally, i.e. even when σ is not in the stable range, as in (2.1), for algebraic arguments γ unconditionally, and for general ones under the condition that the relevant pair of homology/cohomology theories is extended to include $\overline{M}(V,\sigma)$:

$$(3.5) \qquad \langle \tau_{\mathbf{d}} \gamma \rangle_\sigma = \int_{J(V,\sigma,\mathbf{d})} \mathrm{ev}^*(\gamma) = \deg(\mathrm{ev}^*(\gamma) \cap J(V,\sigma,\mathbf{d})).$$

The last equality of (3.3) follows from one of the simplest cases of the projection formula. The standard projection formula in the smooth proper case $f_*(f^*(a)b) = af_*(b)$ has at least three non–equivalent versions for the paired theories A_*, A^* depending on where we put a,b, and even more if we include operational theories (cf. V.8). So to illustrate its first occurrence we will formalize our situation in the following simple lemma:

3.1.1. Lemma. *Let X, Y, Z be three proper DM–stacks, $f = (f_X, f_Y) : Z \to X \times Y$, $z \in A_*(Z)$, $x \in A^*(X)$. Then*

$$(3.6) \qquad \mathrm{pr}_{Y*}(\mathrm{pr}_X^*(x) \cap f_*(z)) = f_{Y*}(f_X^*(x) \cap z).$$

To check this, use $f_X = \mathrm{pr}_X \circ f$, $f_Y = \mathrm{pr}_Y \circ f$ and refer to V.(8.21) and V.(8.24). For applications to (3.3), put $z = J(V,\sigma,\mathbf{d})$, $f = (\mathrm{ev},\mathrm{st})$.

As we have already remarked in 1.3, the main difference between the classes J (or $[\overline{M}]^{virt}$) and I is that J are supported by the stacks of maps about which very little is known except for some simplest cases, whereas I provide an extra structure on the quite explicit spaces. So from this viewpoint it seems wiser to concentrate on I (or even physical correlators) and to treat J only as an auxiliary (and fairly difficult) construction.

On the other hand, the union of all $\overline{M}(V,\sigma)$ endowed with the canonical (forgetful, boundary, and stabilization) morphisms and the tautological classes

$$\psi_j,\ \mathrm{ev}^*(\gamma),\ [\overline{M}]^{virt}$$

looks as a sophisticated relative (over V) version of the stacks of stable curves. The intersection theory of the tautological classes appears to have many stable formal properties extending those of the tautological rings of $\overline{M}_{g,n}$ and largely insensitive to the unstable geometry of the carrier spaces.

We now turn to the elementary properties of the classes $J(V,\sigma,\mathbf{d})$ generalizing the discussion in §1.

3.2. S_n–symmetry. The permutation group of labels S acts on $\overline{M}_{g,S}(V,\beta)$, $V^S \times \overline{M}_{g,S}$, and markings $\mathbf{d}: S \to \mathbf{Z}_{\geq 0}$. Classes J and I are invariant with respect to this action. For the physical correlators (2.1) the usual signs arise formally.

More generally, isomorphisms of $(B(V),\mathbf{d})$-marked graphs induce the isomorphisms of correlators.

This is a part of the functoriality of J with respect to graphs.

3.3. Dimension. If $x \in A_m(X)$ (resp. $A^m(X)$), we write $\dim x = m$ (resp. $\mathrm{codim}\, x = m$). We have

(3.7) $$\dim J(V,\sigma,\mathbf{d}) = \dim I(V,\sigma^s,\mathbf{d}) = \dim(V,\sigma) - \sum_{j \in S_\sigma} d_j,$$

where

$$\dim(V,\sigma) := \left(\chi(\|\sigma\|) - \sum_{v \in V_\sigma} g_v\right)(\dim V - 3)$$

(3.8) $$- \left(\sum_{v \in V_\sigma} \beta_v\right)(\omega_V) + |S_\sigma| - |E_\sigma|$$

is the virtual dimension of the boundary stratum $\overline{M}(V,\sigma)$, $\|\sigma\|$ being the geometric realization of σ.

This follows from (3.1): proper pushforward does not change dimension. Hence

$$\dim(\mathrm{ev}^*(\gamma) \cap J(V,\sigma,\mathbf{d})) = \dim I(V,\sigma^s,\mathbf{d})(\gamma)$$
(3.9) $$= \dim(V,\sigma) - \sum_{j \in S_\sigma} d_j - \mathrm{codim}\,\gamma.$$

In particular, $\langle \tau_\mathbf{d}\gamma\rangle_\sigma$ can be non–zero only if

(3.10) $$\dim(V,\sigma) = \sum_{j \in S_\sigma} d_j + \mathrm{codim}\,\gamma.$$

The remaining part of this section is devoted to the behavior of the classes $J(V,\sigma,\mathbf{d})$ with respect to the forgetful morphisms.

3.4. Commuting forgetful maps. Let $S \cup \{l, m\}$ be a finite set. Consider the diagram

(3.11)
$$\begin{array}{ccc} \overline{M}_{g,S\cup\{l,m\}}(V,\beta) & \xrightarrow{\mu} & \overline{M}_{g,S\cup\{l\}}(V,\beta) \\ \lambda \downarrow & & \downarrow \lambda' \\ \overline{M}_{g,S\cup\{m\}}(V,\beta) & \xrightarrow{\mu'} & \overline{M}_{g,S}(V,\beta) \end{array}$$

in which λ, λ' stably forget l, whereas μ, μ' stably forget m. (The genus expansion parameter denoted also by λ in (2.7) will not appear in this section.) This diagram is commutative. This can be checked by appealing to the explicit projective construction of stably forgetful maps in V.4; cf. [BehM] for more general statements about functoriality with respect to graph morphisms.

However, this diagram generally is not Cartesian. In fact, the morphism of $\overline{M}_{g,S\cup\{l,m\}}(V,\beta)$ to the respective fibered product is a blow up. We will always assume that $\overline{M}_{g,S}(V,\beta)$ is non-empty, so that the other stacks are non-empty as well.

The basic fact about any forgetful morphism, say, λ' is that it is canonically isomorphic to the universal curve morphism

$$C_{g,S}(V,\beta) \to \overline{M}_{g,S}(V,\beta)$$

(see V.4.5). In particular, it is flat, and admits the dualizing sheaf $\omega_{\lambda'}$ compatible with base changes. Moreover, this dualizing sheaf is invertible (cf. V.2.4.1).

Furthermore, λ' possesses canonical sections labeled by S, which we now denote

$$\lambda_j'^{-1} : \overline{M}_{g,S}(V,\beta) \to \overline{M}_{g,S\cup\{l\}}(V,\beta), \; j \in S.$$

For one-vertex graphs with $S = \{1, \ldots, n\}$ we have

(3.12) $$L_{j;g,S} = (\lambda_j'^{-1})^*(\omega_{\lambda'}).$$

The Chern classes of these invertible sheaves can be represented by Cartier divisors and have good functorial and operational properties.

We will apply similar notation to any arrow in (3.11).

3.5. Lemma. *For $j \in S$, denote by D_j the divisor on $\overline{M}_{g,S\cup\{m\}}(V,\beta)$ represented by the j-th section of μ'. Then there is a canonical isomorphism*

(3.13) $$(\mu')^* \circ (\lambda_j'^{-1})^*(\omega_{\lambda'})(D_j) \cong (\lambda_j^{-1})^*(\omega_\lambda).$$

Proof. First of all, we have:

(3.14) $$\lambda_j'^{-1} \circ \mu' = \mu \circ \lambda_j^{-1}$$

as morphisms $\overline{M}_{g,S\cup\{m\}}(V,\beta) \to \overline{M}_{g,S\cup\{l\}}(V,\beta)$. In fact, $\lambda_j'^{-1}$ can be canonically identified with the special boundary morphism

$$b(V,\tau) : \overline{M}(V,\tau) \to \overline{M}_{g,S\cup\{l\}}(V,\beta),$$

where τ is the following two-vertex one-edge marked modular graph: one vertex is labeled by (g,β) and carries the tails $S \setminus \{j\}$, another vertex is labeled $(0,0)$ and carries the tails $\{j,l\}$ (cf. V.4.7.1). One can similarly describe λ_j^{-1}. Then (3.14) becomes a particular case of the statement that forgetful maps commute with the boundary maps; see [BehM].

From (3.14) we see that

$$(3.15) \qquad (\mu')^* \circ (\lambda_j'^{-1})^*(\omega_{\lambda'}) = (\lambda_j^{-1})^* \circ \mu^*(\omega_{\lambda'}).$$

We will now construct an isomorphism

$$(3.16) \qquad \mu^*(\omega_{\lambda'}) \cong \omega_\lambda \left(-\sum_{k \in S} E_k\right),$$

where E_k, $k \in S$, is the boundary divisor corresponding to the two–vertex one–edge marked modular graph whose one vertex is labeled by (g, β) and carries the tails $S \setminus \{k\}$, whereas another vertex is labeled by $(0, 0)$ and carries the tails $\{k, l, m\}$. Below we will use for such a graph the notation $((g, \beta; S \setminus \{k\}), (0, 0; k, l, m))$.

In fact, we have the canonical map $\mu^*(\omega_{\lambda'}) \to \omega_\lambda$. It maps a local section of $\omega_{\lambda'}$ which is a λ'-vertical differential form with singularities to its lift as a λ-vertical form. The lift will have a first order zero exactly on the divisor spanned by those fibers of λ which contain a component $\mathbf{P}^1$ contracted by μ. Looking at λ as the universal curve morphism, we see that the λ-image of this divisor on $\overline{M}_{g, S \cup \{m\}}(V, \beta)$ is the sum of boundary divisors $((g, \beta; S \setminus \{k\}), (0, 0; k, m))$ taken over all $k \in S$. Its inverse image on $\overline{M}_{g, S \cup \{l, m\}}(V, \beta)$ is then the sum of boundary divisors $((g, \beta; S \setminus \{k\}), (0, 0; k, l, m))$ and $((g, \beta; S \setminus \{k\} \cup \{l\}), (0, 0; k, m))$ over all k. The morphism μ stably forgetting m contracts precisely the first summands. This proves (3.16).

Now we want to apply $(\lambda_j^{-1})^*$ to the right hand side of (3.16).

For $k \neq j$, the image of λ_j^{-1} does not intersect with E_k. In fact, after stably forgetting m, they become boundary divisors on $\overline{M}_{g, S}(V, \beta)$ corresponding to the graphs $((g, \beta; S \setminus \{j\}), (0, 0; j, l))$ and $((g, \beta; S \setminus \{k\}), (0, 0; k, l))$ respectively. But these two divisors are different structure sections of the morphism stably forgetting l, considered as the universal curve. Hence they are disjoint. Thus, combining (3.14), (3.15), and (3.16), we have

$$(3.17) \qquad (\mu')^* \circ (\lambda_j'^{-1})^*(\omega_{\lambda'}) \cong (\lambda_j^{-1})^*(\omega_\lambda(-E_j)).$$

To finish the proof of (3.13), it remains to establish the isomorphism

$$(3.18) \qquad (\lambda_j^{-1})^*(\mathcal{O}(-E_j)) \cong \mathcal{O}(-D_j).$$

Now, λ_j^{-1} is the boundary morphism corresponding to the graph

$$((g, \beta; S \setminus \{j\} \cup \{m\}), (0, 0; j, l)),$$

and E_k is a boundary divisor as well. These boundary divisors intersect transversally. The intersection is represented by the boundary morphism to $\overline{M}_{g, S \cup \{l, m\}}(V, \beta)$ corresponding to a three–vertex two–edge marked modular graph. The middle vertex is labeled by $(0, 0; m)$, the remaining two are $(g, \beta; S \setminus \{j\})$ and $(0, 0; j, l)$. Viewed as a boundary morphism to $\overline{M}_{g, S \cup \{m\}}(V, \beta)$ (identified with the j-th section), this stratum becomes the divisor $((g, \beta; S \setminus \{j\}), (0, 0; j, m))$ (because we must stably forget l). This is just D_j.

3.6. Relations in the Chow groups I. Now we put for $j \in S$:

$$\psi_j := c_1((\lambda_j^{-1})^*\omega_\lambda),$$

$$\psi_j' := c_1((\lambda_j'^{-1})^*\omega_{\lambda'}),$$

$$\chi_j := (\mu')^*\psi_j' = c_1((\mu')^* \circ (\lambda_j'^{-1})^*\omega_{\lambda'}).$$

These Chern classes are elements of the operational Chow rings A^1 (see V.8.6). From (3.13) it follows that

(3.19) $$\psi_j = \chi_j + [D_j].$$

Notice that since D_j is a Cartier divisor, it is c_1 of an invertible sheaf, and hence its class $[D_j]$ makes sense in A^1. Each arrow in (3.11) is representable, proper, and flat, hence it defines the functorial pullback and pushforward on the homological Chow groups.

3.6.1. Proposition. *We have*

(3.20) $$\mu'_\bullet\left(\prod_{j\in S}\psi_j^{d_j}\right) = \sum_{k\in S,\, d_k\geq 1}\prod_{j\in S}(\psi_j')^{d_j-\delta_{kj}},$$

where $\mu'_\bullet$ is the proper flat pushforward defined in V.8.9.

Proof. The restriction of $(\lambda_j^{-1})^*\omega_\lambda$ onto D_j is trivial. In fact, the geometric fiber of this sheaf at the point $[(C; x_k\,|\,k \in S \cup \{m\}; f)]$ is $T_{x_j}^*C$. When this point belongs to D_j, C contains the $\mathbf{P}^1$–component with exactly three special points, x_j, x_m, and the intersection point $*$ with the rest of the curve. This component is contracted by f. Hence we can algebraically trivialize our vector bundle over D_j, e.g., by identifying $(x_j, x_m, *)$ with $(0, 1, \infty)$. As a corollary, we have $\psi_j.[D_j] = 0$, so that

(3.21) $$\forall d \geq 1,\; \psi_j^d = \chi_j^d + [D_j].\chi_j^{d-1}.$$

Now, $[D_j].[D_k] = 0$ for $j \neq k$. Hence

$$\mu'_\bullet\left(\prod_{j\in S}\psi_j^{d_j}\right) = \mu'_\bullet\left(\prod_{j\in S,\,d_j\geq 1}(\chi_j^{d_j} + [D_j].\chi_j^{d_j-1})\right)$$

$$= \mu'_\bullet\left(\sum_{k\in S,\,d_k\geq 1}[D_k]\prod_{j\in S}\chi_j^{d_j-\delta_{kj}}\right) = \mu'_\bullet\left(\sum_{k\in S,\,d_k\geq 1}[D_k]\mu'^*\left(\prod_{j\in S}(\psi_j')^{d_j-\delta_{kj}}\right)\right)$$

$$= \sum_{k\in S,\,d_k\geq 1}\mu'_\bullet([D_k])\prod_{j\in S}(\psi_j')^{d_j-\delta_{kj}}.$$

In the last equality we used V.(8.29). Applying the stack version of the simple case of the excess intersection formula ([Ful], Example 6.3.4, p. 105, with $\text{rk}\, E = 1$ and $f = \text{id}$ so that $i^! = i^*$), we see that

$$[D_k] \cap = \mu'^{-1}_{k*} \circ (\mu'^{-1}_k)^*.$$

Moreover, $\mu'_k \circ \mu'^{-1}_k = \text{id}$. It follows that $\mu'_\bullet([D_k]) = \text{id}$ which implies (3.20).

3.7. Relations in the Chow groups II. In the left hand side of (3.20), we can consider more general monomials in ψ–classes, involving $\psi_m = c_1((\lambda_m^{-1})^*(\omega_\lambda))$ as well. We start with the following identity:

3.7.1. Proposition. *If $d_m \geq 1$, we have*

$$(3.22) \qquad \psi_m^{d_m} \prod_{j \in S} \psi_j^{d_j} = \psi_m^{d_m} \prod_{j \in S} \chi_j^{d_j} = \psi_m^{d_m} \mu'^* \left(\prod_{j \in S} \psi_j'^{d_j} \right).$$

For the proof we need the following modification of Lemma 3.5:

3.7.2. Lemma. *We have*

$$(3.23) \qquad \omega_{\mu'}\left(\sum_{j \in S} D_j \right) \cong (\lambda_m^{-1})^*(\omega_\lambda).$$

Proof. We will apply $(\lambda_m^{-1})^*$ to (3.16). First of all, the map

$$\mu \circ \lambda_m^{-1} : \overline{M}_{g,S\cup\{m\}}(V,\beta) \to \overline{M}_{g,S\cup\{l\}}(V,\beta)$$

is a relabeling isomorphism: $x_m \mapsto x_l$, $x_j \mapsto x_j$ for $j \in S$. Hence

$$(\lambda_m^{-1})^* \circ \mu^*(\omega_{\lambda'}) = \omega_{\mu'}.$$

It remains to show that for all $k \in S$

$$(3.24) \qquad (\lambda_m^{-1})^*(\mathcal{O}(-E_k)) \cong \mathcal{O}(-D_k).$$

The argument is similar to that in the proof of (3.18). The morphism λ_m^{-1} is the boundary morphism corresponding to the graph $((g,\beta;S),(0,0;m,l))$, and E_k is a boundary divisor as well. These boundary divisors intersect transversally. The intersection is represented by the boundary morphism to $\overline{M}_{g,S\cup\{l,m\}}(V,\beta)$ corresponding to a three–vertex two–edge marked modular graph. The middle vertex is labeled by $(0,0;k)$, the remaining two are $(g,\beta;S\setminus\{k\})$ and $(0,0;m,l)$. Viewed as a boundary morphism to $\overline{M}_{g,S\cup\{m\}}(V,\beta)$ (identified with the m–th section), this stratum becomes the divisor $((g,\beta;S\setminus\{k\}),(0,0;k,m))$ (because we must stably forget l). This is just D_k.

3.7.3. Deduction of (3.22). From (3.23) we get

$$(3.25) \qquad \psi_m = c_1(\omega_{\mu'}) + \sum_{k \in S}[D_k].$$

The residue map identifies $\omega_{\mu'}(D_j)$ restricted to D_j with $\mathcal{O}_{D_j}$. It follows that $(c_1(\omega_{\mu'}) + [D_j]).[D_j] = 0$ and furthermore $\psi_m.[D_j] = 0$ for all $j \in S$. Now, using (3.24), we get for $d_m \geq 1$:

$$\psi_m^{d_m} \prod_{j \in S} \psi_j^{d_j} = \psi_m^{d_m} \prod_{j \in S, d_j \geq 1} (\chi_j^{d_j} + [D_j].\chi_j^{d_j - 1}) = \psi_m^{d_m} \prod_{j \in S} \chi_j^{d_j}.$$

3.8. Theorem. *a) In the situation of 3.4, let* **d** *be a map* $S \cup \{m\} \to \mathbf{Z}$ *with* $d_m = 0$, $\gamma \in A^*(V^S)$ *(or in* $H^*(V^S)$*), and* $\gamma \otimes 1$ *the pullback of* γ *with respect to the projection* $V^{S \cup \{m\}} \to V^S$. *Then we have*

$$(3.26) \quad \mu'_*(\mathrm{ev}^*(\gamma \otimes 1) \cap J_{g,S \cup \{m\}}(V,\beta,\mathbf{d})) = \sum_{k \in S, d_k \geq 1} \gamma \cap J_{g,S}(V, \beta, \mathbf{d} \,|\, S - \delta_k).$$

Here $\mathbf{d} \,|\, S$ is the restriction of $\mathbf{d}$ to S, and δ_k is the function equal to 1 at k and 0 elsewhere.

b) Similarly, if $d_m = 1$,

$$(3.27) \quad \mu'_*(\gamma \cap J_{g,S\cup\{m\}}(V,\beta,\mathbf{d})) = (2g - 2 + |S|)\gamma \cap J_{g,S}(V,\beta, \mathbf{d} \,|\, S).$$

Proof. Put

$$\alpha = \mathrm{ev}^*(\gamma) \cap J_{g,S}(V,\beta), \quad \gamma \in A^*(V^S).$$

The composition $\mathrm{ev} \circ \mu'$ coincides with the evaluation map followed by projection which we denote

$$\mathrm{ev}' : \overline{M}_{g,S\cup\{m\}}(V,\beta) \to V^S.$$

In view of (1.14), we also have

$$\mu'^*(J_{g,S}(V,\beta)) = J_{g,S\cup\{m\}}(V,\beta).$$

Since the cap multiplication commutes with the flat pullback, we have

$$\mu'^*(\alpha) = \mathrm{ev}'^*(\gamma) \cap J_{g,S\cup\{m\}}(V,\beta).$$

Taking the cap product of (3.20) with α we get

$$\mu'_*\left(\prod_{j \in S} \psi_j^{d_j} \, \mathrm{ev}'^*(\gamma) \cap J_{g,S\cup\{m\}}(V,\beta)\right)$$

$$= \sum_{k \in S, d_k \geq 1} \prod_{j \in S} (\psi'_j)^{d_j - \delta_{kj}} \mathrm{ev}^*(\gamma) \cap J_{g,S}(V,\beta)$$

which is (3.26).

Turning now to (3.27), we first obtain from (3.22):

$$(3.28) \quad \psi_m^{d_m} \prod_{j \in S} \psi_j^{d_j} \cap \mu'^*(\alpha) = \psi_m^{d_m} \cap \mu'^*\left(\prod_{j \in S} \psi_j'^{d_j} \cap \alpha\right).$$

Put $d_m = 1$ here. From (3.25) one sees that ψ_m can be represented as the class of the divisor each component D of which is a closed substack of $\overline{M}_{g,S\cap\{m\}}(V,\beta)$ whose projection to $\overline{M}_{g,S}(V,\beta)$ is flat and finite, of degree equal the intersection index of D with the generic fiber of μ'. Let $i : D \to \overline{M}_{g,S\cap\{m\}}(V,\beta)$ be the closed embedding. Arguing as in the end of the proof of (3.20), we find:

$$\mu'_*\left([D] \cap \prod_{j \in S} \psi_j'^{d_j} \cap \mu'^*(\alpha)\right) = (\mu' \circ i)_*(\mu' \circ i)^*\left(\prod_{j \in S} \psi_j'^{d_j} \cap \alpha\right)$$

$$(3.29) \qquad\qquad = \deg(\mu' \circ i) \prod_{j \in S} \psi_j'^{d_j} \cap \alpha.$$

The degree of $\mu' \circ i$ equals the intersection index of D with any fiber of μ'. Therefore if we replace $[D]$ in the left hand side of (3.29) by ψ_m, the coefficient in the right hand side will become $2g - 2 + |S|$ (cf. (3.25)).

The theorem we have just proved extends the Identity Axiom of quantum cohomology from III.5.3. The next result extends the Divisor Axiom.

3.9. Theorem. *As above, let* $\mathbf{d}$ *be a map* $S \cup \{m\} \to \mathbf{Z}$ *with* $d_m = 0$, $\gamma \in A^*(V^S)$ *(or in* $H^*(V^S)$*),* $\gamma_m \in A^1(V)$ *(or* $H^2(V)$*). Then we have*

$$(3.30) \qquad \mu'_*(\mathrm{ev}^*(\gamma \otimes \gamma_m) \cap J_{g,S\cup\{m\}}(V,\beta,\mathbf{d}))$$
$$= (\gamma_m, \beta) J_{g,S}(V,\beta,\mathbf{d}|S) + \sum_{k\in S, d_k \geq 1} \gamma \cup \mathrm{pr}_k^*(\gamma_m) \cap J_{g,S}(V,\beta,\mathbf{d}|S - \delta_k),$$

where pr_k *is the projection of* V^S *to the* k*-th factor.*

Proof. Put $\mathrm{ev}_j = \mathrm{pr}_j \circ \mathrm{ev}$. Arguing as in the proof of (3.20) and (3.22), we have for $\gamma_m \in A^1(V), \alpha \in A_*(\overline{M}_{g,S\cup\{m\}}(V,\beta))$

$$\mu'_*\left(\mathrm{ev}_m^*(\gamma_m) \prod_{j\in S} \psi_j^{d_j} \cap \mu'^*(\alpha)\right)$$

$$= \mu'_*\left(\mathrm{ev}_m^*(\gamma_m) \prod_{j\in S} (\chi_j^{d_j} + [D_j].\chi_j^{d_j-1}) \cap \mu'^*(\alpha)\right)$$

$$= \mu'_*\left(\mathrm{ev}_m^*(\gamma_m) \prod_{j\in S} \chi_j^{d_j} \cap \mu'^*(\alpha)\right)$$

$$(3.31) \qquad + \mu'_*\left(\sum_{k\in S, d_k \geq 1} \mathrm{ev}_m^*(\gamma_m)[D_k] \prod_{j\in S} \chi_j^{d_j - \delta_{kj}} \cap \mu'^*(\alpha)\right).$$

The first summand in the last expression of (3.31) can be treated as in the previous subsection. We will get an expression similar to the last one in (3.29), with the coefficient (γ_m, β) before the product. This will lead to the first term of (3.30).

Furthermore, identifying the section morphisms with the boundary morphisms as in 3.5, we see that

$$(3.32) \qquad \mathrm{ev}'_m \circ \mu_k'^{-1} = \mathrm{ev}_k,$$

where this time ev'_m means the m-th evaluation morphism of $\overline{M}_{g,S\cup\{m\}}(V,\beta)$.

Hence

$$(3.33) \qquad \mu_k'^{-1*} \circ \mathrm{ev}_m'^*(\gamma_m) = \mathrm{ev}_k^*(\gamma_m).$$

Eliminating $[D_k]$ as in the proof of Proposition 3.6.1, we can therefore rewrite the k-th summand in the last sum in (3.31) in the following way:

$$\mu'_* \circ \mu_{k*}'^{-1} \circ (\mu_k'^{-1})^* \left(\mathrm{ev}_m'^*(\gamma_m) \mu'^* \left(\prod_{j\in S} \psi_j'^{d_j} \cap \alpha\right)\right)$$

$$\tag{3.34} = \mathrm{ev}_k^*(\gamma_m) \prod_{j \in S} \psi_j'^{d_j} \cap \alpha.$$

As above, this will specialize to the k–th summand of (3.30).

3.10. Relations in the Chow group III. In this subsection, we study relations involving the analogs of the Mumford–Morita–Miller classes, modified as in [AC1]. The basic classes are defined in the Chow ring of any space $\overline{M}_{g,S}(V,\beta)$ as a pushforward from the respective universal curve, or, as in diagram (3.11), along any one of the morphisms λ', μ'. We choose the vertical arrows:

$$\tag{3.35} \kappa_d' := \kappa_{d;g,S}(V,\beta) := \lambda'_\bullet \left((c_1(\omega_{\lambda'}) + \sum_{j \in S}[F_j'])^{d+1} \right),$$

where $F_j', j \in S$, denote the sections of λ'. Similarly

$$\tag{3.36} \kappa_d := \kappa_{d;g,S\cup\{m\}}(V,\beta) := \lambda_\bullet \left((c_1(\omega_\lambda) + \sum_{j \in S\cup\{m\}}[F_j])^{d+1} \right),$$

where F_j are the sections of λ.

Now let $\overline{M}$ be the fibered product of the morphisms λ' and μ' in (3.11), p_λ, p_μ the respective projections, and $\nu : \overline{M}_{g,S\cup\{l,m\}}(V,\beta) \to \overline{M}$ the canonical morphism such that $\lambda = p_\lambda \circ \nu$, $\mu = p_\mu \circ \nu$.

3.10.1. Proposition. *Assume that $\nu_* \nu^* = \mathrm{id}$ on $A_*(\overline{M})$. Then*

$$\tag{3.37} \kappa_d \cap = (\mu'^*(\kappa_d') + \psi_m'^d) \cap$$

as operations on $A_*(\overline{M}_{g,S\cup\{m\}}(V,\beta))$.

3.10.2. Lemma. *We have*

$$\tag{3.38} \mu^* \left(\omega_{\lambda'}(\sum_{j \in S} F_j') \right) \cong \omega_\lambda(\sum_{j \in S} F_j).$$

(Notice the absence of F_m on the right hand side.)

Proof. This will follow from (3.16), if we check that

$$\tag{3.39} \mu^*(\mathcal{O}(F_j')) \cong \mathcal{O}(F_j + E_j).$$

But in fact, $\mu^{-1}(F_j') = F_j + E_j$. This can be seen from the identification of the sections with boundary morphisms as in the proof of Lemma 3.5. Geometrically, on a curve over a generic point of F_j' the new point labeled m can be put either on the $(g, \beta; S \setminus \{j\})$–component, which gives F_j, or on the $(0,0;j,l)$–component, which gives E_j.

Proof of (3.37). As in 3.7.3, we have $(c_1(\omega_\lambda) + \sum_{j \in S\{m\}}[F_j]).[F_m] = 0$. Induction on d and (3.38) then show that

$$(c_1(\omega_\lambda) + \sum_{j \in S\cup\{m\}}[F_j])^{d+1} = (c_1(\omega_\lambda) + \sum_{j \in S}[F_j])^{d+1} + c_1(\omega_\lambda)^d.[F_m]$$

$$(3.40) \qquad = \mu^* \left((c_1(\omega_{\lambda'}) + \sum_{j \in S} [F_j''])^{d+1} \right) + c_1(\omega_\lambda)^d \cdot [F_m].$$

Applying $\lambda_\bullet$ we find:

$$(3.41) \qquad \kappa_d = \lambda_\bullet \mu^* \left((c_1(\omega_{\lambda'}) + \sum_{j \in S} [F_j''])^{d+1} \right) + \psi_m'^d.$$

Moreover,
$$\lambda_\bullet \mu^* = p_{\lambda\bullet} \nu_\bullet \nu^* p_\mu^*.$$

Considering the first sum on the right hand side of (3.41) as an operation on the covariant Chow group, we can first omit $\nu_\bullet \nu^*$ (cf. V.(8.32)), then replace $p_{\lambda\bullet} p_\mu^*$ with $\mu'^* \lambda'_\bullet$ (cf. V.(8.30)). In view of (3.35), we will obtain $\mu'^*(\kappa_d') \cap$.

§4. Correlators and boundary maps

4.1. The setting. In this section we will study the relations between the correlators following from the linear relations between the classes of boundary strata in $\overline{M}_{g,n}$.

The resulting relations between the generating series include the Associativity Equations in genus zero, Getzler's relation in genus one ([Ge4]), and many more differential equations whose total potential is not yet explored.

We will first consider the correlators without gravitational descendants. The computations can be readily generalized to the case of modified correlators with descendants, defined by the version of the formula (2.1) which involves Gromov–Witten correspondences and classes $c_1(L_i)$ on $\overline{M}_{g,n}$ rather than virtual fundamental classes and $c_1(L_i)$ on $\overline{M}_{g,n}(V, \beta)$ (see (6.5) below). In §6 and §7 we will explain how to pass back and forth between the modified correlators and the initial ones, and extend the main theorem of this section to the large phase space.

This section extends parts of III.4 to arbitrary genera. In order to stress this, we introduce a version of the notation employed there. Let $I_{g,S;\beta} : H(V)^{\otimes S} \to H^*(\overline{M}_{g,S})$ denote the Gromov–Witten maps. Put $I_{g,S} = \sum_{\beta \in B(V)} q^\beta I_{g,S;\beta}$. The target of this map is the cohomology of $\overline{M}_{g,S}$ with coefficients in the Novikov ring Λ. We now denote the respective correlators by

$$(4.1) \qquad Y_{g,S}\left(\bigotimes_{i \in S} \gamma_i\right) = \int_{\overline{M}_{g,S}} I_{g,S}\left(\bigotimes_{i \in S} \gamma_i\right),$$

where $\gamma_i \in H^*(V)$. In the case $S = \{1, \ldots, n\}$, we can write as well

$$(4.2) \qquad Y_{g,n}(\gamma_1 \otimes \cdots \otimes \gamma_n) = \langle \gamma_1 \ldots \gamma_n \rangle_g := \sum_{\beta \in B(V)} q^\beta \langle \gamma_1 \ldots \gamma_n \rangle_{g,\beta},$$

where the correlators $\langle \ldots \rangle_{g,n}$ are defined by (2.1) with $d_1 = \cdots = d_n = 0$.

More generally, let σ be a stable modular graph with the set of tails (labeled by) S. We then put

$$(4.3) \qquad Y(\sigma)\left(\bigotimes_{i \in S} \gamma_i\right) = \int_{\widetilde{M}(\sigma)} I_{g,S}\left(\bigotimes_{i \in S} \gamma_i\right),$$

where $\widetilde{M}(\sigma)$ was described in V.7.1.4.

4.2. Proposition. *We have*

$$(4.4) \quad Y(\sigma)\left(\bigotimes_{i \in S} \gamma_i\right) = \frac{1}{|\mathrm{Aut}\,\sigma|}\left(\bigotimes_{v \in V_\sigma} Y_{g_v, F_\sigma(v)}\right)\left(\bigotimes_{i \in S} \gamma_i \otimes \Delta^{\otimes E_\sigma}\right).$$

Proof. This is the generalization of the formula III.(4.9). It can be checked by combining V.(7.4) and Proposition 1.2.1 of this chapter.

4.2.1. Corollary. *For any linear relation*

$$(4.5) \quad \sum_j c_j [\widetilde{M}(\sigma_j)] = 0$$

in the Chow ring of $\overline{M}_{g,S}$ *or, less restrictively, for the similar relation in the (co)homology group, we have*

$$(4.6) \quad \sum_j \frac{c_j}{|\mathrm{Aut}\,\sigma_j|}\left(\bigotimes_{v \in V_{\sigma_j}} Y_{g_v, F_{\sigma_j}(v)}\right)\left(\bigotimes_{i \in S} \gamma_i \otimes \Delta^{\otimes E_{\sigma_j}}\right) = 0.$$

4.3. Equations for generating functions.
Following [Ge4] we will now explain how to process any single relation (4.5) into a system of non–linear differential equations for the potentials $\Phi_h, h \leq g$. Here Φ_g is the restriction of F_g from (2.7) to the small phase space:

$$(4.7) \quad \Phi_g(x) = \langle e^\gamma \rangle_g = \sum_{N=0}^\infty \sum_{(a_i) \in A^N} \frac{x^{a_N} \ldots x^{a_1}}{N!} \langle \Delta_{a_1} \ldots \Delta_{a_N} \rangle_g.$$

We denote by $x^a, a \in A$, the coordinates on $H^*(V)$ as in 2.4, $\gamma = \sum_a x^a \Delta_a$.

To devise the more mnemonic notation, we first rewrite (4.7) in the form:

$$(4.8) \quad \Phi_g(x) = \left(\sum_{N=0}^\infty \int_{[\widetilde{M}_{g,N}]} I_{g,N}\right)\left(e^{\sum x^a \Delta_a}\right).$$

By our convention, a pair of summands taken from the two inner sums can provide a non–zero contribution only if their multi–indices match as in (4.7).

Now let σ be any stable (g, n)–modular graph and $\mathbf{a} = (a_1, \ldots, a_n) \in A^n$ an additional labeling of its tails. Put

$$(4.9) \quad \Phi_{\sigma,\mathbf{a}}(x) = \left(\sum_{N=0}^\infty \int_{\pi_N^*[\widetilde{M}(\sigma)]} I_{g,n+N}\right)\left(\Delta_{a_1} \ldots \Delta_{a_n} e^{\sum x^a \Delta_a}\right),$$

where $\pi_N : \overline{M}_{g,n+N} \to \overline{M}_{g,n}$ is the morphism stably forgetting the last N marked points. For the one–vertex graph without tails we get (4.8).

4.3.1. Theorem. *a) For any linear relation*

$$\sum_j c_j [\widetilde{M}(\sigma_j)] = 0 \tag{4.10}$$

in the homology of $\overline{M}_{g,n}$ and any labeling $(a_1, \ldots, a_n)$ (common for all σ_j) we have

$$\sum_j c_j \, \Phi_{\sigma_j, \mathbf{a}}(x) = 0. \tag{4.11}$$

b) $\Phi_{\sigma, \mathbf{a}}(x)$ can be expressed as a polynomial in $\Phi_h(x), h \leq g$, and their derivatives. This polynomial is determined by the combinatorial structure of $(\sigma, \mathbf{a})$ and the metric coefficients g_{ab}.

Proof. a) From (4.5) it follows that for any $N \geq 0$ we have

$$\sum_j c_j \pi_N^*([\widetilde{M}(\sigma_j)]) = 0. \tag{4.12}$$

In view of (4.9), this proves (4.11).

b) For a map $r : \underline{N} = \{1, \ldots, N\} \to V_\sigma$, denote by σ^r the graph having the same vertices, edges, and tails as σ and N extra tails $\{n+1, \ldots, n+N\}$: if $r(i) = v$, the new tail $n+i$ is attached to the vertex v. We have

$$\pi_N^*[\widetilde{M}(\sigma)] = \sum_r \frac{|\text{Aut } \sigma^r|}{|\text{Aut } \sigma|} [\widetilde{M}(\sigma^r)]. \tag{4.13}$$

In fact, in view of V.(7.4) this is equivalent to

$$\pi_N^* b(\sigma)_* [\overline{M}(\sigma)] = \sum_r b(\sigma^r)_* \pi_N^* [\overline{M}(\sigma)]. \tag{4.14}$$

To check the latter equality, we appeal to V.(6.13).

Now input (4.13) into (4.9):

$$\Phi_{\sigma, \mathbf{a}}(x) = \sum_{N=0}^\infty \sum_{r:\underline{N} \to V_\sigma} \frac{|\text{Aut } \sigma^r|}{|\text{Aut } \sigma|} \left(\int_{\widetilde{M}(\sigma^r)} I_{g, n+N} \right) \left(\Delta_{a_1} \ldots \Delta_{a_n} e^{\sum x^a \Delta_a} \right). \tag{4.15}$$

Leaving inside the rightmost brackets only the terms matching the tails $S_{\sigma^r} = \{1, \ldots, n+N\}$ we can rewrite this as

$$\sum_{N=0}^\infty \sum_{r:\underline{N} \to V_\sigma} \frac{|\text{Aut } \sigma^r|}{|\text{Aut } \sigma|} \left(\int_{\widetilde{M}(\sigma^r)} I_{g, n+N} \right) \left(\sum_{(a_{n+i}) \in A^N} \prod_{i=1}^{n+N} \Delta_i \frac{x^{a_{n+N}} \ldots x^{a_{n+1}}}{N!} \right).$$

In view of (4.3) this is the same as

$$\sum_{N=0}^\infty \sum_{r:\underline{N} \to V_\sigma} \frac{|\text{Aut } \sigma^r|}{|\text{Aut } \sigma|} \sum_{(a_{n+i}) \in A^N} Y(\sigma^r) \left(\bigotimes_{i=1}^{n+N} \Delta_{a_i} \right) \frac{x^{a_{n+N}} \ldots x^{a_{n+1}}}{N!}.$$

Applying (4.4) for σ^r, we can replace this by

$$\frac{1}{|\text{Aut } \sigma|} \sum_{N=0}^\infty \sum_{r:\underline{N} \to V_\sigma} \sum_{(a_{n+i})} \left(\bigotimes_{v \in V_\sigma} Y_{g_v, F_{\sigma^r}(v)} \right) \left(\bigotimes_{i=1}^{n+N} \Delta_{a_i} \otimes \Delta^{\otimes E_\sigma} \right) \frac{\prod_{i=N}^1 x^{a_{n+i}}}{N!}. \tag{4.16}$$

Let σ be the one–vertex graph with tails $\{1,\dots,n\}$. Calculating $\Phi_{\sigma,\mathbf{a}}$ first directly from (4.15) and then via (4.16) we obtain

$$(4.17) \quad \Phi_{\sigma,\mathbf{a}} = \partial_{a_1}\dots\partial_{a_n}\Phi_g = \sum_{N=0}^{\infty}\sum_{(a_{n+i})\in A^N} Y_{g,n+N}\left(\bigotimes_{i=1}^{n+N}\Delta_{a_i}\right)\frac{x^{a_{n+N}}\dots x^{a_{n+1}}}{N!}.$$

It must be clear by now that (4.16) for a general σ is the linear combination of monomials in several expressions of the type (4.15). In order to describe it succinctly, we introduce the symmetric differential form

$$(4.18) \quad \Omega_{g,n} = D^n\Phi_g = \frac{1}{n!}\sum_{(a_i)\in A^n} dx^{a_n}\dots dx^{a_1}\partial_{a_1}\dots\partial_{a_n}\Phi_g$$

endowing dx^a by the same parity as x^a. This is the n–th symmetric differential of Φ_g with respect to the canonical flat structure of $H^*(V): \Phi_g(x+dx) = \sum_n D^n\Phi_g$. All such forms constitute a (super)commutative ring which is canonically isomorphic to $\Lambda[H]\otimes\Lambda[[H]]$, $H = H^*(V)$ (the first factor accommodates differentials).

Similarly, for the stable graph σ put

$$(4.19) \quad \Omega_\sigma = \sum_{(a_i)\in A^n} dx^{a_n}\dots dx^{a_1}\Phi_{\sigma,\mathbf{a}}.$$

Now consider the (geometric realization of) the graph σ with the following labeling. Each edge is (arbitrarily) oriented and labeled by the Laplace operator

$$(4.20) \quad \Box = \sum_{a,b\in A}\partial_a g^{ab}\partial_b,$$

where the subscript a (resp. b) is assigned to the outcoming (resp. incoming) flag of the edge. Each vertex v is labeled by the form Ω_{g_v,n_v} where $n_v = |F_\sigma(v)|$. Now form the product of all Ω_{g_v,n_v} and apply to this product the operator $\mathrm{Contr}_{E_\sigma}$ of multiple contraction with respect to the pairs of indices corresponding to all edges of σ and determined by the operator $\Box$ at each edge. The contraction is defined by the rule $\partial_a \dashv dx^b = (-1)^{ab}dx^b \vdash \partial_a = \delta_{ab}$. The result will be $|\mathrm{Aut}\,\sigma|\Omega_\sigma$.

4.4. Examples. *(i) Genus zero.* The simplest relations of the type (4.10) occur in $H_0(\overline{M}_{04})$: any two boundary divisors (points) are equivalent. This leads, of course, to the Associativity Equations for Φ_0.

(ii) Genus one. This case was thoroughly studied in [Ge4]. Similarly to the genus zero case, any two zero–dimensional boundary strata in $\overline{M}_{1,2}$ are equivalent. Relations of the new type appear first in the middle homology $H_4(\overline{M}_{14})$. E. Getzler explicitly produces a relation between some boundary S_4-invariant cycles and calculates the relevant differential equation for Φ_1. Using this differential equation, B. Dubrovin and Y. Zhang in [DZh2] have shown that in the semisimple quantum cohomology case Φ_1 can be explicitly expressed through Φ_0. This expression makes sense for any semisimple Frobenius manifold and thus provides the definition of the genus one potential for it.

§5. The simplest Virasoro constraints

Here we deduce the cases $n = 0, -1$ of Conjecture 2.7.

5.1. Proposition. *Write 1 for $\tau_0 \Delta_0$ where Δ_0 is the fundamental class of V. We have*

$$(5.1) \quad \langle 1\, \tau_{d_1}\gamma_1 \ldots \tau_{d_n}\gamma_n \rangle_{g,\beta} = \sum_{j:\, d_j \geq 1} \langle \tau_{d_1}\gamma_1 \ldots \tau_{d_j-1}\gamma_j \ldots \tau_{d_n}\gamma_n \rangle_{g,\beta}$$

unless $\overline{M}_{g,n}(V,\beta)$ is empty whereas $\overline{M}_{g,n+1}(V,\beta)$ is non-empty. In these exceptional cases the non-vanishing correlators with 1 among their arguments can occur only when $\beta = 0$ and $(g,n) = (0,2)$ or $(1,0)$, and they are

$$(5.2) \quad \langle 1\, \gamma_1 \gamma_2 \rangle_{0,0} = \int_V \gamma_1 \cup \gamma_2,$$

$$(5.3) \quad \langle 1 \rangle_{1,0} = \deg c_\delta(V), \quad \delta = \dim V.$$

Proof. The first statement follows from Theorem 3.8 a) with $S = \{1, \ldots, n\}$, $m = 0$ (cf. (3.26)).

In the exceptional cases (3.26) makes no sense. This cannot occur unless $\beta = 0$, and if $\beta = 0$, this occurs only for $g = 0, n = 2$ and $g = 1, n = 0$ because $\overline{M}_{g,S}(V,0) = V^S \times \overline{M}_{g,S}$. For $g = 0, n = 3$, $\overline{M}_{0,3}$ is a point, $\psi_i = 0$, and the Mapping to a Point Axiom (1.14) shows that $J_{0,3}(V,0) = [V]$. This proves that (3.2) provides the only non-vanishing correlators. The same argument shows more generally that

$$(5.4) \quad \deg\left(\mathrm{ev}^*(\gamma_1 \otimes \gamma_2 \otimes \gamma_3) \cap J_{0,3}(V,0)\right) = \int_V \gamma_1 \cup \gamma_2 \cup \gamma_3$$

and that the similar expressions with $n \geq 4$ vanish.

For $g = 1$, $\overline{M}_{1,1}$ is one-dimensional, and

$$(5.5) \quad J_{1,1}(V,0) = c_\delta(\mathcal{T}_V \boxtimes \omega^{-1}) = c_\delta(V) \boxtimes 1 - c_{\delta-1}(V) \boxtimes c_1(\omega),$$

where ω is the pushforward of the relative dualizing sheaf of the universal curve over $\overline{M}_{1,1}$. This establishes (5.3).

In the physical literature, (5.1) is called *the puncture equation*; cf. [W2].

5.2. String equation and L_{-1}. With F defined by (2.7), directly from (5.1) and (5.2) we obtain:

$$(5.6) \quad \frac{\partial F}{\partial x_0^0} = \sum_{a,d} x_d^a \frac{\partial F}{\partial x_{d-1}^a} + \frac{1}{2\lambda^2} \sum_{a,b} g_{ab} x_0^a x_0^b$$

and

$$(5.7) \quad \left(-\frac{\partial}{\partial x_0^0} + \sum_{a,d} x_d^a \frac{\partial}{\partial x_{d-1}^a} + \frac{1}{2\lambda^2} \sum_{a,b} g_{ab} x_0^a x_0^b\right) e^{F(x)} = 0.$$

(5.6) is called *the string equation*, and (5.7) is equivalent to Conjecture 2.7 for $n = -1$.

5.3. Proposition. *This time write 1 for Δ_0. We have*

(5.8) $\qquad \langle \tau_1 1 \tau_{d_1}\gamma_1 \ldots \tau_{d_n}\gamma_n \rangle_{g,\beta} = (2g - 2 + n) \langle \tau_{d_1}\gamma_1 \ldots \tau_{d_n}\gamma_n \rangle_{g,\beta}$

unless $\beta = 0$ and $(g,n) = (0,2)$ or $(1,0)$. All non-vanishing correlators of this form in the cases $(g,n) = (0,2)$ or $(1,0)$ are

(5.9) $\qquad\qquad\qquad \langle \tau_1 1 \rangle_{1,0} = \dfrac{1}{24} \deg c_\delta(V).$

Proof. The first statement follows from Theorem 3.8 b) (cf. (3.27)).

The exceptional cases are the same as in Proposition 5.1, and the only non-vanishing correlator stems from the first term of (5.5), because

(5.10) $\qquad\qquad\qquad \displaystyle\int_{[\overline{M}_{1,1}]} \psi_1 = \dfrac{1}{24}.$

In the physical literature, (5.8) is called *the dilaton equation* ([W2]).

5.4. Proposition. *Let γ_0 be a divisor class on V or more generally, a class in $H^2(V)$. Then we have*

(5.11) $\begin{aligned}\langle \gamma_0 \tau_{d_1}\gamma_1 \ldots \tau_{d_n}\gamma_n \rangle_{g,\beta} &= (\gamma_0, \beta) \langle \tau_{d_1}\gamma_1 \ldots \tau_{d_n}\gamma_n \rangle_{g,\beta} \\ &+ \sum_{k:\, d_k \geq 1} \langle \tau_{d_1}\gamma_1 \ldots \tau_{d_k-1}(\gamma_0 \cup \gamma_k) \ldots \tau_{d_n}\gamma_n \rangle_{g,\beta}\end{aligned}$

unless $\beta = 0$ and $(g,n) = (0,2)$ or $(1,0)$. In the exceptional cases, all non-vanishing correlators are

(5.12) $\qquad\qquad\qquad \langle \gamma_0 \gamma_1 \gamma_2 \rangle_{0,0} = \displaystyle\int_V \gamma_0 \cup \gamma_1 \cup \gamma_2,$

(5.13) $\qquad\qquad\qquad \langle \gamma_0 \rangle_{1,0} = -\dfrac{1}{24}(\gamma_0, c_{\delta-1}(V)).$

Proof. The first statement follows from Theorem 3.9 (in (3.30) replace S by $\{1,\ldots,n\}$ and γ_m by γ_0).

Moreover, (5.12) is a particular case of (5.4), whereas (5.13) follows from (5.5) and (5.10).

5.5. The dilaton equation and $L_0, \overline{L}_0$. Identities (5.8) and (5.9) readily translate into

(5.14) $\qquad\qquad \lambda \dfrac{\partial F}{\partial \lambda} + \sum s_d^a \dfrac{\partial F}{\partial s_d^a} + \dfrac{1}{24} \deg c_\delta(V) = 0.$

Similarly, for a divisorial class γ_0 define the derivation acting non-trivially only upon q^β: $\partial_{\gamma_0}(q^\beta) := (\gamma_0, \beta) q^\beta$. Put $\gamma_0 \cup \Delta_a = \sum_b G_a^b \Delta_b$. Then from (5.11)–(5.13) we get

(5.15) $\qquad \partial_{\gamma_0} F + \displaystyle\sum_{a,d} G_a^b s_d^a \dfrac{\partial F}{\partial s_{d-1}^b} + \dfrac{1}{2\lambda^2}\sum_{a,b,c} G_a^c g_{bc} x_0^a x_0^b - \dfrac{1}{24}(\gamma_0, c_{\delta-1}(V)) = 0.$

Finally, (2.8) translates into two identities:

$$\frac{3-\delta}{2}\lambda\frac{\partial F}{\partial \lambda} + \partial_{-K_V} F - \sum_{a,d}(d+p_a-1)s_d^a\frac{\partial F}{\partial s_d^a} = 0, \tag{5.16}$$

and similarly with p_a replaced by q_a. K. Hori in [H] remarked that one can get rid of λ- and q-derivatives in (5.16) by adding an appropriate linear combination of (5.12) and (5.13) for $\gamma_0 = -K_V$. The resulting equations will be equivalent to $L_0(e^F) = \overline{L}_0(e^F) = 0$ (cf. (2.11)).

§6. Generalized correlators

6.1. Notation. In this section we prove a theorem which will allow us in principle to calculate the correlators (2.1) via the Gromov–Witten correspondences and intersection theory on $V^n \times \overline{M}_{g,n}$.

Put $\psi_i := c_1(L_{i;g,n}(V,\beta))$.

In the stable range $2g - 2 - n > 0$ we have the absolute stabilization map st: $\overline{M}_{g,n}(V,\beta) \to \overline{M}_{g,n}$, and the respective bundles L_i on $\overline{M}_{g,n}$. Put $\phi_i := \mathrm{st}^*(c_1(L_i))$.

Our generalized correlators, by definition, are:

$$\langle \tau_{d_1,e_1}\gamma_1 \ldots \tau_{d_n,e_n}\gamma_n\rangle_{g,\beta}$$
$$\tag{6.1} := \int_{J_{g,n}(V,\beta)} \psi_1^{d_1}\phi_1^{e_1} \cup \mathrm{ev}_1^*(\gamma_1) \cup \ldots \cup \psi_n^{d_n}\phi_n^{e_n} \cup \mathrm{ev}_n^*(\gamma_n).$$

Since $\overline{M}_{0,2}(V,0) = \emptyset$, we have

$$\langle \tau_{d_1}\gamma_1 \tau_{d_2}\gamma_2\rangle_{0,0} = 0. \tag{6.2}$$

Furthermore, in the stable range we have

$$\langle \prod_{i=1}^n \tau_{d_i,0}\gamma_i\rangle_{g,\beta} = \langle \prod_{i=1}^n \tau_{d_i}\gamma_i\rangle_{g,\beta}.$$

6.2. Theorem. *If $2g - 2 + n > 0$, then for any j with $d_j \geq 1$ we have*

$$\langle \prod_{i=1}^n \tau_{d_i,e_i}\gamma_i\rangle_{g,\beta} = \langle \prod_{i=1}^n \tau_{d_i-\delta_{ij},e_i+\delta_{ij}}\gamma_i\rangle_{g,\beta}$$
$$\tag{6.3} + \sum_{a,\beta_1+\beta_2=\beta} \pm\langle \tau_{d_j-1}\gamma_j\,\tau_0\Delta^a\rangle_{0,\beta_1}\langle \tau_{0,e_j}\Delta_a \prod_{i:\,i\neq j}\tau_{d_i,e_i}\gamma_i\rangle_{g,\beta_2}.$$

Here (Δ_a), (Δ^a) are Poincaré dual bases of $H^*(V)$, and the sign arises from permuting γ_j with γ_i for all $i < j$.

6.2.1. Corollary. *For $g = 0$, $n = 3$, $d_1 \geq 1$ we have:*

$$(6.4) \quad \langle \tau_{d_1}\gamma_1 \tau_{d_2}\gamma_2 \tau_{d_3}\gamma_3 \rangle_{0,\beta} = \sum_{a,\, \beta_1+\beta_2=\beta} \langle \tau_{d_1-1}\gamma_1\, \tau_0 \Delta^a \rangle_{0,\beta_1} \langle \tau_0 \Delta_a\, \tau_{d_2}\gamma_2\, \tau_{d_3}\gamma_3 \rangle_{0,\beta_2}.$$

In fact, $\phi_i = 0$ here, so one should put $e_i = 0$ in (6.3), and the first summand will vanish.

Before proving Theorem 6.2, we will discuss the computational algorithms provided by it. Clearly, relations (6.3) allow us to reduce all the generalized (in particular, the conventional ones) correlators to those with $\beta = 0$, to the conventional ones in the unstable range and to the generalized ones with all $d_i = 0$ in the stable range. Using (6.1) and the projection formula, one can rewrite the latter in the form

$$\langle \tau_{0,e_1}\gamma_1 \ldots \tau_{0,e_n}\gamma_n \rangle_{g,\beta}$$

$$(6.5) \quad := \int_{I_{g,n}(V,\beta)} c_1(\mathrm{pr}_2^*(L_1))^{e_1} \cup \mathrm{pr}_1^*(\gamma_1) \cup \ldots \cup c_1(\mathrm{pr}_2^*(L_n))^{e_n} \cup \mathrm{pr}_1^*(\gamma_n),$$

where this time the integration refers to $V^n \times \overline{M}_{g,n}$, $I = (\mathrm{ev}, \mathrm{st})_* J$ is the Gromov–Witten correspondence, and pr_i are the two projections. Hence the correlators in the stable range with $d_i = 0$ are calculable if we know the full (not just top) Gromov–Witten invariants. We will call the expressions above *the modified ("downstairs") correlators*, in order to distinguish them from the conventional ("upstairs") correlators (2.1).

Notice that for $\beta = 0$ we have $\psi_i = \phi_i$, hence $\tau_{d,e} = \tau_{d+e}$, so that (6.3) gives no new information and is tautologically true because of (6.2). So we will recall what happens in the case $\beta = 0$, $\dim V > 0$ separately.

6.3. The mapping to a point case. Recall that $\overline{M}_{g,n}(V,0)$ is canonically isomorphic to $\overline{M}_{g,n} \times V$, and with this identification,

$$(6.6) \quad J_{g,n}(V,0) = c_G(\mathcal{E} \boxtimes \mathcal{T}_V) \cap [\overline{M}_{g,n} \times V],$$

where $\mathcal{E} = R^1\pi_*\mathcal{O}_C$, $\pi : C \to \overline{M}_{g,n}$ is the universal curve, and $G = g \dim V$. Consider the Chern classes and Chern roots of $\mathcal{E}$ and $\mathcal{T}_V$:

$$c_t(\mathcal{E}) = \prod_{i=1}^{g}(1 + a_i t) = \sum_{i=0}^{g}(-1)^i \lambda_{i;g,n} t^i,$$

where λ_i are Mumford's tautological classes defined as Chern classes of $\pi_*(\omega_\pi)$,

$$c_t(\mathcal{T}_V) = \prod_{j=1}^{\delta}(1 + v_j t) = \sum_{j=0}^{\delta} c_j(V) t^j, \quad \delta = \dim V.$$

Then we get

$$c_G(\mathcal{E} \boxtimes \mathcal{T}_V) = \prod_{i=1}^{g}\prod_{j=1}^{\delta}(a_i \boxtimes 1 + 1 \boxtimes v_j) = \prod_{j=1}^{\delta}\sum_{i=0}^{g}(-1)^i \lambda_{i;g,n} \boxtimes v_j^{g-i}$$

$$= \sum_{(i_1,\ldots,i_\delta)}(-1)^{i_1+\cdots+i_\delta} \lambda_{i_1;g,n} \ldots \lambda_{i_\delta;g,n} \boxtimes v_1^{g-i_1} \ldots v_\delta^{g-i_\delta}$$

(6.7) $\quad = (-1)^G \sum_{0 \leq i_1 \leq \cdots \leq i_\delta \leq g} \lambda_{i_1;g,n} \cdots \lambda_{i_\delta;g,n} \boxtimes m_{g-i_1,\ldots,g-i_\delta}(c_0(V),\ldots,c_\delta(V)).$

Here $m_{g-i_1,\ldots,g-i_\delta}$ is the symmetric function obtained by symmetrization of the obvious monomial in $-v_j$ and expressed via the Chern classes of V.

Furthermore, $L_{i;g,n}(V,0)$ is the lift of $L_{i;g,n}$ wrt the projection $\overline{M}_{g,n} \times V \to \overline{M}_{g,n}$ and ev_i is the projection $\overline{M}_{g,n} \times V \to V$. Hence we get

$$\langle \tau_{d_1}\gamma_1 \ldots \tau_{d_n}\gamma_n \rangle_{g,0}$$

$$= (-1)^G \sum_{0 \leq i_1 \leq \cdots \leq i_\delta \leq g} \left(\int_{\overline{M}_{g,n}} \lambda_{i_1;g,n} \cdots \lambda_{i_\delta;g,n} \psi_{1;g,n}^{d_1} \cdots \psi_{n;g,n}^{d_n} \right.$$

(6.8) $\qquad\qquad \times \left. \int_V m_{g-i_1,\ldots,g-i_\delta}(c_0(V),\ldots,c_\delta(V))\gamma_1 \ldots \gamma_n \right),$

where $\psi_{i;g,n} = c_1(L_{i;g,n})$.

The generalized correlators give nothing new: $\tau_{d,e} = \tau_{d+e}$.

Most of the correlators (6.8) vanish for dimensional reasons. Here is the list of those that may remain.

6.3.1. Proposition. *The correlators (6.8) identically vanish except for the following cases.*

a) $g = 0$, $n \geq 3$, $\sum d_i = n - 3$, $\sum |\gamma_i| = 2\delta$, where $\gamma \in H^{|\gamma|}(V)$, $\delta = \dim V$:

(6.9) $\qquad \langle \tau_{d_1}\gamma_1 \ldots \tau_{d_n}\gamma_n \rangle_{0,0} = \dfrac{(d_1 + \cdots + d_n)!}{d_1! \ldots d_n!} \int_V \gamma_1 \ldots \gamma_n.$

b) $g = 1$, $n \geq 1$, $\sum d_i = n$ *(resp.* $n-1$*)*, $\sum |\gamma_i| = 0$ *(resp. 2):*

(6.10) $\qquad \langle \tau_{d_1}1 \ldots \tau_{d_n}1 \rangle_{1,0} = \deg c_\delta(V) \int_{\overline{M}_{1,n}} \psi_{1;1,n}^{d_1} \cdots \psi_{n;1,n}^{d_n},$

(6.11) $\qquad \langle \tau_{d_1}\gamma \tau_{d_2}1 \ldots \tau_{d_n}1 \rangle_{1,0} = -(c_{\delta-1}(V),\gamma) \int_{\overline{M}_{1,n}} \lambda_{1,1,n} \psi_{1;1,n}^{d_1} \cdots \psi_{n;1,n}^{d_n}$

for $|\gamma| = 2$.

c) $g \geq 2$, $n \geq 0$, $\sum |\gamma_i|/2 \leq \delta \leq 3$, $\sum(d_i + |\gamma_i|/2) = (g-1)(3-\delta) + n$.

In particular, the $g \geq 2$, $\beta = 0$ *correlators vanish for* $\dim V \geq 4$.

Proof. First of all, $\mathcal{E} = \mathcal{E}_{g,n}$ is lifted from $\overline{M}_{\geq 2,0}$, $\overline{M}_{1,1}$, or $\overline{M}_{0,3}$.

For $g = 0$, $\mathcal{E}$ is the zero bundle, and $J_{0,n}(V,0) = [\overline{M}_{0,n} \times V]$. Formula (6.9) follows from this and from the known expression for $g = 0$, $V = $ *a point* correlators:

(6.12) $\qquad \int_{\overline{M}_{0,n}} \psi_{1;0,n}^{d_1} \cdots \psi_{n;0,n}^{d_n} = \dfrac{(d_1 + \cdots + d_n)!}{d_1! \ldots d_n!}.$

For $g = 1$, (6.7) becomes

$$c_\delta(\mathcal{E} \boxtimes \mathcal{T}_V) = c_\delta(V) \boxtimes 1 - c_{\delta-1}(V) \boxtimes \lambda_{1,1,n}$$

from which (6.10) and (6.11) follow.

Finally, for $g \geq 2$ one sees that the virtual fundamental class can be non-zero only if the virtual dimension for $n = 0$ is non-negative, which means that $\dim V \leq 3$. The remaining inequalities follow from the dimension matching.

One can further specialize (6.8) and write formulas similar to (6.9)–(6.11) separately for curves, surfaces, and threefolds, $g \geq 2$. For more details, see [GePa].

6.4. Unstable range case. If $2g - 2 + n \leq 0$, we cannot use the absolute stabilization morphism as in (6.5) because $\overline{M}_{g,n}$ is empty, whereas for $\beta \neq 0$, the stack $\overline{M}_{g,n}(V, \beta)$ may well be non-empty. Always assuming this (otherwise the relevant correlators vanish), we will use instead the forgetful morphism $\overline{M}_{g,n+1}(V, \beta) \to \overline{M}_{g,n}(V, \beta)$ to produce recursion.

6.4.1. Proposition. *All the unstable range correlators can be calculated through the genus zero and one primary ($d_i = 0$) stable range correlators, and the $\beta = 0$ correlators.*

Proof. We will be considering the cases $(g, n) = (0, 2), (0, 1), (0, 0), (1, 0)$ in this order, reducing each in turn to the previously treated ones.

To treat the two-point correlators with, say, $d_1 > 0$, we first use (5.11) and write for some divisor class γ_0 with $(\gamma_0, \beta) \neq 0$:

$$(6.13) \quad \langle \tau_{d_1} \gamma_1 \tau_{d_2} \gamma_2 \rangle_{0,\beta} = \frac{1}{(\gamma_0, \beta)} (\langle \gamma_0 \tau_{d_1} \gamma_1 \tau_{d_2} \gamma_2 \rangle_{0,\beta} - \langle \tau_{d_1-1}(\gamma_0 \cup \gamma_1) \tau_{d_2} \gamma_2 \rangle_{0,\beta} - \langle \tau_{d_1} \gamma_1 \tau_{d_2-1}(\gamma_0 \cup \gamma_2) \rangle_{0,\beta}).$$

The last two terms in (6.13) contain only two-point correlators with smaller sum $d_1 + d_2 - 1$. To the first term we apply (6.4):

$$(6.14) \quad \langle \gamma_0 \tau_{d_1} \gamma_1 \tau_{d_2} \gamma_2 \rangle_{0,\beta} = \sum_{a, \beta_1 + \beta_2 = \beta} \langle \tau_{d_1-1} \gamma_1 \Delta_a \rangle_{0,\beta_1} \langle \Delta^a \gamma_0 \tau_{d_2} \gamma_2 \rangle_{0,\beta_2}.$$

The right hand side contains only two-point correlators with smaller sum $d_1 - 1$ and three-point correlators with maximum one $\tau_d, d \neq 0$. If necessary, we can again apply (6.14) to the three-point correlators there, again reducing the order of the gravitational descendants involved.

Iterating this procedure, we will arrive at the expressions containing only primary correlators. Finally, the two-point primary correlators can be reduced to the three-point stable range ones:

$$(6.15) \quad \langle \gamma_1 \gamma_2 \rangle_{0,\beta} = \frac{1}{(\gamma_0, \beta)} \langle \gamma_0 \gamma_1 \gamma_2 \rangle_{0,\beta}.$$

For later use, we register the following explicit reduction of some two-point correlators to the three-point ones following from (6.13):

$$(6.16) \quad \langle \tau_d \gamma_1 \tau_0 \gamma_2 \rangle_{0,\beta} = \sum_{j=1}^{d+1} (-1)^{j+1} (\gamma_0, \beta)^{-j} \langle \gamma_0 \tau_{d+1-j} \gamma_1 \tau_0(\gamma_0^{j-1} \cup \gamma_2) \rangle_{0,\beta}.$$

Clearly, one can invoke (5.11) in the same way in order to calculate the one-point and zero-point correlators. Alternatively, one can exploit the dilaton equation (5.8).

6.5. Correlators for zero–dimensional V. This case is covered by the Witten–Kontsevich theory and additional relations summarized in [Fab1].

6.6. Proof of (6.3). We now turn to proving Theorem 6.2. The general strategy is straightforward. Consider the morphism of universal curves $\widetilde{\mathrm{st}}: C_{g,n}(V,\beta) \to C_{g,n}$ covering st. It induces the morphism of relative 1–form sheaves $\omega \to \omega(V,\beta)$, at least at the complement of singular points of the fiber. Restricting the latter to the j–th section ($j \in S$ being fixed), we get the morphism $\mathrm{st}^*(L_{j;g,n}) \to L_{j;g,n}(V,\beta)$ on $\overline{M}_{g,n}(V,\beta)$. It is a local isomorphism everywhere except for the points in this stack over which the j–th section lies on the component of fiber which gets contracted by $\widetilde{\mathrm{st}}$. These points constitute the union of boundary strata $\overline{M}(V, \sigma_j(\beta_1,\beta_2))$ where $\sigma_j(\beta_1,\beta_2)$ is the one–edge, two–vertex n–graph with one vertex of genus 0, class β_1, with tail j, and another of genus g, class β_2, with tails $\neq j$. Naively, one would expect that all these boundaries are divisors, and over them sections of $\mathrm{st}^*(L_{j;g,n})$ have an extra zero of the first order. Hence in (6.1) we could replace one factor ψ_j by $\phi_j + \sum_{\beta_1+\beta_2=\beta}[\overline{M}(V,\sigma_j(\beta_1,\beta_2))]$. Then the restriction to the boundary would give (6.3). A more precise reasoning shows that this identity in fact holds after replacing the fundamental classes by the appropriate virtual fundamental classes.

Denote by μ the union of boundary morphisms

$$\mu: \coprod_{\beta_1+\beta_2=\beta} \overline{M}(V,\sigma_j(\beta_1,\beta_2)) \to \overline{M}_{g,n}(V,\beta).$$

Notice that the term corresponding to $\beta_1 = 0$ is empty. The following statement and its proof are due to K. Behrend.

6.6.1. Proposition. *For any $\alpha \in A^*(\overline{M}_{g,n}(V,\beta))$ and $1 \leq j \leq n$ we have*

(6.17)
$$\psi_j \cap (\alpha \cap J_{g,n}(V,\beta)) = \phi_j \cap (\alpha \cap J_{g,n}(V,\beta)) + \mu_*\left(\alpha \cap \sum_{\beta_1+\beta_2=\beta} J(V,\sigma_j(\beta_1,\beta_2))\right).$$

Proof. Consider a cartesian diagram:

(6.18)
$$\begin{array}{ccc} N & \xrightarrow{\mu} & M \\ \downarrow & & \downarrow \\ \mathcal{N} & \xrightarrow{\nu} & \mathcal{M} \end{array}$$

Assume that N, M are DM–stacks, μ is proper, $\mathcal{N}, \mathcal{M}$ are Artin stacks, and ν is a regular local imbedding of codimension one. Denote by $D \subset \mathcal{M}$ the Cartier divisor which is the image of ν. Suppose that $\mathcal{O}_D$ fits into an exact sequence

$$0 \to L' \to L \to \mathcal{O}_D \to 0,$$

where L, L' are line bundles. Then for all $\gamma \in A_*(M)$ we have

(6.19) $$c_1(\pi^*L) \cap \gamma = c_1(\pi^*L') \cap \gamma + \mu_*\nu^!\gamma.$$

This can be checked by unravelling the definitions.

§6. GENERALIZED CORRELATORS

We apply this result to the diagram

(6.20)
$$\begin{array}{ccc} \coprod_{\beta_1+\beta_2=\beta} \overline{M}(V,\sigma_j(\beta_1,\beta_2)) & \xrightarrow{\mu} & \overline{M}_{g,n}(V,\beta) \\ \downarrow & & \downarrow \\ \mathcal{M}_{0,2} \times \mathcal{M}_{g,n} & \xrightarrow{\nu} & \mathcal{M}_{g,n} \end{array}$$

in which the vertical arrows forget V, and the lower arrow is

$$(C; x_j, y_0), (C'; y_1, x_k \mid k \neq j) \mapsto (C \amalg_{y_0=y_1} C'; x_1, \ldots x_n).$$

The local deformation theory shows that it is indeed a local regular immersion of codimension one. Let D be its image. The sheaves L, L' are constructed as follows. Let $\pi : C \to \mathcal{M}_{g,n}$ be the universal curve and $p : C \to C^s$ its stabilization morphism, so that $\pi = \pi^s \circ p$. Denote by $x_j^s : \mathcal{M}_{g,n} \to C^s$ the image of the section x_j. Then we have an exact sequence

$$0 \to x_j^{s*}\omega_{\pi^s} \to x_j^*\omega_\pi \to \mathcal{O}_D \to 0.$$

Now specialize (6.19) to this case and take $\gamma = \alpha \cap J_{g,n}(V,\beta)$. To deduce (6.17), it remains to check that

$$\nu^! J_{g,n}(V,\beta) = \sum_{\beta_1+\beta_2=\beta} J(V, \sigma_j(\beta_1,\beta_2)).$$

This can be proved in the same way as Lemma 10 of [Beh] and Proposition 7.2 of [BehF], because it reflects the pullback property of the virtual fundamental classes under pullback of the obstruction theories.

It remains to deduce (6.3) from (6.17). We choose

(6.21)
$$\alpha = \pm [\psi_j^{d_j-1} \cup \mathrm{ev}_j^*(\gamma_j)] \cup \left[\phi_j^{e_j} \cap \prod_{i:\, i \neq j} \psi_i^{d_i} \cup \mathrm{ev}_i^*(\gamma_i) \right].$$

The regrouping of factors in (6.21) takes into account that we have fixed j. We write for brevity (6.21) as $\alpha = \pm \alpha_0 \cup \alpha_g$ stressing that the first factor comes from the component of genus zero, whereas the second comes from the component of genus g. Since we deal now with cohomology, we need to extrapolate (6.17) correspondingly, or else use the familiar motivic arguments. Taking this point for granted, we see that the first summand in the rhs of (6.17) produces the first summand in the rhs of (6.3). To identify the summands corresponding to a fixed (β_1, β_2), write

(6.22)
$$\begin{aligned} \deg \mu_*(\alpha \cap J(V, \sigma_j(\beta_1,\beta_2))) &= \deg \alpha \cap J(V, \sigma_j(\beta_1,\beta_2)) \\ &= \deg c^*(\alpha_0 \otimes \alpha_g) \cap \Delta^!(J_0 \otimes J_g). \end{aligned}$$

Here $J_0 = J_{02}(V,\beta_1)$, $J_g = J_{g,n}(V,\beta_2)$, Δ is the diagonal of V, and c is the upper arrow in the cartesian square:

$$\begin{array}{ccc} \overline{M}(V, \sigma_j(\beta_1, \beta_2)) & \xrightarrow{c} & \overline{M}_{02}(V,\beta_1) \times \overline{M}_{g,n}(V,\beta_2) \\ \downarrow & & \downarrow \\ V & \xrightarrow{\Delta} & V \times V \end{array}$$

Rewriting (6.22) further, we find

$$\deg \Delta'((\alpha_0 \cap J_0) \otimes (\alpha_g \cap J_g)) = \deg [\Delta] \cdot ((\alpha_0 \cap J_0) \otimes (\alpha_g \cap J_g)).$$

Applying the Künneth formula, we finally obtain (6.3).

§7. Generating functions on the large phase space

7.1. Notation. We return to the setup of 2.4. For further reference, we recall the formula for the genus g components of (2.7):

(7.1)
$$F_g(x) = \langle e^\Gamma \rangle_g = \sum_n \frac{\langle \Gamma^{\otimes n} \rangle_g}{n!} = \sum_{n,(a_1,d_1),\ldots,(a_n,d_n)} \frac{x_{d_n}^{a_n} \cdots x_{d_1}^{a_1}}{n!} \langle \tau_{d_1} \Delta_{a_1} \cdots \tau_{d_n} \Delta_{a_n} \rangle_g.$$

We got rid of the summation over β using the notation (4.2). Now define $F_g^{\text{st}}(x)$ by the same formula in which the last summation is restricted to the stable range of (g,n), that is, $n \geq 3$ for $g = 0$ and $n \geq 1$ for $g = 1$.

We will introduce the generating function $G_g(x)$ for modified correlators by the same formula as F_g^{st} in which every τ_d in the stable range correlators is replaced by $\tau_{0,d}$, where $\tau_{d,e}$ are defined by (6.1):

(7.2)
$$G_g(x) = \sum_{n,(a_1,d_1),\ldots,(a_n,d_n)} \frac{x_{d_n}^{a_n} \cdots x_{d_1}^{a_1}}{n!} \langle \tau_{0,d_1} \Delta_{a_1} \cdots \tau_{0,d_n} \Delta_{a_n} \rangle_g.$$

We will prove first of all that the two functions are connected by the linear change of coordinates of the large phase space.

7.2. Theorem. *We have for all $g \geq 0$*

(7.3)
$$F_g^{\text{st}}(x) = G_g(y),$$

where

(7.4)
$$y_c^b = x_c^b + \sum_{(a,d), d \geq c+1} x_d^a \langle \tau_{d-c-1} \Delta_a \tau_0 \Delta^b \rangle_0.$$

Proof. For $d \geq 1$, define the linear operators

$$U_d : H^*(V, \Lambda) \to H^*(V, \Lambda)$$

by the formula

(7.5)
$$U_d(\gamma) := \sum_a \langle \tau_{d-1} \gamma \, \tau_0 \Delta_a \rangle_0 \Delta^a = \sum_{a,\beta} q^\beta \langle \tau_{d-1} \gamma \, \tau_0 \Delta_a \rangle_{0,\beta} \Delta^a$$

and put $U_0(\gamma) = \gamma$.

The formula (6.3) means that in the stable range and for $d \geq 1$ the correlator of any element of the form

$$\tau_{d,e} \gamma - \tau_{d-1,e+1} \gamma - \tau_{0,e}(U_d(\gamma))$$

with any product of others $\tau_{d_i,e_i} \gamma_i$ vanishes; the same is true for $d = 0$ by the definition of U_0. Hence by induction, in any stable range correlator we can replace any expression $\tau_{d,0}\gamma$ by $\sum_{j=0}^{d}\tau_{0,j}(U_{d-j}(\gamma))$ without changing the value of the correlator. In particular,

$$F_g^{st}(x) = \sum_n \frac{1}{n!} \langle \prod_{i=1}^n \sum_{a_i,d_i} x_{d_i}^{a_i}\tau_{d_i}\Delta_{a_i}\rangle_g$$

$$= \sum_n \frac{1}{n!} \langle \prod_{i=1}^n \sum_{a_i}^{d_i} x_{d_i,a_i} \sum_{j_i=0}^{d_i} \tau_{0,j_i}(U_{d_i-j_i}(\Delta_{a_i}))\rangle_g$$

(7.6)
$$= \sum_n \frac{1}{n!} \langle \prod_{i=1}^n \sum_{c_i,b_i} y_{c_i}^{b_i}\tau_{0,c_i}\Delta_{b_i}\rangle_g = G_g(y).$$

To obtain the last equality in (7.6), use (7.5) in order to represent each sum in the correlator product as a linear combination of terms $\tau_{0,c}\Delta_b$. The straightforward calculation of coefficients furnishes (7.4).

Remark. The operator T defined by $y = T(x)$ is a linear transformation of the large phase space with coefficients in Λ defined entirely in terms of genus zero two–point correlators. It is invertible, because (7.4) shows that it is the sum of identity and the operator which strictly raises the gravitational weight c. Hence we may define the corrected version of $G_g(x)$ by $\widetilde{G}_g(x) := F_g(T^{-1}(x))$. Equivalently, we can extend the modified correlators to the unstable range keeping the natural functional equations.

One can also use these formulas in order to give independent meaning to the symbols $\tau_{0,d}$ as linear operators on $\Lambda \otimes \mathcal{H}(V)$.

7.3. Expressing T through the three–point primary correlators. The correlators $\langle \ldots \rangle_g$ are Λ-polylinear functions on $\Lambda \otimes \mathcal{H}(V)$. We will write simply $\langle \ldots \rangle$ when $g = 0$. Setting in (6.14) $d_2 = 0$, multiplying by q^β and summing, we obtain:

(7.7)
$$\langle \gamma_0\, \tau_d\gamma_1\, \gamma_2 \rangle = \sum_a \langle \tau_{d-1}\gamma_1\, \Delta_a\rangle\langle \Delta^a\, \gamma_0\, \gamma_2\rangle.$$

Put

(7.8)
$$\gamma_0 \cdot \gamma_2 := \sum_a \Delta_a \langle \Delta^a \gamma_0\, \gamma_2\rangle$$

(this is essentially the product in "small" quantum cohomology where the structure constants are the third derivatives of the genus zero potential restricted to H^2).

Then we can rewrite (7.7) as

(7.9)
$$\langle \gamma_0\, \tau_d\gamma_1\, \gamma_2\rangle = \langle \tau_{d-1}\gamma_1\, \gamma_0 \cdot \gamma_2\rangle.$$

Now let l be any linear function on $H_2(V,\Lambda)$. It defines the derivation $\partial_l : \Lambda \to \Lambda$, $\partial_l q^\beta := l(\beta)\, q^\beta$. We extend it to formal series over Λ coefficientwise. If γ_0 is an ample divisor class considered as a linear function on H_2, we write ∂_{γ_0} for this derivation, as in 5.5. Turning now to the equation (6.16), multiply it by q^β and sum over all β. The left hand side of (6.16) vanishes for $\beta = 0$, and the right hand side does not make sense, so we get:

$$\langle \tau_d \gamma_1 \, \gamma_2 \rangle = \sum_{j=1}^{d+1} (-1)^{j+1} \partial_{\gamma_0}^{-j} [\langle \gamma_0 \, \tau_{d+1-j} \gamma_1 \, \tau_0(\gamma_0^{j-1} \cup \gamma_2) \rangle$$
$$- \langle \gamma_0 \, \tau_{d+1-j} \gamma_1 \, \tau_0(\gamma_0^{j-1} \cup \gamma_2) \rangle_{0,0}].$$

To interpret this, notice that since $(\gamma_0, \beta) \neq 0$ for all algebraic effective non–zero 2–homology classes on V, $\partial_{\gamma_0}^{-1} F$ makes sense for any series F whose coefficients are correlators not involving the $\beta = 0$ ones. As the result of this "integration" we take the series again not involving the $\beta = 0$ terms.

Actually, in view of (6.9), the $\beta = 0$ terms vanish unless $j = d + 1$. Separating this summand and replacing the remaining triple correlators with the help of (7.9), we get the following result.

7.3.1. Proposition. *The matrix coefficients of T can be expressed inductively through the triple primary correlators, that is, Gromov–Witten invariants, of genus zero: for $d \geq 1$*

$$\langle \tau_d \gamma_1 \, \gamma_2 \rangle = \sum_{j=1}^{d} (-1)^{j+1} \partial_{\gamma_0}^{-j} \langle \tau_{d-j} \gamma_1 \, \gamma_0 \cdot (\gamma_0^{j-1} \cup \gamma_2) \rangle$$
(7.10) $$+ (-1)^d \partial_{\gamma_0}^{-(d+1)} [\langle \gamma_0 \, \gamma_1 \, \gamma_0^d \cup \gamma_2 \rangle - \langle \gamma_0 \, \gamma_1 \, \gamma_0^d \cup \gamma_2 \rangle_{0,0}].$$

7.4. Gravitational descendants for the Frobenius manifolds of qc–type. Now let M be a formal Frobenius manifold over $\operatorname{Spec} \Lambda$ whose space of flat vector fields is a free Λ–module denoted H. Without any additional hypotheses, we can define its modified correlators with gravitational descendants. They are polylinear functions on the large phase space $\bigoplus_{d \geq 0} H[d]$ where $H[d]$ are copies of the space H identified with the help of the shift operator $\tau : H[d] \to H[d+1]$. To define these correlators explicitly, consider the respective genus zero Cohomological Field Theory given by $I_n^M : H^{\otimes n} \to H^*(\overline{M}_{0,n}, \Lambda)$, $n \geq 3$, satisfying the usual properties and put

(7.11) $$\langle \overline{\tau}_{d_1} \Delta_{a_1} \ldots \overline{\tau}_{d_n} \Delta_{a_n} \rangle := \int_{\overline{M}_{0,n}} I_n^M(\Delta_{a_1} \otimes \cdots \otimes \Delta_{a_n}) \psi_1^{d_1} \ldots \psi_n^{d_n},$$

where we write $\overline{\tau}_d$ for what was $\tau_{0,d}$ in the context of quantum cohomology, and $\psi_i = c_1(L_i)$ on $\overline{M}_{0n}$.

The generating function for these correlators, the modified potential on our formal large phase space, is

(7.12) $$G^M(x) = \sum_{n \geq 3, (a_i, d_i)} \frac{x_{d_n}^{a_n} \ldots x_{d_1}^{a_1}}{n!} \langle \overline{\tau}_{d_1} \Delta_{a_1} \ldots \overline{\tau}_{d_n} \Delta_{a_n} \rangle.$$

In order to define the unmodified correlators, we need the operator T. It can be constructed, if M is assumed to be of qc–type: see the discussion in III.5.4. In view of Claim III.5.4.3, we can define also the (unmodified) two argument correlators $\langle \tau_d \gamma_1 \gamma_2 \rangle$ by the inductive formula (7.10):

$$\langle \tau_d \gamma_1 \gamma_2 \rangle = \sum_{j=1}^{d} (-1)^{j+1} \partial_\delta^{-j} \langle \tau_{d-j} \gamma_1 \delta \cdot (\delta^{j-1} \cup \gamma_2) \rangle$$
(7.13)
$$+ (-1)^d \partial_{\gamma_0}^{-(d+1)} [\langle \delta \gamma_1 \delta^d \cup \gamma_2 \rangle - \langle \delta \gamma_1 \delta^d \cup \gamma_2 \rangle_0].$$

Here $\delta \in H^2$ is an arbitrary (say, generic) element such that $(\delta, \beta) \neq 0$ for all $\beta \in B \setminus \{0\}$ and the operator ∂_δ^{-1} divides $\langle \ldots \rangle_\beta$ by (δ, β).

Furthermore, put

$$y_c^b = x_c^b + \sum_{(a,d), d \geq c+1} x_d^a \langle \tau_{d-c-1} \Delta_a \Delta^b \rangle.$$

Then the large phase space potential of M is, by definition, $F^M(x) := G^M(y)$, and the unmodified correlators with gravitational descendants of M in the stable range are defined as the coefficients of F^M:

(7.14)
$$F^M(x) = \sum_{n \geq 3, (a_i}^{d_i)} \frac{x_{d_n}^{a_n} \ldots x_{d_1}^{a_1}}{n!} \langle \tau_{d_1} \Delta_{a_1} \ldots \tau_{d_n} \Delta_{a_n} \rangle.$$

7.5. Differential equations following from the linear relations between the boundary classes. We will now extend Theorem 4.3.1 to the large phase space. Working with modified correlators makes this extension quite straightforward.

Generalizing (4.18) define the symmetric differential forms on the appropriate completion of $\mathcal{H}(V)$:

(7.15)
$$D^n G_g = \frac{1}{n!} \sum_{(a_i, d_i) \in (A \times \mathbf{Z}_{\geq 0})^n} dx_{d_n}^{a_n} \ldots dx_{d_1}^{a_1} \partial_{d_1, a_1} \ldots \partial_{d_n, a_n} G_g.$$

Now for any stable modular (g, n)-graph σ put

(7.16)
$$P_\sigma = \text{Contr}_{E_\sigma} \left(\prod_{v \in V_\sigma} D^{n_v} G_{g_v, n_v} \right).$$

The contraction operator Contr_{E_σ} was described at the end of subsection 4.3. In the context of the large phase space it is defined by essentially *the same* Laplace operator (4.20):

(7.17)
$$\Box = \sum_{a,b \in A} \partial_{0,a} g^{ab} \partial_{0,b}$$

and the contraction rules

$$\partial_{0,a} \dashv dx_d^b = (-1)^{ab} dx_d^b \vdash \partial_{0,a} = \delta_{0,d} \delta_{ab}.$$

For each multi-index $(d_1, a_1), \ldots, (d_n, a_n)$, the coefficient of $dx_{d_1}^{a_1} \ldots dx_{d_n}^{a_n}$ in P_σ is the differential polynomial in G_h for $h \leq g$. We now have the following extension of Theorem 4.3.1:

7.5.1. Theorem. *For any linear relation*
$$\sum_j c_j [\widetilde{M}(\sigma_j)] = 0$$
in the homology of $\overline{M}_{g,n}$ we have:

(7.18) $$\sum_j \frac{c_j}{|\operatorname{Aut} \sigma_j|} P_{\sigma_j} = 0.$$

Proof. The proof follows the same pattern as that of Theorem 4.3.1. We will sketch the necessary modifications.

Generalizing (4.1), for any map $\mathbf{d} : S \to \mathbf{Z}_{\geq 0}$ put

(7.19) $$Y_{g,S,\mathbf{d}} \left(\bigotimes_{i \in S} \gamma_i \right) = \int_{\overline{M}_{g,S}} I_{g,S} \left(\bigotimes_{i \in S} \gamma_i \right) \prod_{i \in S} \psi_i^{d_i}.$$

Similarly extending (4.3), for a stable (g, S)–modular graph σ and a map $\mathbf{d} : S \to \mathbf{Z}_{\geq 0}$ put

(7.20) $$Y(\sigma, \mathbf{d}) \left(\bigotimes_{i \in S} \gamma_i \right) = \int_{\widetilde{M}(\sigma)} I_{g,S} \left(\bigotimes_{i \in S} \gamma_i \right) \prod_{i \in S} \psi_i^{d_i}.$$

The respective extension of (4.4) reads

(7.21) $$Y(\sigma, \mathbf{d}) \left(\bigotimes_{i \in S} \gamma_i \right) = \frac{1}{|\operatorname{Aut} \sigma|} \left(\bigotimes_{v \in V_\sigma} Y_{g_v, F_\sigma(v), \mathbf{d}_v} \right) \left(\bigotimes_{i \in S} \gamma_i \otimes \Delta^{\otimes E_\sigma} \right),$$

where $\mathbf{d}_v$ is the restriction of $\mathbf{d}$ to the subset of tails at v extended by zero to the remaining flags, that is, halves of edges. The geometric meaning of this extension is evident: when a curve degenerates, sections become distributed among its irreducible components. This distribution induces the respective splitting of $\prod_{i \in S} \psi_i^{d_i}$ so that

$$b(\sigma)^* \left(\prod_{i \in S} \psi_i^{d_i} \right) = \prod_{v \in V_\sigma} \prod_{i \in S_\sigma(v)} \psi_i^{d_i}.$$

The remaining special points of normalizations of these components (those that become singular) come without descendants, i.e. have $d_i = 0$. In the final count, this is why the contraction operator is of the form (7.17).

We now pass to the generating functions. We start with extending the Gromov–Witten operators to the large phase space: define $\overline{I}_{g,n} : \Lambda \otimes \mathcal{H}(V)^{\otimes n} \to H^*(\overline{M}_{g,n}, \Lambda)$ by

$$\overline{I}_{g,n}(\tau_{0,d_1} \Delta_{a_1} \otimes \cdots \otimes \tau_{0,d_n} \Delta_{a_n}) = I_{g,n}(\Delta_{a_1} \otimes \cdots \otimes \Delta_{a_n}) \prod_{i=1}^n \psi_i^{d_i}.$$

Here we interpret $\tau_{0,d}$ as linear operators and notice that $\tau_{0,d} \Delta_a$ form a basis of $\Lambda \otimes \mathcal{H}(V)$; cf. Remark in 7.2.

The averaged exponential Φ_g (see (4.7)) is replaced by G_g from (7.2). Putting $\Gamma = \sum_{a,d} x_d^a \tau_{0,d} \Delta_a$ we have an analog of (4.8):

$$G_g(x) = \left(\sum_{N=0}^{\infty} \int_{[\widetilde{M}_{g,N}]} \overline{I}_{g,N} \right) (e^\Gamma).$$

In order to extend (4.9), consider an extended labeling $(\mathbf{a}, \mathbf{d}) : \{1, \ldots, n\} \to A \times \mathbf{Z}_{\geq 0}$ and put (omitting the tensor product signs)

(7.22) $$G_{\sigma, \mathbf{a}, \mathbf{d}}(x) = \sum_{N=0}^{\infty} \int_{\pi_N^*[\widetilde{M}(\sigma)]} I_{g,n+N} \left(\tau_{0,d_1} \Delta_{a_1} \ldots \tau_{0,d_n} \Delta_{a_n} e^\Gamma \right).$$

The remaining part of the proof faithfully follows the calculations starting with (4.15).

7.5.2. Example. In genus zero, we get the extension of the Associativity Equations 0.(0.1) to the large phase space which has the following form. Put $G = G_0$. Then for any $(a_1, d_1), \ldots, (a_4, d_4)$

(7.23)
$$\sum_{e,f} \partial_{d_1, a_1} \partial_{d_2, a_2} \partial_{0, e} G \, g^{ef} \, \partial_{0, f} \partial_{d_3, a_3} \partial_{d_4, a_4} G$$
$$= (-1)^{a_2 a_3} \sum_{e,f} \partial_{d_1, a_1} \partial_{d_3, a_3} \partial_{0, e} G \, g^{ef} \, \partial_{0, f} \partial_{d_2, a_2} \partial_{d_4, a_4} G.$$

Notice that this does not endow the large phase space with a structure of the Frobenius manifold. At most, G can be considered as a family of Frobenius potentials on the small phase space depending on the parameters x_d^a with $d \geq 1$. When these parameters vanish, we get the usual quantum cohomology potential.

7.6. On the geometric meaning of the large phase space. We have introduced the large phase space rather formally, as $\mathcal{H} = H[t]$ where H is the cohomology space, or the space of flat vector fields of a general (say, simply connected) Frobenius manifold M.

Looking for its geometric interpretation, we can think either about the geometry of M, or of V and of $\overline{M}_{g,n}(V, \beta)$ (when dealing with quantum cohomology).

From the viewpoint of geometry of M, $\mathcal{H}$ appears as a version of the parametrized loop space of M. B. Dubrovin uses this interpretation as the heuristic tool in several papers; see for example [D2], p. 322.

From the viewpoint of algebraic geometry, $\mathcal{H}$ looks like the (co)homolgy of $V \times B\mathbf{G}_m$ where $B\mathbf{G}_m$ in turn replaces the non–existent space $\overline{M}_{02}$. (Notice that the stack of prestable curves $\mathcal{M}_{02}$ contains $B\mathbf{G}_m$ but is considerably larger.)

Both interpretations must be refined in order to clarify the geometric origin of several structures essential for the understanding of the generating functions for correlators. Here is a brief list of these structures.

(i) The operator T. It seems that in the context of Frobenius geometry, it is a version of the monodromy operator for one of the structure connections, and in the context of quantum cohomology, a version of the Gromov–Witten correspondence in $A_*(V^2 \times B\mathbf{G}_m)$ or in some related group.

(ii) *The non-linear projection* $u: \mathcal{H} \to H$. This operator is defined by
$$u^*(x^a) = \sum_b g^{ab} \partial_{00}\partial_{0b} F_0. \tag{7.24}$$

A result of [DijW] shows that by lifting the second derivatives of Φ from H via u we obtain the same derivatives of F_0: cf. [Ge5], Theorem 14. There is a version of this theorem for genus 1 and conjecturally, for any genus.

(iii) *Virasoro operators.* They present a certain problem already in the cases $M = $ *a point* and $V = $ *a point*. A sound geometric interpretation must include the geometric description of the whole Sugawara construction and of the representation space of the Virasoro algebra generated by the highest weight wector e^F. Conjecturally, it is a (co)homology or even a motive and the Virasoro operators act via correspondences.

Generally, it seems that the Virasoro generators introduced in 2.5 are some twisted combinations of the "vertical" Virasoro operators of the $V = $ *a point* case and the "horizontal" Virasoro operators $L_n = E^{\circ n}$ associated with the classical or Katz's Euler vector field. For a precise but not strong enough statement to this effect, see [Ge6], Theorem 6.6, which reproduces in the context of quantum cohomology the more general result from [DZh3]. Again, the problem of the geometric origin of these operators remains wide open.

Bibliography

[AbV1] D. Abramovich, A. Vistoli. *Complete moduli for fibered surfaces.* Preprint math.AG/9804097.

[AbV2] D. Abramovich, A. Vistoli. *Complete moduli for families over semistable curves.* Preprint math.AG/9811059.

[AC1] E. Arbarello, M. Cornalba. *Combinatorial and algebro-geometric cohomology classes on the moduli spaces of curves.* Journ. Alg. Geom., 5 (1996), 705–749.

[AC2] E. Arbarello, M. Cornalba. *Calculating cohomology groups of moduli spaces of curves via algebraic geometry.* Preprint math.AG/9803001.

[AL1] D. Arinkin, S. Lysenko. *Isomorphisms between moduli spaces of $SL(2)$-bundles with connections on $\mathbf{P}^1 \setminus \{x_1, \ldots, x_4\}$.* Math. Res. Lett., 4 (1997), 181–190.

[AL2] D. Arinkin, S. Lysenko. *On the moduli of $SL(2)$-bundles with connections on $\mathbf{P}^1 \setminus \{x_1, \ldots, x_4\}$.* Int. Math. Res. Notices, 19 (1997), 983–999.

[AP] D. Arinkin, A. Polishchuk. *Fukaya category and Fourier transform.* Preprint math.AG/9811023.

[AGV] V. Arnold, S. Gusein-Zade, A. Varchenko. *Singularities of differentiable maps*, vols. I, II. Birkhäuser, Boston, 1985 and 1988.

[Ar] M. Artin. *Versal deformations and algebraic stacks.* Inv. Math., 27 (1974), 165–189.

[AM] P. S. Aspinwall, D. R. Morrison. *Topological field theory and rational curves.* Comm. Math. Phys., 151 (1993), 245–262.

[BarK] S. Barannikov, M. Kontsevich. *Frobenius manifolds and formality of Lie algebras of polyvector fields.* Int. Math. Res. Notices, 4 (1998), 201–215.

[Bar] S. Barannikov. *Extended moduli spaces and mirror symmetry in dimensions > 3.* Preprint Math. AG/9903124.

[Bara] V. Baranovski. *Moduli of sheaves on surfaces and action of the oscillator algebra.* Preprint math.AG/9811092.

[Ba1] V. Batyrev. *Dual polyhedra and the mirror symmetry for Calabi-Yau hypersurfaces in toric varieties.* Journ. Alg. Geom., 3 (1994), 493–535.

[Ba2] V. Batyrev. *Variation of the mixed Hodge structure of affine hypersurfaces in algebraic tori.* Duke Math. J., 69 (1993), 349–409.

[Ba3] V. Batyrev. *Quantum cohomology ring of toric manifolds.* Astérisque, 218 (1993), 9–34.

[Ba4] V. Batyrev. *Mirror symmetry and canonical flat coordinates on moduli spaces of Calabi-Yau manifolds via the formal deformation theory.* Preprint, 1993.

[Ba5] V. Batyrev. *Quantum cohomology rings of toric manifolds.* Astérisque, 218 (1993), 9–34.

[BaBo1] V. Batyrev, L. Borisov. *On Calabi–Yau complete intersections in toric varieties.* In: Proc. of Int. Conf. on Higher Dimensional Complex Varieties (Trento, June 1994), ed. by M. Andreatta, De Gruyter, 1996, 39–65.

[BaBo2] V. Batyrev, L. Borisov. *Dual cones and mirror symmetry for generalized Calabi–Yau manifolds.* In: Mirror Symmetry II, ed. by S. T. Yau, 1996, 65–80.

[BaBo3] V. Batyrev, L. Borisov. *Mirror duality and string–theoretic Hodge numbers.* Inv. Math., 126:1 (1996), 183–203.

[BaS] V. Batyrev, D. van Straten. *Generalized hypergeometric functions and rational curves on Calabi–Yau complete intersections in toric varieties.* Comm. Math. Phys., 168 (1995), 493–533.

[Beau] A. Beauville. *Quantum cohomology of complete intersections.* Preprint alg–geom/9501008.

[Beh] K. Behrend. *Gromov–Witten invariants in algebraic geometry.* Inv. Math., 127 (1997), 601–617.

[BehF] K. Behrend, B. Fantechi. *The intrinsic normal cone.* Inv. Math., 128 (1997), 45–88.

[BehM] K. Behrend, Yu. Manin. *Stacks of stable maps and Gromov–Witten invariants.* Duke Math. J., 85:1 (1996), 1–60.

[BeG] A. Beilinson, V. Ginzburg. *Infinitesimal structure of moduli spaces of G-bundles.* Int. Math. Res. Notices, 4 (1992), 63–74.

[BelP] P. Belorousski, R. Pandharipande. *A descendent relation in genus 2.* Preprint math.AG/9803072.

[BerCOV] M. Bershadsky, S. Cecotti, H. Ooguri, C. Vafa. *Kodaira–Spencer theory of gravity and exact results for quantum string amplitudes.* Comm. Math. Phys., 165 (1994), 311–427.

[BertTh] A. Bertram, M. Thaddeus. *On quantum cohomology of symmetric product of an algebraic curve.* Preprint math.AG/9803026.

[BesIZ] D. Bessis, C. Itzykson, J. B. Zuber. *Quantum field theory techniques in graphical enumeration.* Adv. in Appl. Math., 1 (1980), 109–157.

[BiCPP] G. Bini, C. de Concini, M. Polito, C. Procesi. *On the work of Givental relative to Mirror Symmetry.* Preprint math.AG/9805097.

[BlVar] B. Blok, A. Varchenko. *Topological conformal field theories and the flat coordinates.* Int. J. Math. Phys., A7 (1992), 1467.

[Bor1] L. Borisov. *On Betti numbers and Chern classes of varieties with trivial odd cohomology groups.* Preprint alg-geom/9703023.

[Bor2] L. Borisov. *Vertex algebras and mirror symmetry.* Preprint math.AG/9809094.

[Br] J.-L. Brylinski. *A differential complex for Poisson manifolds.* J. Diff. Geom., 28 (1988), 93–114.

[COGP] Ph. Candelas, X. C. de la Ossa, P. S. Green, L. Parkes. *A pair of Calabi–Yau manifolds as an exactly soluble superconformal theory.* Nucl. Phys. B, 359 (1991), 21–74.

[CaoZh1] H.-D. Cao, J. Zhou. *Frobenius manifold structure on Dolbeault and mirror symmetry.* Preprint math.DG/9805094.

[CaoZh2] H.-D. Cao, J. Zhou. *Identification of two Frobenius manifolds in mirror symmetry.* Preprint math.DG/9805095.

[CaH1] L. Caporaso, J. Harris. *Enumerating rational curves: the rational fibration method.* Preprint alg-geom/9608023.

[CaH2] L. Caporaso, J. Harris. *Parameter spaces for curves on surfaces and enumeration of rational curves.* Preprint alg-geom/9608024.

[CaH3] L. Caporaso, J. Harris. *Counting plane curves of any genus.* Preprint alg-geom/9608025.

[CK] E. Cattani, A. Kaplan. *Degenerating variations of Hodge structures.* Astérisque, 179–180 (1989), 67–96.

[Ce] S. Cecotti. $N = 2$ *Landau–Ginzburg vs. Calabi–Yau σ-models: non-perturbative aspects.* Int. J. of Mod. Phys. A, 6:10 (1991), 1749–1813.

[C–F] I. Ciocan–Fontanine. *Quantum cohomology of flag varieties.* Int. Math. Res. Notes, 6 (1995), 263–277.

[De1] P. Deligne. *Équations différentielles á points singuliers réguliers.* Springer Lecture Notes in Math., 163 (1970).

[De2] P. Deligne. *Local behavior of Hodge structures at infinity.* In: Mirror Symmetry II, ed. by B. Greene and S. T. Yau, AMS–International Press, 1996, 683–699.

[DGMS] P. Deligne, Ph. Griffiths, J. Morgan, D. Sullivan. *Real homotopy theory of Kähler manifolds.* Inv. Math., 29 (1975), 245–274.

[DeM] P. Deligne, D. Mumford. *The irreducibility of the space of curves of given genus.* Publ. Math. IHES, 36 (1969), 75–109.

[DFI1] P. Di Francesco, C. Itzykson. *A generating function for fatgraphs.* Preprint hep-th/9212108.

[DFI2] P. Di Francesco, C. Itzykson. *Quantum intersection rings.* In: The Moduli Space of Curves, ed. by R. Dijkgraaf, C. Faber, G. van der Geer, Progress in Math., vol. 129, Birkhäuser, 1995, 149–163.

[DFIZ] P. Di Francesco, C. Itzykson, J.-B. Zuber. *Polynomial averages in the Kontsevich model.* Preprint hep-th/9206090.

[DijVV1] R. Dijkgraaf, H. Verlinde, E. Verlinde. *Loop equations and Virasoro constraints in non–perturbative two-dimensional quantum gravity.* Nucl. Phys. B, 348 (1991), 435–456.

[DijVV2] R. Dijkgraaf, H. Verlinde, E. Verlinde. *Notes on topological string theory and 2D quantum gravity.* Preprint PUTP–1217, 1990.

[DijW] R. Dijkgraaf, E. Witten. *Mean field theory, topological field theory, and multimatrix models.* Nucl. Phys. B, 342 (1990), 486–522.

[DoM] R. Donagi, E. Markman. *Cubics, integrable systems, and Calabi–Yau threefolds.* In: Proc. of the Conf. in Alg. Geometry dedicated to F. Hirzebruch, Israel Math. Conf. Proc., 9 (1996).

[Dor] Ch. Doran. *Picard–Fuchs uniformization: modularity of the mirror map and mirror-moonshine.* Preprint math.AG/9812162.

[D1] B. Dubrovin. *Integrable systems in topological field theory.* Nucl. Phys. B, 379 (1992), 627–689.

[D2] B. Dubrovin. *Geometry of 2D topological field theories.* In: Springer LNM, 1620 (1996), 120–348.

[D3] B. Dubrovin. *Painlevé equations in 2D topological field theories.* In: Painlevé property, One Century Later, Cargèse, 1996. Preprint math.AG/9803107.

[D4] B. Dubrovin. *Flat pencils of metrics and Frobenius manifolds.* Preprint math.AG/9803106.

[D5] B. Dubrovin. *Painlevé transcendents in two-dimensional topological field theory.* Preprint math.AG/9803107.

[D6] B. Dubrovin. *Geometry and analytic theory of Frobenius manifolds.* Proc. ICM Berlin 1998, vol. II, 315–326. Preprint math/9807034.

[DZh1] B. Dubrovin, Y. Zhang. *Extended affine Weyl groups and Frobenius manifolds.* Comp. Math., 111 (1998), 167–219.

[DZh2] B. Dubrovin, Y. Zhang. *Bihamiltonian hierarchies in 2D topological field theory at one-loop approximation.* Preprint hep-th/9712232.

[DZh3] B. Dubrovin, Y. Zhang. *Frobenius manifolds and Virasoro constraints.* Peprint math.AG/9808048.

[EdG1] D. Edidin, W. Graham. *Localization in equivariant intersection theory and the Bott residue formula.* Am. Journ. Math., 120 (1998), 619–636.

[EdG2] D. Edidin, W. Graham. *Equivariant intersection theory.* Inv. Math., 131 (1998), 595–634.

[EHX1] T. Eguchi, K. Hori, Ch.-Sh. Xiong. *Gravitational quantum cohomology.* Int. J. Math. Phys., A12 (1997), 1743–1782. hep–th/9605225.

[EHX2] T. Eguchi, K. Hori, Ch.-Sh. Xiong. *Quantum cohomology and Virasoro algebra.* Phys. Lett. B, 402 (1997), 71–80. Preprint hep–th/9703086.

[EJX] T. Eguchi, M. Jinzenji, Ch.-Sh. Xiong. *Quantum cohomology and free field representation.* Nucl. Phys. B, 510 (1998), 608–622. hep-th/9709152.

[EX] T. Eguchi, Ch.-Sh. Xiong. *Quantum Cohomology at Higher Genus: Topological Recursion Relations and Virasoro Conditions.* Adv. Theor. Math. Phys. 2(1998), 219–229. Preprint hep-th/9801010.

[Fab1] C. Faber. *Algorithms for computing intersection numbers on moduli spaces of curves, with an application to the class of the locus of Jacobians.* Preprint, 1997.

[Fab2] C. Faber. *A conjectural description of the tautological ring of the moduli spaces of curves.* Preprint, 1996, http://www.math.okstate.edu/preprint/1997.html.

[FabP] C. Faber, R. Pandharipande. *Hodge integrals and Gromov-Witten theory.* Preprint math.AG/9810173.

[FGK] S. Fomin, S. Gelfand, A. Postnikov. *Quantum Schubert Polynomials.* Preprint, 1996.

[F] R. Fuchs. *Über lineare homogene Differentialgleichungen zweiter Ordnung mit im endlich gelegene wesentlich singulären Stellen.* Math. Ann., 63 (1907), 301–321.

[Fu] K. Fukaya. *Morse homotopy, A^∞-categories, and Floer homologies.* In: Proc. of the 1993 GARC Workshop on Geometry and Topology, ed. by H. J. Kim, Lecture Notes Ser., 18, Seoul Nat. Univ., 1993.

[FuO] K. Fukaya, K. Ono. *Arnold conjecture and Gromov–Witten invariant.* Preprint, 1996.

[Ful] W. Fulton. *Intersection Theory.* Springer, 1984.

[FulMPh] W. Fulton, R. MacPherson. *A compactification of configuration spaces.* Ann. of Math., 139 (1994), 183–225.

[FulP] W. Fulton, R. Pandharipande. *Notes on stable maps and quantum cohomology.* In: Proc. of Symposia in Pure Math., Algebraic Geom., Santa Cruz 1995 (ed. by J. Kollár, R. Lazarsfeld, D. Morrison), vol. 62, part 2, 45–96. Preprint alg-geom/9608011.

[G] B. Gambier. *Sur les équations différentielles du second ordre et du prémier degré dont l'intégrale générale est à points critiques fixes.* CR Acad. Sci. Paris, 142 (1906), 266–269.

[Gep1] D. Gepner. *On the spectrum of 2D conformal field theory.* Nucl. Phys. B, 287 (1987), 111–126.

[Gep2] D. Gepner. *Fusion rings and geometry.* Comm. Math. Phys., 141 (1991), 381–411.

[Ger] M. Gerstenhaber. *The cohomology structure of an associative ring.* Ann. of Math., 79 (1963), 267–288.

[Ge1] E. Getzler. *Operads and moduli spaces of genus zero Riemann surfaces.* In: The Moduli Space of Curves, ed. by R. Dijkgraaf, C. Faber, G. van der Geer, Progress in Math. vol. 129, Birkhäuser, 1995, 199–230.

[Ge2] E. Getzler. *Resolving mixed Hodge modules on configuration spaces.* Preprint MPI 96-144.

[Ge3] E. Getzler. *The semi-classical approximation for modular operads.* Preprint MPI 96-145.

[Ge4] E. Getzler. *Intersection theory on $\overline{M}_{1,4}$ and elliptic Gromov–Witten invariants.* Journ. AMS, 10 (1997), 973–998. alg-geom/9612004.

[Ge5] E. Getzler. *Topological recursion relations in genus 2.* Preprint math.AG/9801003.

[Ge6] E. Getzler. *The Virasoro conjecture for Gromov–Witten invariants.* Preprint math.AG/9812026.

[GeJ] E. Getzler, J.D.S. Jones. *Operads, homotopy algebra, and iterated integrals for double loop spaces.* Preprint, 1994.

[GeK1] E. Getzler, M. M. Kapranov. *Cyclic operads and cyclic homology.* In: Geometry, Topology, and Physics for Raoul, ed. by B. Mazur, Internat. Press, Cambridge, MA, 1995, 167–201.

[GeK2] E. Getzler, M. M. Kapranov. *Modular operads.* Compositio. Math., 110 (1998), 65–126. Preprint dg-ga/9408003.

[GePa] E. Getzler, R. Pandharipande. *Virasoro constraints and the Chern classes of the Hodge bundle.* Nucl. Phys. B, 530 (1998), 701–714. Preprint math.AG/9805114.

[Gil] H. Gillet. *Intersection theory on algebraic stacks and Q-varieties.* J. Pure Appl. Algebra, 34 (1984), 193–240.

[GiK] V. A. Ginzburg, M. M. Kapranov. *Koszul duality for operads.* Duke Math. J., 76:1 (1994), 203–272.

[Giv1] A. Givental. *Homological geometry I: Projective hypersurfaces.* Selecta Math., new ser. 1:2 (1995), 325–345.

[Giv2] A. Givental. *Equivariant Gromov-Witten invariants.* Int. Math. Res. Notes, 13 (1996), 613–663.

[Giv3] A. Givental. *Stationary phase integrals, quantum Toda lattices, flag manifolds and the mirror conjecture.* Preprint, 1996.

[Giv4] A. Givental. *Homological geometry and mirror symmetry.* In: Proc. of the ICM, Zürich 1994, Birkhäuser, 1995, vol. 1, 472–480.

[Giv5] A. Givental. *A mirror theorem for toric complete intersections.* Preprint alg-geom/9702016.

[Giv6] A. Givental. *Elliptic Gromov-Witten invariants and the generalized mirror conjecture.* Preprint math.AG/9803053.

[Giv7] A. Givental. *The mirror formula for quintic threefolds.* Preprint math. AG/9807070.

[GivK] A. Givental, B. Kim. *Quantum cohomology of flag manifolds and Toda lattices.* Comm. Math. Phys., 168 (1994), 609–641.

[GoM] W. Goldman, J. Millson. *The deformation theory of representations of fundamental groups of compact Kähler manifolds.* Publ. Math. IHES, 86 (1988), 43–96.

[GolLO] V. Golyshev, V. Lunts, D. Orlov. *Mirror symmetry for abelian varieties.* Preprint math.AG/9812003.

[GP] L. Göttsche, R. Pandharipande. *The quantum cohomology of blow-ups of* P^2 *and enumerative geometry.* Preprint, 1997.

[GrPa] T. Graber, R. Pandharipande. *Localization of virtual classes.* Preprint, 1997.

[Gre] B. Greene. *Constructing mirror manifolds.* In: Mirror Symmetry II, ed. by B. Greene and S. T. Yau, AMS–International Press, vol. 1, Amer. Math. Soc., Providence, RI, 1996, 29–69.

[Groj] I. Grojnowski. *Instantons and affine algebras. I. The Hilbert scheme and vertex operators.* Math. Res. Lett., 3:2 (1996), 275–291.

[Gro] M. Gromov. *Pseudoholomorphic curves in symplectic manifolds.* Inv. Math., 82 (1985), 307–447.

[GD] A. Grothendieck, J. Dieudonné. *Eléments de Géométrie Algébrique (EGA).* I, Springer Verlag, 1971; II, Publ. Math. IHES, 8, 1961; III, Publ. Math. IHES, 11, 1961, 17, 1963; IV, Publ. Math. IHES, 20, 1964; 24, 1965; 28, 1966; 32, 1967.

[GroW] M. Gross, P. M. H. Wilson. *Mirror symmetry via 3-tori for a class of Calabi-Yau threefolds.* Preprint alg-geom/9608004.

[G-ZV] S. Gusein-Zade, A. Varchenko. *Verlinde algebras and the intersection form on vanishing cycles.* Selecta Math., New. Ser., 3 (1997), 79–97.

[Gu] D. Guzzetti. *Stokes matrices and monodromy for the quantum cohomology of projective spaces*. Preprint SISSA 87/98/FM.

[HL] R. Hain, E. Looienga. *Mapping class groups and moduli spaces of curves*. Preprint alg-geom/9607004.

[HaZ] J. Harer, D. Zagier. *The Euler characteristic of the moduli space of curves*. Inv. Math., 85 (1986), 457–485.

[Har] J. Harnad. *Dual isomonodromic deformations and moment maps to loop algebras*. Comm. Math. Phys., 166 (1994), 337–365.

[Ha1] R. Hartshorne. *Residues and duality*. Springer LN in Math., 20, 1966.

[Ha2] R. Hartshorne. *Algebraic geometry*. Springer, 1977.

[He] C. Hertling. *Classifying spaces for polarized mixed Hodge structures and for Brieskorn lattices*. Preprint, 1997.

[HeMa] C. Hertling, Yu. Manin. *Weak Frobenius manifolds*. Int. Math. Res. Notes, (1999). Preprint math.QA/9810132.

[H1] N. Hitchin. *Poncelet polygons and the Painlevé equations*. In: Geometry and Analysis, ed. by S. Ramanan, Oxford University Press, Bombay, 1995, 151–185.

[H2] N. Hitchin. *Twistor spaces, Einstein metrics and isomonodromic deformations*. J. Diff. Geom., 3 (1995), 52–134.

[H3] N. Hitchin. *Frobenius manifolds (notes by D. Calderbank)*. Preprint, 1996.

[H4] N. Hitchin. *The moduli space of special Lagrangian submanifolds*. Preprint dg-ga/9711002.

[H] K. Hori. *Constraints for topological strings in $D \geq 1$*. Nucl. Phys. B, 439 (1995), 395–420. Preprint hep-th/9411135.

[HLY1] S. Hosono, B. H. Lian, S. T. Yau. *GKZ-generalized hypergeometric systems in mirror symmetry of Calabi-Yau hypersurfaces*. Preprint alg-geom/9511001.

[HLY2] S. Hosono, B. H. Lian, S. T. Yau. *Maximal degeneracy points of GKZ systems*. Preprint alg-geom/9603014.

[Il] L. Illusie. *Complexe cotangent et déformations I,II*. Springer LN in Math., 239, 283 (1971).

[I] C. Itzykson. *Counting rational curves on rational surfaces*. Preprint Saclay T94/001.

[IZu] C. Itzykson, J.-B. Zuber. *Combinatorics of the modular group II: the Kontsevich integrals*. Int. J. Mod. Phys., A7 (1992), 5661–5705. Preprint hep-th/9201001.

[JM] M. Jimbo, T. Miwa. *Monodromy preserving deformation of linear ordinary differential equations with rational coefficients II*. Physica, 2D (1981), 407–448.

[KabKi] A. Kabanov, T. Kimura. *Intersection numbers and rank one cohomological field theories in genus one*. Preprint alg-geom/9706003.

[KacS] V. Kac, A. Schwarz. *A geometric interpretation of the partition function of 2D gravity*. Phys. Lett. B, 257, no. 3–4 (1991), 329–334.

[K] S. Katz. *Virasoro constraints on Gromov-Witten invariants*. Handwritten notes of the talk in the Mittag–Leffler Inst., May 1997.

[Ka1] R. Kaufmann. *The intersection form in $H^*(\overline{M}_{0n})$ and the explicit Künneth formula in quantum cohomology*. Int. Math. Res. Notices, 19 (1996), 929–952.

[Ka2] R. Kaufmann. *The tensor product in the theory of Frobenius manifolds.* Preprint MPI 98-60, Bonn.

[Ka3] R. Kaufmann. *The geometry of moduli spaces of pointed curves, the tensor product in the theory of Frobenius manifolds, and the explicit Künneth formula in quantum cohomology.* Bonner Math. Schriften, Nr. 312, Bonn, 1998.

[KaMZ] R. Kaufmann, Yu. Manin, D. Zagier. *Higher Weil-Petersson volumes of moduli spaces of stable n-pointed curves.* Comm. Math. Phys., 181 (1996), 763–787.

[Ke] S. Keel. *Intersection theory of moduli spaces of stable n-pointed curves of genus zero.* Trans. AMS, 330 (1992), 545–574.

[Ki1] B. Kim. *Quantum cohomology of partial flag manifolds and a residue formula for their intersection pairing.* Int. Math. Res. Notes, 1 (1995), 1–16.

[Ki2] B. Kim. *Quantum cohomology of flag manifolds G/B and Toda lattices.* Preprint alg–geom/9607001.

[KirMa] A. N. Kirillov, T. Maeno. *Quantum double Schubert polynomials, quantum Schubert polynomials and Vafa-Intriligator formula.* Preprint 1997.

[Kn1] F. Knudsen. *Projectivity of the moduli space of stable curves, II: the stacks $M_{g,n}$.* Math. Scand., 52 (1983), 161–199.

[Kn2] F. Knudsen. *The projectivity of the moduli space of stable curves III: The line bundles on $M_{g,n}$ and a proof of projectivity of $\overline{M}_{g,n}$ in characteristic 0.* Math. Scand., 52 (1983), 200–212.

[Knut] D. Knutson. *Algebraic spaces.* Springer LN in Math., 203 (1971).

[Ko1] M. Kontsevich. *Intersection theory on moduli spaces and matrix Airy function.* Comm. Math. Phys., 147:1 (1992), 1–23.

[Ko2] M. Kontsevich. *Feynman diagrams and low-dimensional topology.* Proc. of the first European Congr. of Math. (Paris, 1992), vol. II, Birkhäuser, 1994, 97–121.

[Ko3] M. Kontsevich. *Formal (non-)commutative differential geometry.* In: The Gelfand Mathematical Seminars, 1990–92, ed. by L. Corwin, I. Gelfand, J. Lepowsky. Birkhäuser, Boston, 1993, 173–187.

[Ko4] M. Kontsevich. A_∞-*algebras in mirror symmetry.* Bonn MPI Arbeitstagung talk, 1993.

[Ko5] M. Kontsevich. *Homological algebra of Mirror Symmetry.* Proceedings of the ICM (Zürich, 1994), vol. I, Birkhäuser, 1995, 120–139. Preprint alg-geom/9411018.

[Ko6] M. Kontsevich. *Mirror symmetry in dimension 3.* Séminaire Bourbaki, n° 801, Juin 1995.

[Ko7] M. Kontsevich. *Enumeration of rational curves via torus actions.* In: The Moduli Space of Curves, ed. by R. Dijkgraaf, C. Faber, G. van der Geer, Progress in Math. vol. 129, Birkhäuser, 1995, 335–368.

[KM1] M. Kontsevich, Yu. Manin. *Gromov-Witten classes, quantum cohomology, and enumerative geometry.* Comm. Math. Phys., 164:3 (1994), 525–562.

[KM2] M. Kontsevich, Yu. Manin. *Relations between the correlators of the topological sigma-model coupled to gravity.* Comm. Math. Phys., 196 (1998), 385–398.

[KMK] M. Kontsevich, Yu. Manin (with Appendix by R. Kaufmann). *Quantum cohomology of a product.* Inv. Math., 124 (1996), f. 1–3, 313–340.

[Kos] B. Kostant. *Flag manifold quantum cohomology, the Toda lattice, and the representation with highest weight ρ.* Selecta Mathematica, New Ser., 2:1 (1996), 43–91.

[Kosz] J.-L. Koszul. *Crochet de Schouten–Nijenhuis et cohomologie.* In: "Elie Cartan et les mathématiques d'aujourd'huis", Astérisque (1985), 251–271.

[Kr] A. Kresch. *Cycle groups for Artin stacks.* Preprint math.AG/9810166.

[Ku] V. S. Kulikov. *Mixed Hodge structures and singularities.* Cambridge Univ. Press, 1998.

[LaM–B] G. Laumon, L. Moret–Bailly. *Champs algébriques.* Preprint Orsay 92/42 (1992).

[Le] M. Lehn. *Chern classes of tautological sheaves on Hilbert schemes of points on surfaces.* Preprint math.AG/9803091.

[LiT1] J. Li, G. Tian. *Virtual moduli spaces and Gromov–Witten invariants of algebraic varieties.* J. Amer. Math. Soc., 11 (1998), 119–174. alg-geom/9602007.

[LiT2] J. Li, G. Tian. *Virtual moduli cycles and Gromov–Witten invariants of general symplectic manifolds.* Preprint alg-geom/9608032.

[LZ] B. H. Lian, G. Zuckerman. *New perspectives on the BRST-algebraic structure of string theory.* Comm. Math. Phys., 154 (1993), 613–646. hep–th/9211072.

[LiLY] B. H. Lian, K. Liu, S.-T. Yau. *Mirror principle I.* Asian J. of Math., vol. I, no. 4 (1997), 729–763.

[LibW] A. Libgober, J. Wood. *Uniqueness of the complex structure on Kähler manifolds of certain homology type.* J. Diff. Geom., 32 (1990), 139–154.

[LS] S. Lichtenbaum, M. Schlessinger. *The cotangent complex of morphism.* Trans. AMS, 128 (1967), 41–70.

[LiuT] X. Liu, G. Tian. *Virasoro constraints for quantum cohomology.* Preprint math.AG/9806028.

[Lo1] E. Looienga. *Intersection theory on Deligne–Mumford compactification (after Witten and Kontsevich).* Sém. Bourbaki, 768, March 1993.

[Lo2] E. Looijenga. *On the tautological ring of M_g.* Inv. Math., 121 (1995), 411–419.

[Lo3] E. Looijenga. *Cellular decompositions of compactified moduli spaces of pointed curves.* In: The Moduli Space of Curves, ed. by R. Dijkgraaf, C. Faber, G. van der Geer, Progress in Math. vol. 129, Birkhäuser, 1995, 369–399

[Mal1] B. Malgrange. *Déformations de systèmes différentielles et microdifférentielles.* In: Séminaire de l'ENS 1979–1982, Progress in Math. 37, Birkhäuser, Boston (1983), 353–379.

[Mal2] B. Malgrange. *La classification des connections irrégulieres à une variable.* ibid., 381–399.

[Mal3] B. Malgrange. *Sur les déformations isomonodromiques. I. Singularités régulières.* ibid., 401–426.

[Mal4] B. Malgrange. *Sur les déformations isomonodromiques. II. Singularités irrégulières.* ibid., 427–438.

[MalSV] F. Malikov, V. Schechtman, A. Vaintrob. *Chiral de Rham complex.* Preprint math.AG/980341.

[Ma1] Yu. Manin. *Rational points of algebraic curves over functional fields.* AMS Translations, ser. 2, vol. 50 (1966), 189–234.

[Ma2] Yu. Manin. *Gauge Field Theory and Complex Geometry.* Springer Verlag, 1988, 2nd edition 1997.

[Ma3] Yu. Manin. *Problems on rational points and rational curves on algebraic varieties.* In: Surveys of Diff. Geometry, vol. II, ed. by C. C. Hsiung, S. -T.Yau, Int. Press (1995), 214–245.

[Ma4] Yu. Manin. *Generating functions in algebraic geometry and sums over trees.* In: The Moduli Space of Curves, ed. by R. Dijkgraaf, C. Faber, G. van der Geer, Progress in Math. vol. 129, Birkhäuser, 1995, 401–418.

[Ma5] Yu. Manin. *Sixth Painlevé equation, universal elliptic curve, and mirror of $\mathbf{P}^2$.* AMS Transl. (2), vol. 186 (1998), 131–151. Preprint alg–geom/9605010.

[Ma6] Yu. Manin. *Three constructions of Frobenius manifolds: a comparative study.* Preprint, accepted for Atiyah's Festschrift, math.QA/9801006.

[MM] Yu. Manin, S. Merkulov. *Semisimple Frobenius (super)manifolds and quantum cohomology of $\mathbf{P}^r$.* Topological Methods in Nonlinear Analysis, 9:1 (1997), 107–161 (Ladyzhenskaya's Festschrift).

[MaZo] Yu. Manin, P. Zograf, *Invertible cohomological field theories and Weil-Petersson volumes.* Preprint math.AG/9902051.

[Mat] O. Mathieu. *Harmonic cohomology classes of symplectic manifolds.* Comm. Math. Helvetici, 70 (1995), 1–9.

[Maz] M. Mazzocco. *Picard and Chazy solutions to the Painlevé VI equation.* Preprint math.AG/9901054.

[MS] D. McDuff, D. Salamon. *J–holomorphic curves and quantum cohomology.* Univ. LN series, vol. 6. AMS, Providence, Rhode Island, 1994.

[Me1] S. Merkulov. *Formality of canonical symplectic complexes and Frobenius manifolds.* Int. Math. Res. Notes, 14 (1998), 727–733.

[Me2] S. Merkulov. *Strong homotopy algebras of a Kähler manifold.* Preprint math.AG/9809172.

[MirS1] S.–T. Yau, ed. *Essays on Mirror Manifolds.* International Press Co., Hong Kong, 1992.

[MirS2] B. Greene, S. T. Yau, eds. *Mirror Symmetry II*, AMS–International Press, Amer. Math. Soc., Providence, RI, 1996.

[Mor] Sh. Mori. *Projective manifolds with ample tangent bundles.* Ann. of Math., 110 (1979), 593–606.

[Mo1] D. Morrison. *Mirror symmetry and rational curves on quintic threefolds: a guide for mathematicians.* J. AMS, 6 (1993), 223–247.

[Mo2] D. Morrison. *Compactifications of moduli spaces inspired by mirror symmetry.* Astérisque, vol. 218 (1993), 243–271.

[Mul1] M. Mulase. *Asymptotic analysis of a Hermitian matrix integral.* Int. J. of Math., 6:6 (1995), 881–892.

[Mul2] M. Mulase. *Lectures on the asymptotic expansion of a Hermitian matrix integral.* Preprint math-ph/9811023.

[MulP] M. Mulase, M. Penkava. *Ribbon graphs, quadratic differentials on Riemann surfaces, and algebraic curves defined over* $\overline{\mathbf{Q}}$. Preprint math-ph/9811024.

[Mu1] D. Mumford. *Lectures on curves on an algebraic surface.* Annals of Math. Studies 59, Princeton Univ. Press, 1966.

[Mu2] D. Mumford. *Abelian varieties.* Oxford Univ. Press, 1970.

[Mu3] D. Mumford. *Toward an enumerative geometry of the moduli spaces of curves.* In: Arithmetic and Geometry, ed. by M. Artin and J. Tate, Birkhäuser, Boston, 1983, 271–326.

[Na1] H. Nakajima. *Heisenberg algebra and Hilbert schemes of points on projective surfaces.* Ann. Math., 145 (1997), 379–388. Preprint alg-geom/9507012.

[Na2] H. Nakajima. *Lectures on Hilbert schemes of points on surfaces.* Preprint http://www.kusm.kyoto-u.ac.jp/ nakajima/TEX.html.

[N] N. Nitsure. *Moduli of semistable logarithmic singularities.* Journ. of the AMS, 6:3 (1993), 597–609.

[Od] T. Oda. *K. Saito's period map for holomorphic functions with isolated critical points.* In: Adv. Studies in Pure Math., 10 (1987), Algebraic Geometry, Sendai, 1985, 591–648.

[O1] K. Okamoto. *Isomonodromic deformation and Painlevé equations, and the Garnier system.* J. Fac. Sci. Univ. Tokyo, Sect. IA Math., 33 (1986), 575–618.

[O2] K. Okamoto. *Studies in the Painlevé equations I. Sixth Painlevé equation PVI.* Annali Mat. Pura Appl., 146 (1987), 337–381.

[O3] K. Okamoto. *Sur les feuilletages associés aux équation du second ordre à points critiques fixes de P. Painlevé. Espaces de conditions initiales.* Japan J. Math., 5:1 (1979), 1–79.

[Pa1] R. Pandharipande. *Intersections of* $\mathbf{Q}$-*divisors on Kontsevich's moduli space* $\overline{M}_{0,n}(\mathbf{P}^r,d)$ *and enumerative geometry.* Preprint, 1995.

[Pa2] R. Pandharipande. *The canonical class of* $\overline{M}_{0,n}(\mathbf{P}^r,d)$ *and enumerative geometry.* Preprint, 1995.

[Pa3] R. Pandharipande. *Hodge integrals and degenerate contributions.* Preprint math.AG/9811140.

[Pa4] R. Pandharipande. *Rational curves on hypersurfaces (after A. Givental).* Preprint math.AG/9806133.

[Pea] G. Pearlstein. *Variations of mixed Hodge structure, Higgs fields, and quantum cohomology.* Preprint math.AG/9808106.

[P1] R. C. Penner. *Perturbative series and the moduli space of Riemann surfaces.* J. Diff. Geo., 27 (1988), 35–53.

[P2] R. C. Penner. *The decorated Teichmüller space of punctured surfaces.* Comm. Math. Phys., 113 (1987), 299–339.

[P3] R. C. Penner. *Calculus on moduli space.* In: Geometry of Group Representations, AMS Contemp. Math., 74 (1988), 277–293.

[P4] R. C. Penner. *Integration over the moduli spaces of Riemann surfaces.* In: Proc. of Superstrings TAMU 1989, Adv. Series in Math. Physics, World Scientific, 1989, 346–353.

[P5] R. C. Penner. *Weil–Petersson volumes.* J. Diff. Geo., 35 (1992), 559–608.

[P6] R. C. Penner. *The Poincaré dual of the Weil–Petersson Kähler form.* Comm. Anal. Geo., 1 (1993), 43–70.

[Po1] A. Polishchuk. *Massey and Fukaya products on elliptic curves.* Preprint math.AG/9803017.

[Po2] A. Polishchuk. *Homological mirror symmetry with higher products.* Preprint math.AG/9901025.

[PoZ] A. Polishchuk, E. Zaslow. *Categorical mirror symmetry: the elliptic curve.* Adv. Theor. Math. Phys., 2 (1998), 443–470. Preprint math.AG/980119.

[Ros] M. Rosellen. *Hurwitz spaces and Frobenius manifolds.* Preprint, 1998.

[R] Y. Ruan. *Topological sigma model and Donaldson type invariants in Gromov theory.* Duke Math. J., 83 (1996), 461–500.

[RT1] Y. Ruan, G. Tian. *A mathematical theory of quantum cohomology.* J. Diff. Geo., 42 (1995), 259–367.

[RT2] Y. Ruan, G. Tian. *Higher genus symplectic invariants and sigma model coupled with gravity.* Inv. Math. 130 (1997), 455–516. alg-geom/9601005.

[Sa] C. Sabbah. *Frobenius manifolds: isomonodromic deformations and infinitesimal period mappings.* Preprint, 1996.

[S1] K. Saito. *Period mapping associated to a primitive form.* Publ. Res. Inst. Math. Sci. Kyoto Univ., 19 (1983), 1231–1264.

[S2] K. Saito. *Primitive forms for a universal unfolding of a function with an isolated critical point.* Journ. Fac. Sci. Univ. Tokyo, Sec. IA, 28:3 (1982), 775–792.

[Sch] A. J. Scholl. *Classical motives.* In: Motives, Proceedings of Symposia in Pure Mathematics Vol. 55, Part I. American Mathematical Society, Providence, RI, 1994, 163–187.

[Se] J. Segert. *Frobenius manifolds from Yang–Mills instantons.* Math. Res. Lett., 5 (1998), 327–344.

[Si] B. Siebert. *Gromov–Witten invariants for general symplectic manifolds.* Preprint dg-ga/9608005.

[St] J. Stasheff. *Deformation theory and the Batalin–Vilkovisky master equation.* In: Deformation Theory and Symplectic Geometry, eds. D. Sternheimer et al., Kluwer, 1997, 271–284.

[StYZ] A. Strominger, S.-T.Yau, E. Zaslow. *Mirror symmetry is T-duality.* Nucl. Phys. B, 479 (1996), 243–259.

[T] G. Tian. *Quantum cohomology and its associativity.* In: Current Developments in Math., Int. Press, 1995, 231–282.

[Va] C. Vafa. *Topological mirrors and quantum rings.* In: Essays on Mirror Manifolds, ed. by S.-T. Yau, International Press, Hong Kong, 1992, 96–119.

[vdeL–M] J. W. van de Leur, R. Martini. *The construction of Frobenius manifolds from KP tau-functions.* Preprint solv-int/9808008.

[Vis] A. Vistoli. *Intersection theory on algebraic stacks and on their moduli spaces.* Inv. Math., 93 (1989), 613–670.

[Voi1] C. Voisin. *Symétrie miroir.* Panoramas et synthèses, 2 (1996), Soc. Math. de France.

[Voi2] C. Voisin. *Variations of Hodge structure of Calabi–Yau threefolds.* Quaderni della Scuola Norm. Sup. di Pisa, 1998.

[V] A. Voronov. *Topological field theories, string backgrounds, and homotopy algebras.* Preprint, 1993.

[W1] E. Witten. *On the structure of the topological phase of two-dimensional gravity.* Nucl. Phys. B, 340 (1990), 281–332.

[W2] E. Witten. *Two-dimensional gravity and intersection theory on moduli space.* Surveys in Diff. Geom., 1 (1991), 243–310.

[Wo] S. Wolpert. *On the homology of the moduli spaces of stable curves.* Ann. of Math., 118 (1983), 491–523.

[Za] E. Zaslow. *Solitons and helices: the search for a Math-Physics bridge.* Comm. Math. Phys., 175 (1996), 337–375.

[Zo1] P. Zograf. *The Weil–Petersson volume of the moduli spaces of punctured spheres.* In: Contemporary Mathematics, 150 (1993), ed. by R. M. Hain and C. F. Bödigheimer, Amer. Math. Soc., Providence, RI, 367–372.

[Zo2] P. Zograf. *Weil–Petersson volume of moduli spaces of curves and the genus expansion in two-dimensional gravity.* Preprint math.AG/9811026.

[Zu] J.-B. Zuber. *Graphs and reflection groups.* Comm. Math. Phys., 179 (1996), 265–294.

Subject Index

III.8.2.1 means Chapter III, §8, subsection 8.2.1; 0.3.1 refers to the Introduction.

A–model, 0.1
Absolute stabilization, V.1.7, V.4.6
Abstract correlation functions, III.1.3
Admissible metric, I.5.4
Affine flat structure on supermanifold, I.1.2
Algebraic space, V.5.5.3
Algebraic stack, V.5.5
Artin stacks, V.5.5
Associative pre–Frobenius manifold, I.1.3
Associativity (WDVV) equations, 0.2, I.1.3.1
Atlas, V.5.5

B–model, 0.1
Boundary morphism of moduli stacks, V.4.7

Cartesian square of groupoids, V.4.3.1
Central operations, V.8.7
Chern classes, V.8.6
Class of prestable map, V.1.4.1
Classical linear operad, IV.1.1.1
Classifying groupoid, V.3.2.5
Co–identity, I.2.1.4
Cohomological correlators, VI.2.2
Cohomological Field Theory (CohFT), 0.3.1, III.4.1
Combinatorial type of prestable curve, III.2.5
Combinatorial type of prestable map, V.1.5
Complete Cohomological Field Theory, III.4.5
Configuration space, IV.4.3
Correlation functions of CohFT, III.4.1
Cycles on schemes, V.6.1
Cycles on DM–stacks, V.7.1.1
Cyclic $Comm_\infty$–algebra, III.1.2
Cyclic operad, IV.2.6

d–spectrum of a Frobenius manifold, III.4.10.4
Darboux–Egoroff's equations, I.3.4
Degeneration Axiom for GW–invariants, III.5.2, VI.1.2

Deligne–Mumford (DM) stacks, V.5.5
Differential Gerstenhaber–Batalin–Vilkovyski (dGBV) algebra, III.9.5
Dilaton equation, V.5.3
Dimension Axiom for GW–invariants, III.5.3
Direct sum diagram (of Saito's frameworks), III.8.5
Direct sum of singularities, III.8.6
Divisor Axiom for GW–invariants, III.5.3
Dual modular graph of prestable curve, III.2.5
Dualizing sheaf, V.1.1

Edge (of graph), III.2.1
Effectivity Axiom for GW–invariants, III.5.2, VI.1.2
Equivalence groupoid, V.5.5.2
Euler field, I.2.2.1
Euler field in quantum cohomology, I.4.4, III.5.3.4
Evaluation morphism, V.4.2.2
Excess intersection formula, V.8.8
Extended structure connection, I.2.5.1

F–algebra, I.5.5
F–manifold, I.5.1
Flag (of graph), III.2.1
Flat families, V.2.1, V.2.2
Flat functions and forms, I.1.3.1
Flat pullback, V.6.1.5, V.7.1.5
Formal Laplace transform, II.1.3
Formal Frobenius manifold, 0.4.1, III.1.1
Formal Frobenius manifold of qc–type, III.5.4.1
Frobenius manifold, 0.4.1, I.1.3

Generalized correlators, VI.6.1
Gepner's Frobenius manifolds, III.8.4.1
Gerstenhaber–Batalin–Vilkovyski (GBV) algebra, III.9.4
Gluing along pairs of sections, V.2.3
Good monomials, III.3.5.1
Graph, III.2.1
Gravitational descendants, VI.2.1
Gravity algebra, III.1.9
Gromov–Witten (GW) invariants, III.5
Gromov–Witten (GW) correspondences, III.5, VI.1.3
Groupoid, V.3.2
Groupoid of prestable curves, V.3.2.1
Groupoid of prestable maps, V.3.2.2
Groupoid of universal curves, V.3.2.3
Gysin maps, V.6.2, V.7.2
Gysin pushforward for operational Chow groups, V.8.9.2

Hamiltonian structure of Schlesinger's equations, II.2.4
Homological Chow groups of schemes, V.6.1.3

Homological Chow groups of DM–stacks, V.7.1.3
Homological Chow groups of Artin stacks, V.7.3

Identity Axiom for GW–invariants, III.5.3
Identity on pre–Frobenius manifold, I.2.1
Induced Frobenius structure, I.1.7

Landin transform, II.5.5.6
Large phase space, VI.2.4
Legendre–type transformation, I.5.4.2
Local regular imbedding (l.r.i.) of stacks, V.7.2
Logarithmic CohFT of rank one, III.6.1.4

Mapping to a Point Axiom for GW–invariants, III.5.3, VI.1.2
Markl's operad, IV.2.5
Maurer–Cartan equations, III.9.1
Metric potential, I.3.3
Modified dualizing sheaf, V.1.2
Modular graph, III.2.4
Moduli space $\overline{M}_{0,n}$, 0.3
Morphism of groupoids, V.4.2, V.4.3
Morphism of operads, IV.1.4
Motivic correlators, VI.2.2
Mumford classes, III.6.2

Normal bundle, V.6.2.1
Normal cone, V.6.2.1
Normalized CohFT of rank one, III.6.1.5
Novikov ring, III.5.2.1

Operational Chow groups, V.8.1
Orientation classes, V.8.1.1, V.8.1.2

Painlevé VI equation, II.5.4
Partition function, IV.3.2.2
Perturbation series, V.3
Potential, 0.2
Potential of qc–type, III.5.4.1
Potential pre–Frobenius manifold, I.1.3
Pre–Frobenius manifold, I.1.3
Prestable curve, III.2.1
Prestable map, V.1.3.1
Primary fields, VI.2.1
Projection formulas, V.8.5, V.8.9.4
Proper pushforward, V.6.1.4, V.7.1.4, V.8.3
Proper pushforward for operational Chow groups, V.8.9.1
Pullback for operational Chow groups, V.8.4
Puncture equation, VI.5.1
Pushforward of operational Chow groups, V.8.9

Quantum cohomology, 0.1, III.5
Quantum cohomology of projective spaces, II.4

Rank one CohFT, III.6.1
Rational equivalence of cycles on schemes, V.6.1.2
Rational equivalence of cycles on DM–stacks, V.7.1.2
Representable morphism of groupoids, V.5.3
Rotation coefficients, I.3.4

Saito's framework, III.8.2.1
Schlesinger's equations, III.2.3.1
Second structure connection, II.1.1
Semisimple Euler field, I.2.4
Semisimple (pre–)Frobenius manifold, I.3.1
Singularities of meromorphic connections, II.2.1
Small quantum multiplication, III.5.4.2
S_n-covariance Axiom for GW–invariants, III.5.2, VI.1.2
Special initial conditions, II.3.4
Special coordinates of tame semisimple germ, III.7.1.1
Special solutions to Schlesinger's equations, II.3.1.1
Species of algebras, IV.6.1.4
Spectral cover of Frobenius manifold, III.8.1
Spectrum of Frobenius manifold, I.2.4
Split identity, I.2.1.3, III.7.5.8
Split semisimple (pre–)Frobenius manifold, I.3.1
Stabilization morphism of moduli stacks, V.4.4
Stabilization of prestable curve, V.1.6
Stabilization of prestable map, V.1.7
Stable curve, III.2.5
Stable map, V.1.3.2
Stable modular graph, III.2.4
Stack, V.5.1
Standard weight, IV.3.2.1
Strictly special solutions to Schlesinger's equations, II.3.1.3
String equation, VI.5.2
Structure connection of pre–Frobenius manifold, I.1.4
Supermanifold, I.1.1.1

Tame semisimple germ of Frobenius manifold, III.7.1
Tame singularity of a connection, II.2.1
Tau–function of solution, II.2.3.2
Tensor product of formal Frobenius manifolds, III.4.4, III.4.10, III.6.6
Tensor product of analytic Frobenius manifolds, III.7
Tensor product diagram, III.7.5.9
Theta–divisor of Schlesinger's equations, II.2.3
Twisted Frobenius manifold, I.5.4.2

Unfolding singularities, III.8.4

Versal deformation of meromorphic connection, II.2.2

Virasoro constraints, VI.2.5, VI.5
Vertex (of graph), III.2.1
Virtual dimension, V.1.9
Virtual fundamental class, VI.1.1
Virtual Poincaré polynomial, IV.4.1

WDVV-equations (= Associativity Equations) I.1.3.1, I.1.9
Weak Euler field, I.5.4.2
Weak Frobenius manifold, I.5.3
Weight of weak Euler field, I.5.4.2
Weight of Euler field, III.4.10
Weight of identity, III.4.10
Weil–Petersson volumes, III.6.4
Wick's lemma, IV.3.4.2

图字：01-2019-0852号

Frobenius Manifolds, Quantum Cohomology, and Moduli Spaces, by Yuri I. Manin,
first published by the American Mathematical Society.
Copyright © 1999 by the American Mathematical Society. All rights reserved.
This present reprint edition is published by Higher Education Press Limited Company under authority of the American Mathematical Society and is published under license.
Special Edition for People's Republic of China Distribution Only. This edition has been authorized by the American Mathematical Society for sale in People's Republic of China only, and is not for export therefrom.

本书最初由美国数学会于1999年出版，原书名为*Frobenius Manifolds, Quantum Cohomology, and Moduli Spaces*，作者为Yuri I. Manin。
美国数学会保留原书所有版权。
原书版权声明：Copyright ©1999 by the American Mathematical Society。
本影印版由高等教育出版社有限公司经美国数学会独家授权出版。
本版只限于中华人民共和国境内发行。本版经由美国数学会授权仅在中华人民共和国境内销售，不得出口。

Frobenius 流形、量子上同调和模空间

Frobenius Liuxing Liangzi
Shangtongdiao he Mokongjian

图书在版编目 (CIP) 数据

Frobenius 流形、量子上同调和模空间：英文 / (俄罗斯) 尤里·曼宁 (Yuri I. Manin) 著 . -- 影印本 . -- 北京：高等教育出版社，2019.5
书名原文：Frobenius Manifolds, Quantum Cohomology, and Moduli Spaces
ISBN 978-7-04-051702-6

Ⅰ.①F… Ⅱ.①尤… Ⅲ.①流形—研究—英文 ②量子—上同调—研究—英文③模（数学）—空间—研究—英文 Ⅳ.① O189.3 ② O189.22 ③ O153.3
中国版本图书馆 CIP 数据核字 (2019) 第 064621 号

策划编辑	李华英	责任编辑	李华英
封面设计	张申申	责任印制	尤 静

出版发行	高等教育出版社	开本	787mm×1092mm 1/16
社址	北京市西城区德外大街4号	印张	20.25
邮政编码	100120	字数	520千字
购书热线	010-58581118	版次	2019 年 5 月第 1 版
咨询电话	400-810-0598	印次	2019 年 5 月第 1 次印刷
网址	http://www.hep.edu.cn	定价	135.00元
	http://www.hep.com.cn		
网上订购	http://www.hepmall.com.cn		
	http://www.hepmall.com		
	http://www.hepmall.cn		
印刷	北京新华印刷有限公司		

本书如有缺页、倒页、脱页等质量问题，请到所购图书销售部门联系调换
版权所有 侵权必究
[物 料 号 51702-00]

郑重声明

高等教育出版社依法对本书享有专有出版权。任何未经许可的复制、销售行为均违反《中华人民共和国著作权法》，其行为人将承担相应的民事责任和行政责任；构成犯罪的，将被依法追究刑事责任。为了维护市场秩序，保护读者的合法权益，避免读者误用盗版书造成不良后果，我社将配合行政执法部门和司法机关对违法犯罪的单位和个人进行严厉打击。社会各界人士如发现上述侵权行为，希望及时举报，本社将奖励举报有功人员。

反盗版举报电话	(010) 58581999 58582371 58582488
反盗版举报传真	(010) 82086060
反盗版举报邮箱	dd@hep.com.cn
通信地址	北京市西城区德外大街 4 号 高等教育出版社法律事务与版权管理部
邮政编码	100120

美国数学会经典影印系列

1	**Lars V. Ahlfors**, Lectures on Quasiconformal Mappings, Second Edition	9787040470109
2	**Dmitri Burago, Yuri Burago, Sergei Ivanov**, A Course in Metric Geometry	9787040469080
3	**Tobias Holck Colding, William P. Minicozzi II**, A Course in Minimal Surfaces	9787040469110
4	**Javier Duoandikoetxea**, Fourier Analysis	9787040469011
5	**John P. D'Angelo**, An Introduction to Complex Analysis and Geometry	9787040469981
6	**Y. Eliashberg, N. Mishachev**, Introduction to the h-Principle	9787040469028
7	**Lawrence C. Evans**, Partial Differential Equations, Second Edition	9787040469356
8	**Robert E. Greene, Steven G. Krantz**, Function Theory of One Complex Variable, Third Edition	9787040469073
9	**Thomas A. Ivey, J. M. Landsberg**, Cartan for Beginners: Differential Geometry via Moving Frames and Exterior Differential Systems	9787040469172
10	**Jens Carsten Jantzen**, Representations of Algebraic Groups, Second Edition	9787040470086
11	**A. A. Kirillov**, Lectures on the Orbit Method	9787040469103
12	**Jean-Marie De Koninck, Armel Mercier**, 1001 Problems in Classical Number Theory	9787040469998
13	**Peter D. Lax, Lawrence Zalcman**, Complex Proofs of Real Theorems	9787040470000
14	**David A. Levin, Yuval Peres, Elizabeth L. Wilmer**, Markov Chains and Mixing Times	9787040469943
15	**Dusa McDuff, Dietmar Salamon**, J-holomorphic Curves and Symplectic Topology	9787040469936
16	**John von Neumann**, Invariant Measures	9787040469974
17	**R. Clark Robinson**, An Introduction to Dynamical Systems: Continuous and Discrete, Second Edition	9787040470093
18	**Terence Tao**, An Epsilon of Room, I: Real Analysis: pages from year three of a mathematical blog	9787040469004
19	**Terence Tao**, An Epsilon of Room, II: pages from year three of a mathematical blog	9787040468991
20	**Terence Tao**, An Introduction to Measure Theory	9787040469059
21	**Terence Tao**, Higher Order Fourier Analysis	9787040469097
22	**Terence Tao**, Poincaré's Legacies, Part I: pages from year two of a mathematical blog	9787040469950
23	**Terence Tao**, Poincaré's Legacies, Part II: pages from year two of a mathematical blog	9787040469967
24	**Cédric Villani**, Topics in Optimal Transportation	9787040469219
25	**R. J. Williams**, Introduction to the Mathematics of Finance	9787040469127
26	**T. Y. Lam**, Introduction to Quadratic Forms over Fields	9787040469196

27 **Jens Carsten Jantzen**, Lectures on Quantum Groups

28 **Henryk Iwaniec**, Topics in Classical Automorphic Forms

29 **Sigurdur Helgason**, Differential Geometry, Lie Groups, and Symmetric Spaces

30 **John B. Conway**, A Course in Operator Theory

31 **James E. Humphreys**, Representations of Semisimple Lie Algebras in the BGG Category O

32 **Nathanial P. Brown, Narutaka Ozawa**, C*-Algebras and Finite-Dimensional Approximations

33 **Hiraku Nakajima**, Lectures on Hilbert Schemes of Points on Surfaces

34 **S. P. Novikov, I. A. Taimanov, Translated by Dmitry Chibisov**, Modern Geometric Structures and Fields

35 **Luis Caffarelli, Sandro Salsa**, A Geometric Approach to Free Boundary Problems

36 **Paul H. Rabinowitz**, Minimax Methods in Critical Point Theory with Applications to Differential Equations

37 **Fan R. K. Chung**, Spectral Graph Theory

38 **Susan Montgomery**, Hopf Algebras and Their Actions on Rings

39 **C. T. C. Wall, Edited by A. A. Ranicki**, Surgery on Compact Manifolds, Second Edition

40 **Frank Sottile**, Real Solutions to Equations from Geometry

41 **Bernd Sturmfels**, Gröbner Bases and Convex Polytopes

42 **Terence Tao**, Nonlinear Dispersive Equations: Local and Global Analysis

43 **David A. Cox, John B. Little, Henry K. Schenck**, Toric Varieties

44 **Luca Capogna, Carlos E. Kenig, Loredana Lanzani**, Harmonic Measure: Geometric and Analytic Points of View

45 **Luis A. Caffarelli, Xavier Cabré**, Fully Nonlinear Elliptic Equations

46 **Teresa Crespo, Zbigniew Hajto**, Algebraic Groups and Differential Galois Theory

47 **Barbara Fantechi, Lothar Göttsche, Luc Illusie, Steven L. Kleiman, Nitin Nitsure, Angelo Vistoli**, Fundamental Algebraic Geometry: Grothendieck's FGA Explained

48 **Shinichi Mochizuki**, Foundations of p-adic Teichmüller Theory

49 **Manfred Leopold Einsiedler, David Alexandre Ellwood, Alex Eskin, Dmitry Kleinbock, Elon Lindenstrauss, Gregory Margulis, Stefano Marmi, Jean-Christophe Yoccoz**, Homogeneous Flows, Moduli Spaces and Arithmetic

50 **David A. Ellwood, Emma Previato**, Grassmannians, Moduli Spaces and Vector Bundles

51 **Jeffery McNeal, Mircea Mustaţă**, Analytic and Algebraic Geometry: Common Problems, Different Methods

52 **V. Kumar Murty**, Algebraic Curves and Cryptography

53 **James Arthur, James W. Cogdell, Steve Gelbart, David Goldberg, Dinakar Ramakrishnan, Jiu-Kang Yu**, On Certain L-Functions

54 **Rick Miranda,** Algebraic Curves and Riemann Surfaces

55 **Hershel M. Farkas, Irwin Kra,** Theta Constants, Riemann Surfaces and the Modular Group

56 **Fritz John,** Nonlinear Wave Equations, Formation of Singularities

57 **Henryk Iwaniec, Emmanuel Kowalski,** Analytic Number Theory

58 **Jan Malý, William P. Ziemer,** Fine Regularity of Solutions of Elliptic Partial Differential Equations

59 **Jin Hong, Seok-Jin Kang,** Introduction to Quantum Groups and Crystal Bases

60 **V. I. Arnold,** Topological Invariants of Plane Curves and Caustics

61 **Dusa McDuff, Dietmar Salamon,** J-Holomorphic Curves and Quantum Cohomology

62 **James Eells, Luc Lemaire,** Selected Topics in Harmonic Maps

63 **Yuri I. Manin,** Frobenius Manifolds, Quantum Cohomology, and Moduli Spaces

64 **Bernd Sturmfels,** Solving Systems of Polynomial Equations

65 **Liviu I. Nicolaescu,** Notes on Seiberg-Witten Theory